BIOFLUID MECHANICS · 2

BIOFLUID MECHANICS · 2

Edited by

Daniel J. Schneck
Virginia Polytechnic Institute and State University
Blacksburg, Virginia

Springer Science+Business Media, LLC

Library of Congress Cataloging in Publication Data

Mid-Atlantic Conference on Biofluid Mechanics, 2d, Virginia Polytechnic Institute and State University, 1980.
Biofluid mechanics.

Includes index.
1. Hemodynamics – Congresses. 2. Rheology (Biology) – Congresses. 3. Body fluid flow – Congresses. 4. Biomedical engineering – Congresses. I. Schneck, Daniel J. II. Title.
QP105.M54 1980 599.01'01'532 80-10092
ISBN 978-1-4757-4612-9 ISBN 978-1-4757-4610-5 (eBook)
DOI 10.1007/978-1-4757-4610-5

Proceedings of the Second Mid-Atlantic Conference on Biofluid Mechanics, held at Virginia Polytechnic Institute and State University, Blacksburg, Virginia, May 5-7, 1980

Conference Chairman: Daniel J. Schneck
Conference Co-Chairman: Donald L. Vawter

This Conference was sponsored in part by a grant from the National Science Foundation

Originally published by Plenum Press, New York in 1980
Softcover reprint of the hardcover 1st edition 1980

PREFACE

The Department of Engineering Science and Mechanics at Virginia Polytechnic Institute and State University sponsored the First Mid-Atlantic Conference on Bio-Fluid Mechanics, which was held in Blacksburg, Virginia during the period 9-11 August 1978. Some 40 life-scientists, engineers, physicians and others who share a common interest in the advancement of basic and applied knowledge in bio-fluid mechanics gathered at the Donaldson Brown Center for Continuing Education to hear 25 papers presented in seven technical sessions. At the conclusion of the conference, those present decided unanimously that its success warranted having at least one more -- and that it was conceptually a sound idea to plan it on a biennial basis for late spring. Hence, the second Mid-Atlantic Conference on Bio-Fluid Mechanics took place at Virginia Tech on May 4-6, 1980.

This volume documents the Proceedings of the second conference. It contains full texts of 23 contributed papers, 2 guest lectures and 1 invited seminar. The papers are grouped according to subject matter, beginning with 3 in the area of respiration, followed by 1 in kidney dialysis, 1 in reproduction, 1 in joint lubrication, 1 in prosthetic fluidics, 2 in zoology, and ending with 14 in the general field of cardiovascular dynamics. Of the latter, 5 deal with the subject of heart valves, 2 concern themselves with the microcirculation, 6 address vascular system hemodynamics and 1 covers some aspects of blood rheology. The guest presentations are of a review nature, reflecting upon Bio-engineering and the Real World, a Historical Perspective of Control of the Output of the Heart, and 1980 Updates in Optical and Thermal Anemometry.

Examining the table of contents of this book, one quickly gains an appreciation for the breadth and diversity of ways in which the laws and principles of fluid mechanics can be applied to solving problems in biology and medicine.

This, in fact, was the theme of the conference. Its objective was to provide a forum for the interchange of current developments in the field, and to encourage more dialogue among investigators concerned with problems in biofluid mechanics. Hence, sincere thanks go to all of the authors and guest lecturers who contributed their time, effort and manuscripts to the Conference and to this volume. I also acknowledge gratefully the encouragement and support of Dr. Daniel Frederick, Head of the Department of Engineering Science and Mechanics, and Dr. Paul E. Torgersen, Dean of the College of Engineering at Virginia Tech. A special thanks goes to Dr. Richard F. Harshberger, Associate Director of Program Development, and his staff at the Donaldson Brown Center for Continuing Education, for their gracious hosting of this Conference; and to the National Science Foundation, for their generous financial contribution. Last, but certainly not least, I wish to express my appreciation to all those who served as session chairmen, and to Ms. Patricia Ellen Holcomb, who put in sometimes long, many times tedious, but always pleasant hours of expert secretarial and editorial assistance to help put this volume together. Indeed, it takes many dedicated people putting in a great deal of hard work to make a venture such as this successful.

Daniel J. Schneck
Editor and
Conference Chairman

May, 1980
Virginia Polytechnic Institute
and State University
Blacksburg, Virginia

CONTENTS

BIOENGINEERING AND THE REAL WORLD

Alvin H. Sacks, Ph.D.

Institute for Medical Research

751 So. Bascom Ave., San Jose, California

In comparison to the "good old days" in which engineering problems were well defined and the solutions depended upon well established techniques, the engineer in bioengineering does indeed face the real world today. There seem to be two aspects in which bioengineering differs substantially from the classical engineering of the past. First, the technical problems are often vaguely defined, the measurements are hard to come by and difficult to repeat, and the techniques are imprecise. As a result, the conclusions appear to be considerably more open to debate. But the second, and perhaps the more difficult, aspect in which bioengineering differs from classical engineering is the fact that multidisciplinary teams are almost essential to substantial progress. Here the engineer is forced to enter the real world of personal relationships with people who come from a substantially different background and training. I suspect that those of us who went into engineering or into physics, mathematics, or chemistry tend to be those who relate to inanimate objects and to equations more easily than we do to people, whereas the physician is almost the opposite. He usually relates very well to people, especially those who come for his advice, and frequently he does not relate well to mathematics or physics. Unfortunately, one characteristic the engineer seems to share with the physician is competetiveness.

In the early days of bioengineering, about 20 years ago, we saw many attempts by engineers to solve what they called biological problems on their own (I was one of those). The result was what we might call grand solutions to trivial problems. About the same time, the physician began to introduce basic electronic equipment into his laboratory and began to see himself as an engineer. There were indeed also some early tentative efforts to form teams, but too often the engineer played the role of an assistant who was asked to design a piece of equipment to make a specific measurement, only to find later that the results could not be interpreted. Today we are beginning to see some real collaboration, and that in itself raises some interesting problems. This paper will look at some of the problems of bioengineering as opposed to classical engineering, with particular emphasis on the personal problems of working together in multidisciplinary teams. In order to illustrate some of the specific problems concerned with the latter, we will consider the special problems that arise in connection with the development of an instrument intended for commercial sale, an increasingly common occurence in the practice of bioengineering.

TECHNICAL PROBLEMS: Even the choice of the problem to be attacked depends strongly upon the viewpoint of the invesigators, i.e., their speciality, their aptitudes, and their impressions of what is important and what is possible, as well as upon the ability of the interdisciplinary members of the team to get along with one another. Only through an open discussion involving considerable patience and mutual respect does one arrive at a truly interesting and significant problem for which there is some hope of a useful solution. The success or failure of a bioengineering project depends most strongly upon the proper selection of the problem. After that is settled, the methods of experiment and analysis are often a stumbling block for bioengineering teams, since the methods familiar to the engineer are quite unfamiliar and therefore uncomfortable for the physician or physiologist, and vice versa. From the engineer's viewpoint, an acquaintance with dimensional analysis, for example, would seem to be a prerequisite, and yet it is customarily absent in the life sciences. The engineer also is accustomed to systems whose performance is the same from moment to moment and from day to day, a situation which is not encountered in the real world of physiology, in which statistical methods are an absolute necessity. Finally, the interpretation of results depends strongly upon the backgrounds of the investigators, and it is here that the diversity of view point can be tremendously advantageous, if indeed it is permitted to flow in the presence of a good working relationship.

WORKING TOGETHER: Even in the earliest days of bioengineering, we all knew of the potential tremendous advantages of working in bioengineering teams consisting of both engineers and physicians or physiologists, but I don't believe that any of us foresaw the problems involved in such a working relationship. It has been my observation that so long as the role of each investigator is clearly defined and does not impinge on the role of other investigators within the project, then the personal problems are minimal, and mutual respect appears to flow easily. However, if a physician begins to see himself as an engineer, or worse yet, if an engineer begins to see himself as a physiologist, then there is potential trouble. Since electrical engineers may fall easily into the position of experts on computers or electronics, and since mechanical engineers may fall easily into the category of expert equipment designers, the conflicts in those areas appear to be minimal, although some physicians do seem to fancy themselves as experts in those fields. For the fluid mechanics specialist, there is room for considerable conflict because of the fact that his field overlaps many areas in which the physician feels that he should be the expert, namely in the understanding of blood flow, red cell behavior, sound generation, blood pressure measurement, and the like. Here, as in any personal relationship, a little humility can do wonders.

With regard to conflict within a team, it may be useful to refer to a chart which is sometimes used in psychology to investigate those parts of our personalities which affect realationships with other people. This chart is called the Johari window and is shown in the sketch below. It seems clear that the more of our characteristics that are in the open, to be seen by both parties, the less chance there will be for conflict. On the other hand, the ones that are blind or hidden, and certainly those that are closed, are bound to be sources of trouble. We might tend to think that we are very open and have few things concealed in the closed box, but those are the ones that lead to the outbursts or upsets that later even we ourselves do not understand. Characteristics that are either blind or hidden are clearly sources of trouble. Any fears we might have (perhaps unknown to ourselves) which are unintentionally challenged, such as the fear of being shown up, put down, or outdone, can be a serious problem for both parties in the bioengineering setting.

	You see	You don't see
I see	Open	Hidden
I don't see	Blind	Closed

Johari Window: A look at our character traits

Assuming that these relationships can be worked out between the engineer and the physician, we can represent the types of bioengineering that can be expected to come out of different types of collaboration in the form of a chart (shown below) in which both the engineering and physician's side have been divided into two categories, namely, practical and research. (Note that the term "physician" might be replaced by any professional from the life sciences.) The first collaborations we saw in the early days of bioengineering were largely in the clinical or applied areas, and the main developments were largely those of automation (an example of grand solutions to trivial problems?). Later, we began to see the emergence of new devices and new techniques, some of which fell into the same category, such as the electronic stethoscope, which yielded additional but unintelligible information. Only very recently have we seen the collaboration of research physicians or research physiologists with research engineers and the consequent emergence of new concepts in the area of bioengineering. Perhaps the two areas of personal traits which contribute most frequently to the breakdown of bioengineering teams are the tendencies on the part of both engineers and physicians to employ put-downs and to compete for credit. Unfortunately, both of these traits are politely encouraged in both professional fields, as can be seen at many technical conferences. How often have we heard or read the phrase--"Unfortunately, in the work of Dr. Jones,---".

	Development Engineer	Research Engineer
Clinical MD	Automation	New Devices
Research MD	New Techniques	New Concepts

Categories of Bioengineering Collaborations

INSTRUMENT DEVELOPMENT, A SPECIAL PROBLEM. Perhaps the largest area of potential conflict in bioengineering teams is the same one that breaks up many marriages, namely a conflict over money. Since many of you are likely to become involved in the development of instruments which might be sold commercially, I should like to outline briefly the story of one instrument development in which I have been involved and some of the problems that have arisen out of it. The technical problem to be studied was selected nearly 20 years ago, namely the question of what it is that the physician really measures when he takes your blood pressure by the standard auscultatory technique, and what are the errors involved in that measurement. In applying the auscultatory technique, the physician applies a pneumatic cuff to encircle the upper arm, pressurizes the cuff by means of a squeeze bulb to a point where the pulse is obliterated at the radial artery, and then listens with a stethoscope at the brachial artery just below the cuff as the cuff pressure is gradually reduced. During the deflation of the cuff, he hears through the stethoscope the onset of a faint but distinct tapping sound (one per beat), which then proceeds to grow louder, to change in character, diminish or fade, and finally disappear. These are the five phases of the Korotkoff sounds. He takes two readings, one at the first appearance of the sound and one at the final disappearance. These two readings are denoted as the systolic and diastolic blood pressure. We undertook to study this technique by constructing a large scale, dynamically similar model of the brachial artery with a simulated artifical cuff by which we succeeded in producing the entire range of Korotkoff sounds. We then proceeded to take direct and indirect blood pressure readings and made comparisons over a wide range of fluids, vessel elasticity, fluid viscosity and pressure levels, as well as pulse rates. We studied the generation of Korotkoff sounds as well as the errors in auscultatory measurements and the influence of the various parameters.

It did, of course, occur to me during this project and afterward that there might be some simpler, more straight-forward method of obtaining the same measurements. However, it was quite by chance years later that a physician asked me if I would be interested in designing a device for home use with which one could measure his or her own blood pressure. It was shortly after that question was raised that ideas began to come together in rapid succession, and the development of a new instrument was underway. Over the next few months, I realized that (1) I did not need a stethoscope to measure my own blood pressure, since I could feel sensations which very clearly indicated systolic and diastolic pressures, (2) I could feel these sensations equally well at the wrist or even in the fingers, (3) inflating a cuff on the finger would require such a small volume of air that one could in fact eliminate the pressure gage by using the perfect gas law to calibrate the volume change, and (4) adjusting for various finger sizes could be accomplished by some sort of pop-out valve to zero the calibrated pressure gage. With these four elements, a new instrument was gradually fabricated from available parts and is shown in the figure below.

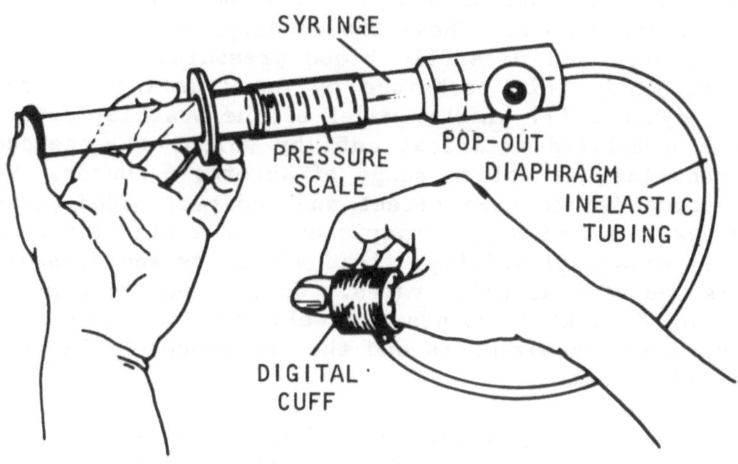

Early model of a device for self measurement of blood pressure

At this stage of development, the technical problems were solved in principle, although there remained many details to be worked out. On the other hand, the more difficult problems involved with relationships were just beginning. It was, for example, at this point that I realized that I had an employment agreement with my employer which did not seem to cover this particular circumstance, since the device was developed at home, outside of any project, and on my own time. It therefore became the subject of a long (and sometimes uncomfortable) negotiation during which ownership was determined. One of the first lessons for me here was that the object of negotiation is not to win or to prove you are right, but rather to arrive at a mutually acceptable agreement. There was then the question of the contributions of various individuals with whom I had discussed the project at various stages and received comments or suggestions. It is indeed surprising to find the value that is later placed on suggestions, mostly by the people who made them. One of these situations actually became the subject of a year-long negotiation which involved two attorneys and resulted in a lengthy written agreement.

Since blood pressure measurement has recently become a subject of considerable attention, and since the device shown in the figure is clearly a simple and inexpensive one, it soon became apparent that considerable amounts of money might be involved. Somehow, this factor seems to alter one's perspective, and at this point one needs to consider what is important, what is ethical, what one wants from such a development, and to what ends one is willing to go, both in terms of effort and confrontations, in order to get it. Some of these answers seem to lie in the "closed" section of the Johari window. It was my impression up to that time that if I should ever actually come up with a salable invention, I would simply file for a patent, sell it to a company, and proceed to collect royalties. The reality was that the patent application permitted me to enter into a series of negotiations with both companies and individuals which I personally found most uncomfortable, largely because I did not have these important and very personal questions satisfactorily answered in my own mind. I was ill-prepared for my first corporate negotiation, which nevertheless resulted in an option-license agreement. After that time of course, one must continue to pursue the patent application, which is customarily denied on the first go-round on the grounds that (a) it will not work, and (b) it has already been done. In my own case, there was a visit to the patent examiner's office with a demonstration to prove that the device did in fact work. Thanks to the involvement of several attorneys and the subsequent arguments and revisions of the patent, this patent has taken nearly five years from initial filing to issue.

During this entire process, of course, the inventor is likely to become involved in such questions as product design (the design has now changed considerably), and written instructions for the user. The individual interpretations of the same instructions verge on the unbelieveable, and I was astounded to see how heavily customer acceptance depended upon the appearance and color of the device! All in all, the area of commercialization of patented devices in the biomedical field involves a lot more than I had bargained for and should not be entered into lightly. On the other hand, the entire experience can be both rewarding and educational and has taught me perhaps more than I wanted to know about both myself and the realities of patents, negotiations, and product development.

One major distinction between the product or business world of bioengineering and the academic or research world is in the area of secrecy. It is obviously advantageous to keep progess as secret as possible on a product, whereas research is of little value to the researcher if it is not published. Therefore, one must decide in advance whether monetary gains and the satisfaction of a product brought successfully to market are more important to him or her than publishing an original piece of research. In either case, one must decide how he will behave when personal ethics may require relinquishing at least a share of the end reward that he thought was so important. This may well be the most important challenge to the bioengineer in the real world, where we find we must face our greed, our competitiveness, and our defensiveness, as well as that of our collaborators. One of the rewards of negotiation (and one of the traumas) is that it seems to open up the Johari window, thereby shedding some light on who and what we really are.

There are indeed new challenges in bioengineering, not the least of which is a new awareness of the feelings that lie hidden in each of us.

A NEW VIEW OF THE DYNAMIC MECHANISMS OF MAXIMUM EXPIRATORY GAS FLOW IN THE LUNGS[1]

WILLIS G. DOWNING, JR.

UNIVERSITY OF SOUTHERN CALIFORNIA

LOS ANGELES, CALIFORNIA 90007[2]

INTRODUCTION

The physical nature of the dynamic mechanisms of impedance to gas flow in the lungs has been investigated experimentally and modeled mathematically for many years. This has been particularly true of the mechanisms believed responsible for the fixed upper limits to gas volume flow rates during the effort-independent portion of a maximal exhalation. In 1958, Hyatt, Schilder, and Fry described the relationship between maximum expiratory flow and degree of lung inflation (1) and developed the concept of the maximal expiratory flow-volume curve. Their experimental results showed that over the lower half of the vital capacity, during a maximal exhalation, the flow was effort-independent. Since that time, the effort-independent portion of the maximum expiratory flow-volume curve has been reported to reach as high as 90% of the vital capacity for some subjects (2).

[1]This paper describes part of the work done by the author toward the doctoral degree at USC. This research was supported through funds provided by the National Institutes of Health, Institute of General Medical Sciences Grants GM 47860 and GM 01724.

[2]The author is now with the VA Medical Center, Sepulveda, California 91343.

Three different major physical mechanisms have been proposed as primarily responsible for the effort-independent portion of the maximum expiratory flow-volume curve. Two of these mechanisms have been described as occurring at sites along the airway tubes within the tracheobronchial tree. These mechanisms have been called "dynamic airway compression" and "flow limitation at wave speed." In contrast, the third mechanism, "selective merging" has been described as occurring at the airway bifurcating junctions within the tracheobronchial tree. These three mechanisms will be briefly described below, along with a few objections to or limitations of each.

Dynamic Airway Compression

Einthoven described dynamic airway compression and first noted its importance to an understanding of some mechanisms of obstructive airway disease (3). Since that time, several different but complementary descriptions of lung mechanics have been developed to describe and explain the phenomenon of dynamic airway compression (1,4-6). Dynamic airway compression is described as occurring at sites along air passageways where the gas pressure inside the air passageway, P_2 is equal to or less than the pressure bearing inward from outside the airway. The inward bearing pressure is considered to be approximately equal to P_{pl}, the pleural pressure. Airway closure could occur if airway compression continued. Figure 1a illustrates this mechanism.

Mead described the start of dynamic airway compression at or near the site of an "Equal Pressure Point," EPP (5). This point occurred at a local airway site where the pleural pressure outside the airway equaled the gas pressure inside the airway. Mead's model of lung mechanics during maximal expiratory flow implies a fixed gas flow resistance upstream from the EPP at a given lung volume. This resistive pressure drop is exactly equal to the elastic pressure (P_{el}) of the lung upstream from the EPP. P_{el} ($P_{alv} - P_{pl}$) was the upstream driving pressure for Mead's analysis.

Pride, Permutt, et al. described the relationship between dynamic airway compression and the limiting gas volume flow rate at a given lung volume in terms of a "waterfall effect." They reasoned that, at and below some critical transmural pressure, $P'_{tm}(P_2-P_{pl})$, a waterfall effect was created at the point of airway constriction. Driving pressure from the alveoli up to the waterfall was considered to be equal to $P_{el}-P'_{tm}$. When P_{tm}, the transmural pressure, was decreased below P'_{tm}, a waterfall effect was created in which relatively small downstream pressure variations, with respect to the pleural pressure, could not change the instantaneous volume expiratory flow rate. This flow rate would be determined entirely by the upstream driving pressure and the upstream resistance to the waterfall, both of which were presumed constant at a given lung volume.

One of the difficulties with the concept of dynamic airway compression and its relation to the waterfall effect was described by Pride and Permutt, et al. (6). They pointed out that, if volume flow rate remains constant at a given lung volume at a site of local airway compression, the local gas velocity must increase at that site as airway compression increases. The result is a Bernoulli effect, creating a local pressure decrease. This, in turn, causes further compression of the airway. The result is a positive feedback sequence of events, stopped only by 1) possible rigidity of the airway, 2) decrease in expiratory flow rate due to inordinately increasing resistance, and/or 3) momentary airway collapse. The result is a very complex and unstable situation, difficult to characterize. The analysis becomes even more complex when these effects are considered to occur throughout the lung air passageways.

Flow Limitation at Wave Speed

Recently, Dawson and Elliott have proposed "flow limitation at wave speed" as the determinant mechanism responsible for upper limits to expiratory gas flow found during a forceful exhalation (7,8). A theoretical basis for wave speed analysis of blood flow has been described (9-11). The transfer of energy by the moving wave between the elastic vessel walls radially and the axial inertia of the moving fluid could be likened to the movement of a pulse wave of electrical energy along an electric transmission line. Analogous wave speed equation forms could be used for the mechanical and electrical models. In applying the concept of wave speed to blood flow, the analysis is simplified by the fact that both the vessel walls and the moving blood are primarily liquids with similar densities and compressibilities.

Dawson and Elliott extended what had been a nearly single medium analysis to two dissimilar mediums, applying the concept of wave speed limitation to compliance of the membrane of the airway walls and the inertia of the moving gas within the airways. They described the flow limitation at wave speed, postulated as occurring in airways, as analogous to flow limitation occurring at the narrowest cross-section of a converging-diverging nozzle when the fluid flowing through the nozzle reached mach 1 (the wave speed of sound for the fluid). Flow limitation at wave speed was described by Dawson and Elliott as resulting in a "waterfall effect" similar to that used to describe maximal flow ascribed to dynamic airway compression by Pride, Permutt, et al. (6). For the mechanism of flow limitation at wave speed, the waterfall effect is created because a downstream disturbance cannot be propagated upstream faster than the wave speed (which has already been reached in the opposite direction). This is illustrated in Figure 1b. The concept of flow limitation at wave speed placed an upper limit to local expiratory gas velocity which would result in a halt to the positive feedback of the Bernoulli effect. Airway collapse, then, would not take place once wave speed velocity was reached.

The theory, as developed by Dawson and Elliott, assumed steady flow at a given lung volume, neglecting transient inertial effects. They also neglected frictional effects upstream from the "choke point." Elliott and Dawson tested the wave speed theory in excised dog trachea (12). Winter, et al. transmitted sinusoidal pressure waves down dog trachea in vivo (13). Neither of the two tests was performed both in vivo and under the pressure stress distribution found during a maximal exhalation. The mechanical states of the lung air passageways and alveoli in vitro can differ from in vivo states both because of intrinsic differences in tissue chemical and physical conditions and because of different physical stresses around and throughout the lungs. It has been found that the effective compliance of the lungs during a maximal exhalation is markedly reduced below that found during a slow exhalation (14). This may imply marked changes in the mechanical characteristics of the lung air passageways during maximal exhalation with an increase in possible wave speed in air passageways, including the trachea. Thus, the theory as developed so far does not conclusively locate the site or sites of maximal flow limitation or show that this mechanism is sufficient alone to explain maximal expiratory flow.

Selective Merging

In 1971, C.A. Jacobs proposed an alternative hypothesis to "dynamic airway compression" (15), saying that the mechanisms primarily responsible for maximal expiratory flow occurred at airway junctions. One mechanism was flow impedance due to linear and angular acceleration of gas flow at airway junctions, as total conducting cross-sectional area and gas flow direction change, moving from the alveoli to the trachea. The second and major effect Jacobs called "selective merging." This occurs at an airway junction when flow is blocked from one upstream "daughter tube" into the downstream "mother tube." Flow may even be forced backward up one daughter tube, creating pendelluft. Selection of stream direction and source has been demonstrated in the study of fluidic circuit models (16-18). Jacobs described a number of possible causes for this selection in vivo, but primarily unequal gas pressure from adjacent daughter tubes at the bifurcating junctions due to different upstream pressures. This mechanism is illustrated in Figure 1c. His hypothesis was that the extent of selective merging increased in response to increasing pleural pressure. As the extent of selective merging increases, the effective cross-sectional conducting area for unidirectional gas flow is decreased. This, in turn, increases frictional resistance and changes inertial effects in the lung air passageways.

Jacobs developed a computerized mathematical model of maximal expiratory flow based on his hypothesis, with plausible results. He also cited physiological evidence for sequential filling and emptying of gas during inspiration and slow expiration and for modification of the expiratory pattern during maximal expiratory flow as evidence of selective merging switching effects. It is interesting to note that Jacobs' model predicted that over 93% of expiratory resis-

a) ILLUSTRATION OF DYNAMIC AIRWAY COMPRESSION

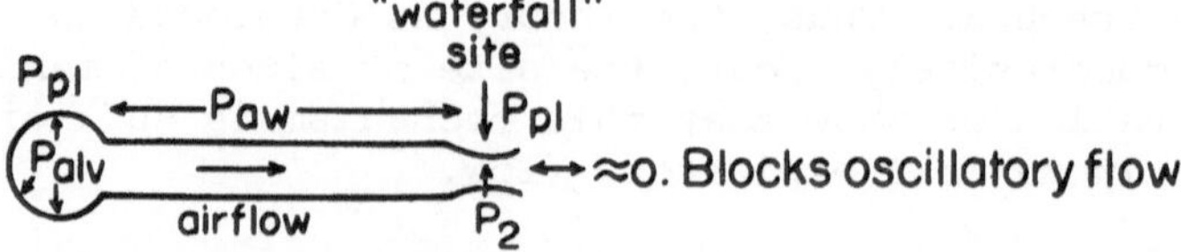

1. P_{pl} = pleural pressure
2. P_{alv} = alveolar pressure
3. $P_{el} = P_{alv} - P_{pl}$ = elastic pressure
4. P_2 = lateral intraluminal pressure
5. $P_{tm} = P_2 - P_{pl}$ = transmural pressure

• *maximum volume flow rate reached above some critical P_{tm}, for a given lung volume.*

b) ILLUSTRATION OF FLOW LIMITATION AT WAVE SPEED

minimum wave speed site

P_{alv} P_1 $P_E \leftrightarrow \approx o$. Blocks oscillatory flow

P_2

$P_1 = P_{critical} > P_2$ at wave speed

$P_E = P_{exit}$

P_1, P_2 = local intraluminal pressures

• *maximum unidirectional gas particle velocity and fixed cross-sectional area at flow limiting site for a given lung volume.*

c) ILLUSTRATION OF SELECTIVE MERGING

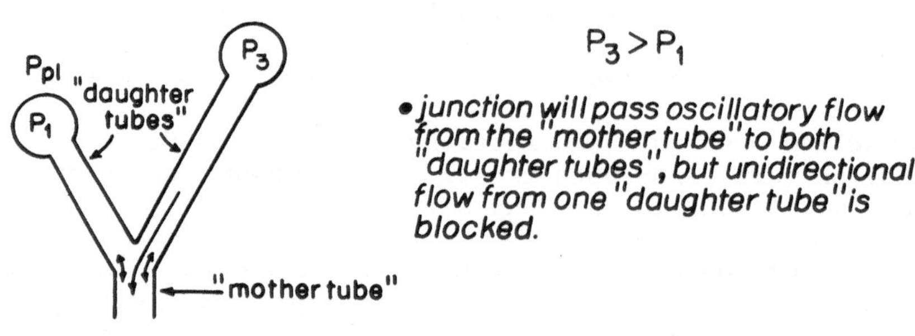

P3
Ppl
"daughter tubes"
P1
"mother tube"
P3 > P1
• junction will pass oscillatory flow from the "mother tube" to both "daughter tubes", but unidirectional flow from one "daughter tube" is blocked.

Figure 1. Three Expiratory Gas Flow Limiting Mechanisms

tance to unidirectional flow during maximal expiratory flow occurs between branch level 6 in the lung air passageways and the mouth for normal subjects. It is also interesting to note that, if complete selective merging were postulated from the trachea to branch level 6, the velocity of the exhaled gas at branch level 6 during a maximal exhalation could reach mach 1, the wave speed of sound in air (19, p. 179). These observations suggest that a combination of dynamic airway compression, selective merging, and flow limitation at wave speed could be postulated as the primary causes of maximum expiratory flow, working in concert.

While Jacobs demonstrated the plausibility of selective merging in his computerized mathematical model and cited experimental physiological evidence from the literature to support it, he did not perform any new in vivo experiments to test the hypothesis. Using his mathematical model, Jacobs calculated that selective merging caused 55% of the increase in expiratory air flow resistance during a maximal exhalation. Thus, he did not exclude the possibility of other concurrent expiratory gas flow limiting mechanisms, although he believed that selective merging was the primary mechanism.

In Vivo Test for Selective Merging

A rationale was needed for experiments to distinguish between selective merging and the other two gas flow limiting mechanisms (i.e., dynamic airway compression and flow limitation at wave speed). It was noted that dynamic airway compression and flow limitation at wave speed both shared in common a waterfall effect in which downstream pressure variations (e.g., at a subject's mouth) would not be transmitted upstream past the waterfall. There was no waterfall for selective merging, but instead a blocking or slowing of flow from one converging gas stream by a higher pressure and velocity flow of the other converging gas stream at a lung passageway bifurcating junction. The waterfall effect of the first two mechanisms would predict high resistance to relatively low amplitude downstream pressure variations, but selective merging would not.

Using this rationale, a hypothesis was formulated that selective merging was not generally present throughout the lungs during a maximal exhalation for normal subjects. To test this hypothesis experimentally, it was decided to apply a small amplitude oscillatory pressure variation at the mouths of normal human subjects during both maximal and slow exhalations. It would be expected that resistance to unidirectional flow during a maximal exhalation would be greatly increased over resistance to unidirectional flow during a slow exhalation for all three flow limiting mechanisms. If selective merging were generally present throughout a maximal exhalation, it would be expected that resistance to superimposed sinusoidal flow would not increase significantly in comparison to the increase in resistance to unidirectional flow.

Henceforth, "unidirectional expiratory gas flow" will be designated $\dot{V}_{LE}$: "superimposed sinusoidal gas flow" will be designated $\dot{V}_{L\sim}$.

METHODS

Instrument Arrangement

Figure 2 shows a block diagram of the basic instruments used for experimentation and their arrangement. A special dynamic pulmonary impedance unit was needed to perform this experiment. The unit was designed and built to ensure low gas flow impedance, good impedance matching and laminar gas flow through the expiratory pneumotachograph. The expiratory pneumotachograph was a No. 4 Fleisch.

The new system electronically compensates the transducer outputs from both the pneumotachograph and esophageal balloon system for frequency vs. magnitude and phase up to about 30 Hz (19).

Experimental Procedure

Eight normal human subjects were tested using the experimental procedure that follows. Reference to Figure 2, a block diagram of the testing apparatus, will be useful.

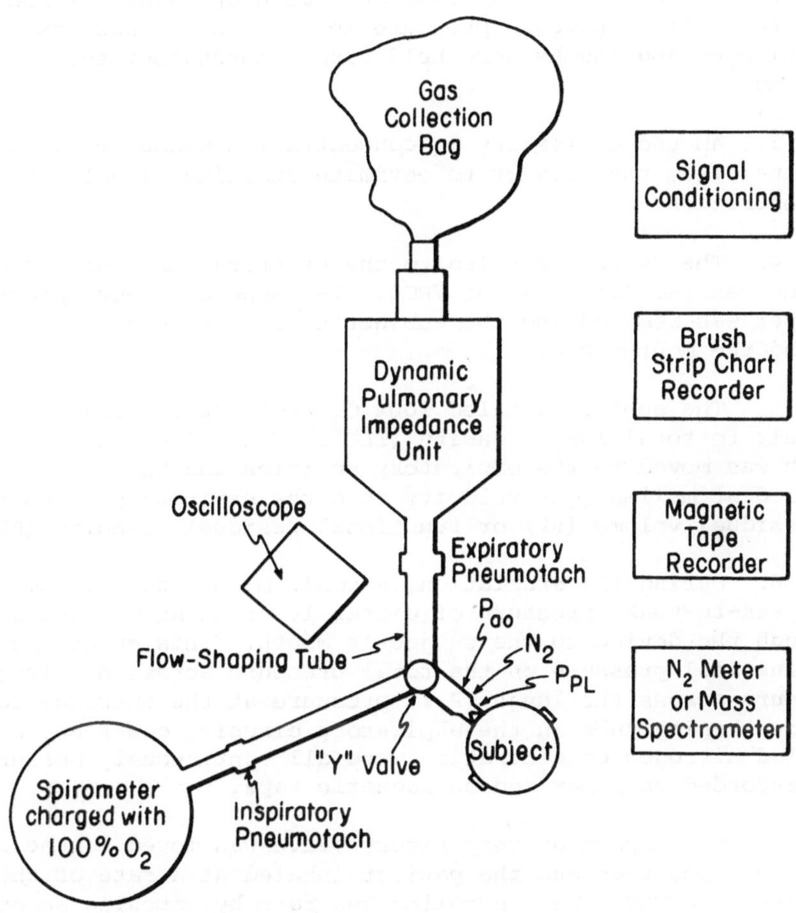

Figure 2. Block Diagram of Experiment

1. This is an optional step. Before starting an experiment, all air was flushed from the dynamic pulmonary impedance unit and replaced with 100% O_2 from the spirometer.

2. The subject first swallowed an esophageal balloon, used to estimate pleural pressure in the lungs. His nose was clamped and cheeks were held firmly throughout the testing.

3. An end-expiratory N_2 concentration measurement was obtained from the subject to estimate starting alveolar N_2 concentration.

4. The subject exhaled to the starting lung reference volume desired (e.g., RV or FRC). The apparatus mouthpiece stopper was removed and the subject's mouth placed firmly around the mouthpiece.

5. The subject inhaled 100% O_2 from the inspiratory circuit to total lung capacity (TLC). The valve near his mouth was moved to the expiratory position and the subject exhaled at maximal gas velocity into the expiratory circuit to residual volume (RV) or functional residual capacity (FRC).

6. During the exhalation, a small magnitude (3.75cm H_2O peak-to-peak) pressure of either 10 or 12 Hz was applied through the device to the subject's mouth. This superimposed a sinusoidal pressure on the total pressure across his lungs. Pressure across the lungs (P_L), pressure at the entrance to a flow-shaping tube in the expiratory circuit, gas flow, and exhaled nitrogen concentration were all continuously measured and recorded on paper and on magnetic tape.

7. The apparatus respiratory valve was moved to the inspiratory position and the patient inhaled at a rate of about .5 ℓ/sec. to TLC. He controlled the rate by watching an oscilloscope trace. Inspiratory flow rate was recorded, but oscillatory pressure was not applied at the mouth during inspiration.

8. The valve was next moved to the expiratory position and the subject exhaled at a slow rate of about .5 ℓ/sec. Again, oscillatory pressure was applied at the subject's mouth during the exhalation.

9. Alternate maximal and slow exhalations were continued in this manner, usually for a total of 6-8 breaths. This constituted one experimental "run."

10. 4-8 "runs" were done per subject (steps 3-9). An average starting lung volume was calculated using average starting and ending N_2 lung concentrations for the subject, end-experiment N_2 concentration in the expiratory circuit and total gas volume exhaled.

Data Collection and Processing

The testing transducer electronic signals corresponding to four measured variables were signal conditioned continuously during the experiments before recording on analog magnetic tape. The analog signals were converted to digital at the rate of 100 pps and edited on a PDP 11-45 digital computer. These data were then stored on a disk for further processing on an IBM 360/44 digital computer. Expiratory single flow data were differentiated to obtain the rate of change of flow and integrated to obtain exhaled gas volume as a function of time. For single exhalation analysis, digital bandpass filters, centered at 10 or 12 Hz, were used to separate superimposed sinusoidal pressure-flow data from the underlying unidirectional pressure-flow data.

Multiple linear regression was applied to the measured and derived data variables from maximal and slow exhalations using the IBM 360/44 digital computer. REGRE, the multiple linear regression computer program from version 3 of the IBM Scientific Subroutine Package was used; it was modified somewhat for this research. The basic equation for the regression analysis is shown below:

$$\hat{P}_{pl} - \hat{P}_{ao} = \hat{P}_L = (1/\hat{C}_L)V_L + \hat{R}_L\dot{V}_L + \hat{I}_L\ddot{V}_L + \hat{a}_o \quad (1)$$

where

$\hat{P}_{pl} - \hat{P}_{ao} = \hat{P}_L$ = the pressure across the lungs, pharynx and mouth, calculated.

V_L = exhaled gas volume

$\dot{V}_L$ = gas volume flow rate

$\ddot{V}_L$ = derivative of $\dot{V}_L$

$\hat{R}_L$ = estimated lung resistance

$\hat{C}_L$ = (lung tissue + gas) compliance, estimated

$\hat{I}_L$ = (lung tissue + gas) inertance, estimated

$\hat{a}_o$ = the intercept

$\hat{R}_L$, $(\hat{1}/C_L)$, and $\hat{I}_L$ were adjusted to minimize $\sum_j^M (P_{L\ \text{observed}j} - \hat{P}_{Lj})^2$. Multiple linear regression was applied sequentially over successive tenth volumes of exhaled breath for each of 3 subjects to identify $\hat{R}_L$, $(\hat{1}/C_L)$ and $\hat{I}_L$. This process was applied to both unidirectional pressure-flow data and superimposed sinusoidal pressure-flow data sequentially over successive tenth volumes of exhaled breath.

A small correction was applied to the single breath data for gas compressibility in the expiratory flow-shaping tube. This flow correction = $C_g\ (dP_{tube}/dt)$. $C_g \simeq .0006$ ℓ/cm H_2O.

Lung pressure and expiratory flow data were ensemble-averaged using 3-6 slow and 3-6 fast exhalations per subject. The ensemble-averaged data were smoothed with 20 Hz low pass digital filtering. Multiple linear regression was then applied to the smoothed data in the same manner used for single exhalations.

The process of parameter identification was applied to the data of 3 of the 8 subjects.

RESULTS

Typical strip chart recordings of P_L and corresponding expiratory flow are shown in Figure 3 for maximal and slow single exhalations. This shows that sinusoidal flow was superimposed on maximal as well as slow exhalations. This was true for every subject tested. For this subject, sinusoidal flow was reduced in magnitude during the maximal exhalation compared to a slow exhalation, even though superimposed sinusoidal pressure was the same for the two breaths. For other subjects, superimposed sinusoidal flow changed very little with change in underlying expiratory flow rate at corresponding lung volumes.

(a) SLOW EXHALATION

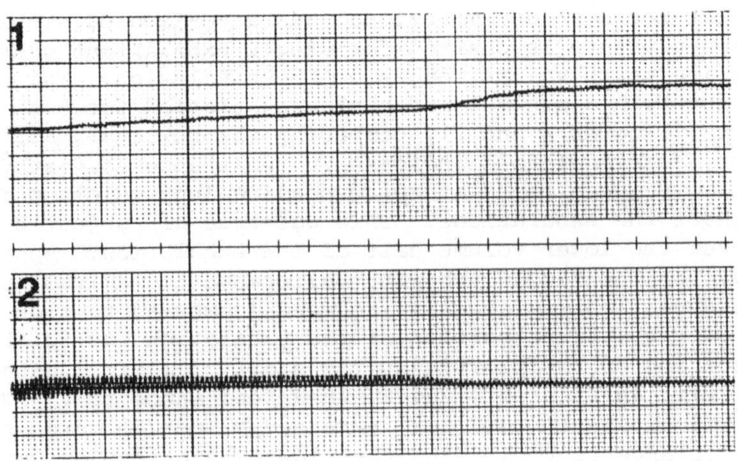

(b) MAXIMAL EXHALATION

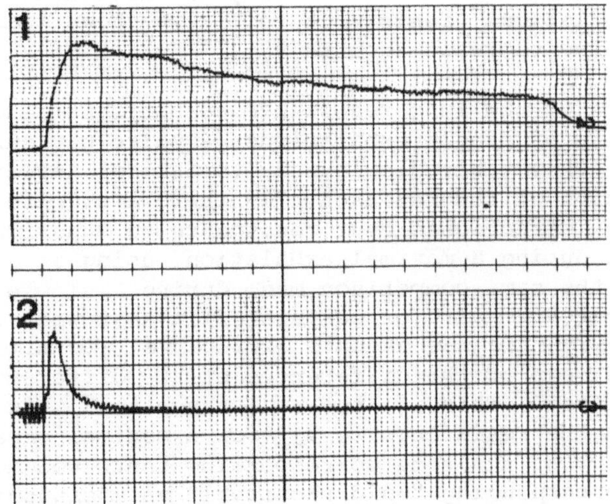

Figure 3. Typical Strip Chart Recordings for Single Exhalations
1. Pressure Across Lungs.
2. Expiratory Gas Flow.

Figures 4 and 5 show some results of single breath analysis of the lung resistance parameter. The resistance parameter for both unidirectional and superimposed sinusoidal flow-pressure is charted for single exhalations over successive tenth volume ranges of the exhalation. This is shown for both a maximal and a slow exhalation for both subjects (Figures 4 and 5), comparing the results. For both figures, a several-fold increase in resistance to fast $\dot{V}_{LE}$ over slow $\dot{V}_{LE}$ was found at corresponding lung volume points throughout the exhalations. This was true of subject 2 only over the lower volume half of the exhalation. Apparently, for the second subject, resistance to maximal unidirectional flow was actually reduced (compared to slow flow) over the first portion of the exhalation due to the wide opening of the glottis at the start of maximal exhalation (20). In Figures 4 and 5, lung resistance to $\dot{V}_{L\sim}$ during a maximal exhalation did not increase as much as corresponding resistance to $\dot{V}_{LE}$, using the slow exhalations as controls. Note, also, that in both figures the separation between resistance to $\dot{V}_{LE}$ and resistance to $\dot{V}_{L\sim}$ during the maximal exhalation starts even before the effort-independent portion of the exhalation is reached. In contrast, throughout most of the slow exhalations, resistance to $\dot{V}_{LE}$ is shown nearly identical in value to resistance to $\dot{V}_{L\sim}$.

Table 1 makes the lung resistance comparisons described above for 6 pairs of single exhalations (2 pairs per subject) statistically. This table displays a statistical answer to a central question of the research, "Is there a significant resistance to unidirectional over resistance to superimposed sinusoidal flow during a maximal exhalation, using as a control reference the same comparison made during a slow exhalation?" The matched-pair (or correlated) t-test was used to analyze the significance of this difference. To make this comparison, the mean of the differences of the differences is first calculated. That is, the resistance to $\dot{V}_{L\sim}$ is first subtracted from the resistance to $\dot{V}_{LE}$ for a maximal exhalation of a subject at the same corresponding tenth volume ranges. The corresponding "unidirectional" vs. "sinusoidal" resistance differences for the slow exhalation are used as a control and are, in turn, subtracted from resistance differences for a maximal exhalation at corresponding tenth volume ranges for the same subject. The subtraction process to obtain the differences in parameter values was started either at the volume tenth where peak maximal flow occurred

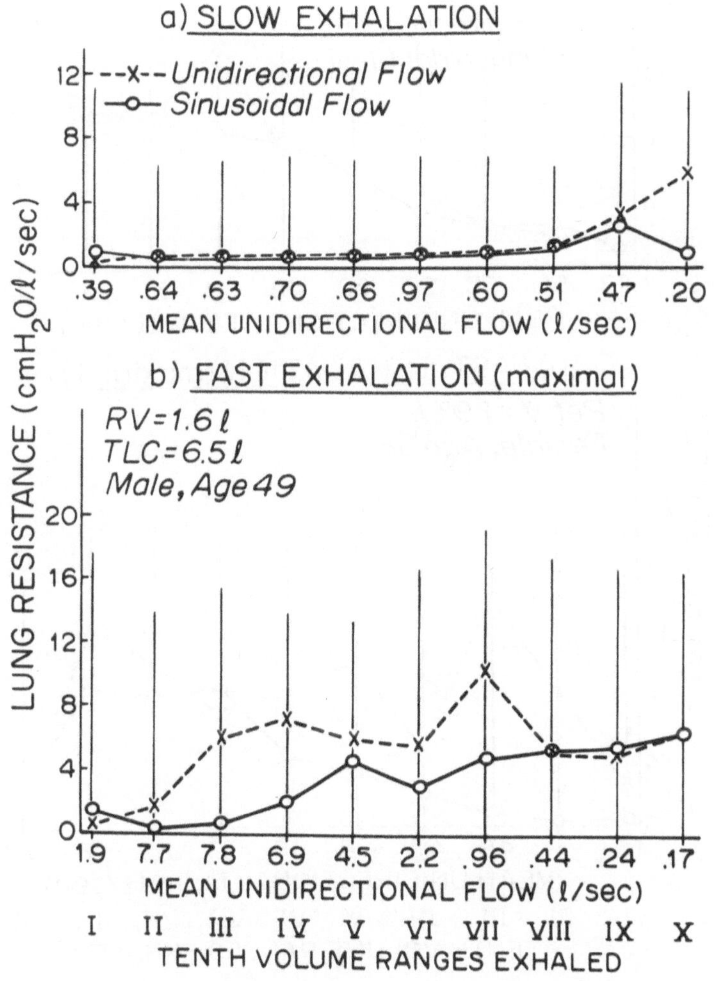

Figure 4. Subject 1 Dynamic Lung Resistance versus Gas Volume Exhaled and Mean Unidirectional Flow

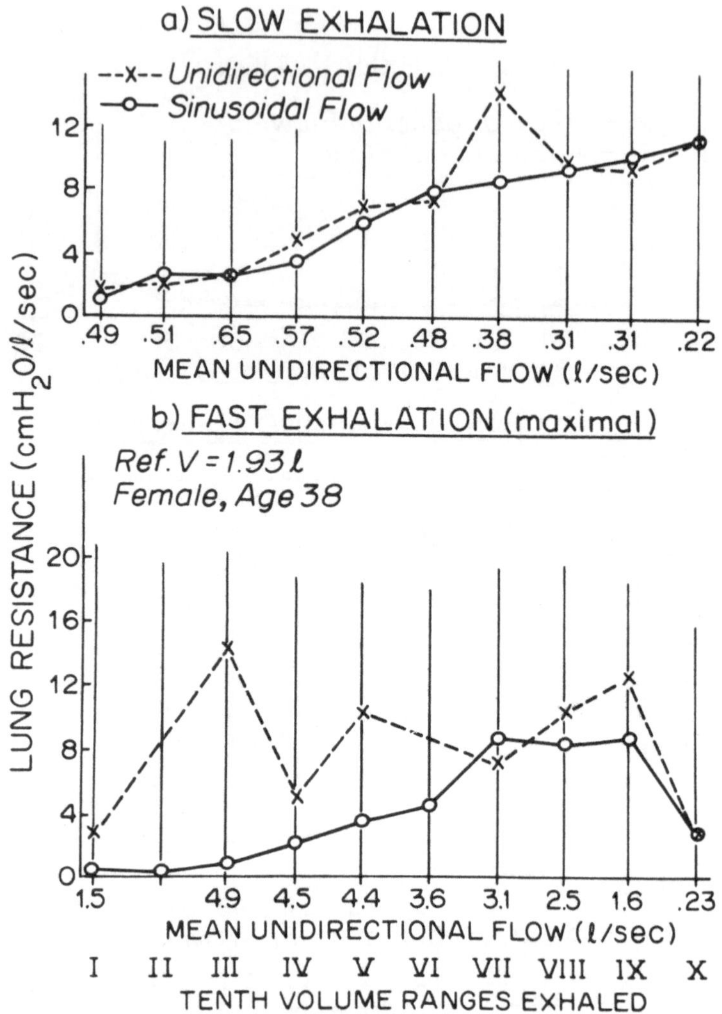

Figure 5. Subject 3 Dynamic Lung Resistance versus Gas Volume Exhaled and Mean Unidirectional Flow

Table 1

Significance of the Mean of Differences in Lung Resistance†

Subject No.	Expirations Compared	P Value of R_L Mean Difference	Power of Results at α=0.1	Number of Data Points	Value of t (Test Statistic)	H_0
1	S1R1E3 vs S1R1E2	0.05	0.667	9	1.856	rejected
1	S1R1E5 vs S1R1E4	<0.0035	>0.99	6	4.900	rejected
1	S1R1E3,S1R1E5 vs S1R1E2,S1R1E4	0.0017	0.99	15	3.825	rejected
2	S2R1E3 vs S2R1E4	0.266		8	0.698	accepted
2	S2R2E6 vs S2R2E1	0.083	0.6	5	1.739	rejected
2	S2R1E3,S2R2E6 vs S2R1E4,S2R2E1	0.062	0.685	13	1.676	rejected
3	S3R3E5 vs S3R3E2	<0.006	0.967	6	3.860	rejected
3	S3R5E5 vs S3R5E4	<0.068	>0.6	8	1.716	rejected
3	S3R3E5,S3R5E5 vs S3R3E2,S3R5E4	0.005	0.937	14	3.009	rejected
All	6 "fast" vs 6 "slow"	<0.0005	>0.9995	42	4.862	rejected

† $\{(\text{Resistance to Unidirectional Flow}) - (\text{Resistance to Sinusoidal Flow})\}_{\text{maximal exhalation}} - \{\text{Corresponding Difference}\}_{\text{slow exhalation}}$

or earlier if resistance to $\dot{V}_{LE}$ increased over resistance to $\dot{V}_{L\sim}$. The rest of the data points were used for comparison except where 1) parameter values were omitted because of low t values or low multiple correlation coefficient or 2) obvious relaxation by a subject from a maximal exhalation occurred.

The mean of the differences was calculated, between the exhalation limits just described, and analyzed for significance using a standard level of significance, $\alpha=0.1$. Examining Table 1, we conclude that, for 5 out of 6 of the maximal exhalations H_0, the null hypothesis is rejected and a significant increase in resistance to $\dot{V}_{LE}$ over resistance to $\dot{V}_{L\sim}$ occurs during a maximal exhalation. Further, when all the resistance data points for a single subject are used, H_0 is rejected for all 3 subjects. When all the data points for all 3 subjects are used, the P value is over 2 orders of magnitude less than 0.1, the level of significance.

"Sinusoidal" inertance and compliance values were considered separately inaccurate because of the problem of "multicollinearity" (21). $\ddot{V}_{L\sim}$ and $V_{L\sim}$ associated with the inertance and compliance parameters were highly correlated for a single frequency. However, the estimation process for these parameters was done because it resulted in a more accurate estimation of lung resistance.

Figure 6 compares lung compliance for $\dot{V}_{LE}$ estimated for 6 ensemble-averaged maximal and 6 ensemble-averaged slow exhalations for subject 1. The same comparison is made for subject 3 using compliance values averaged from 2 exhalations (2 maximal and 2 slow). A very significant decrease in measured lung compliance throughout the maximal exhalations compared to lung compliance throughout the slow exhalations is noted. This result was also found for subject 2. The pattern of dynamic change of lung compliance throughout exhalation varied somewhat from subject to subject, however.

Values for lung inertance during V_{LE} were obtained for the 3 subjects. The lung inertance values exhibited erratic change during many of the maximal exhalations. Inertance values shown in Table 2 for the 3 subjects compare well with literature values for slow flow at FRC.

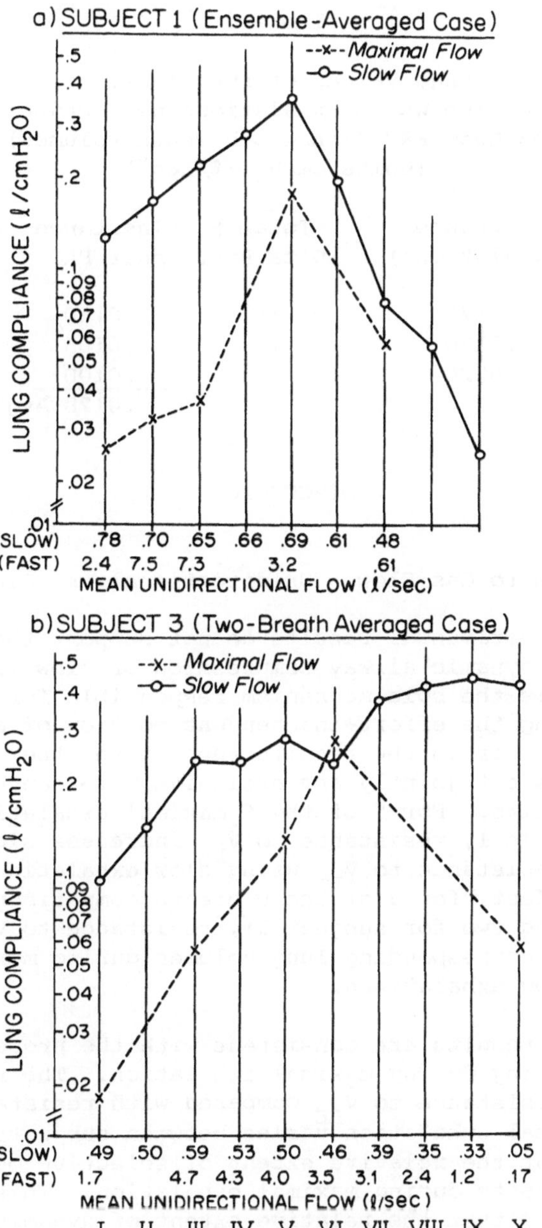

Figure 6. Dynamic Lung Compliance versus Gas Volume Exhaled and Mean Unidirectional Flow

Table 2

Average Values of Lung Inertance over Two Whole Exhalations per Subject and Near FRC (7th & 8th Tenth Volumes) (Units cm $H_2O/\ell/sec^2$)

Subj. #	Inertance (2 Exhalations)	Total # Data Pts.	Inertance near FRC	# Data Pts. at FRC
1	.0175	15	.0180	4
2	.0170	11	.0105	3
3	.0226	12	.0100	3
			.0128 Avg.	

DISCUSSION

Resistance to Gas Flow - Unidirectional vs. Sinusoidal

The experimental results do not support the thesis that either dynamic airway compression or flow limitation at wave speed is the sole mechanism responsible for upper limits to $\dot{V}_{LE}$ during the effort-independent portion of a maximal exhalation. Nor do the results support the thesis that these 2 mechanisms act jointly and exclusively to cause these upper flow limits. For 5 of the 6 maximal exhalations analyzed in Table 1, resistance to $\dot{V}_{LE}$ increases significantly more than resistance to $\dot{V}_{L\sim}$ using slow exhalations as controls. In fact, for 3 of the 6 breath comparisons (one for subject 2 and two for subject 3), resistance to $\dot{V}_{L\sim}$ is nearly the same at corresponding lung volumes during maximal and compared slow exhalations.

These results are consistent with the presence of selective merging during maximal exhalation. The relative increase of resistance to $\dot{V}_{LE}$ compared with resistance to $\dot{V}_{L\sim}$ during maximal exhalation varies between subjects. This suggests that the relative extent of selective merging varies between subjects during maximal exhalation. This, in turn, also suggests that the relative extent of dynamic airway compression and/or flow limitation at wave speed also varies between subjects.

The increase in resistance to maximal $\dot{V}_{LE}$ over resistance to $V_{L\sim}$ is noted in both Figures 4 and 5 to start even before peak expiratory flow is reached. Thus, it seems reasonable to infer that selective merging begins at the very start of a maximal exhalation even before the effort-independent region is reached. Pride, Permutt, et al. suggested that the waterfall effect due to dynamic airway compression began at sequential times and different locations in the lungs until it was generally present across the total cross-section of expiratory flow path (6, p.656). For all the single exhalations analyzed, resistance to $\dot{V}_{L\sim}$ did not increase noticeably during maximal exhalation before peak $\dot{V}_{LE}$ was reached. This implies that flow limitation at wave speed is not significantly present before peak maximal flow is reached and that the waterfall effect associated with dynamic airway compression is also absent or minimal during that time. This is consistent with all 3 theories.

Lung Compliance Decrease During Maximal Exhalation

The precipitous drop in lung compliance "seen" by $\dot{V}_{LE}$ during a maximal exhalation is illustrated in Figure 6 and was found for subject 2 as well. Lung compliance was lower at the start of exhalation at TLC for both maximal and slow exhalations. For a slow exhalation this is probably due to loss of compliance of the lung tissue when extended. However, the lung tissue is extended to the same degree at TLC at the start of a maximal exhalation and measured lung compliance is lower then than at the start of a slow exhalation. Something else must be contributing to this compliance loss during maximal exhalation.

Any of the 3 major mechanisms of gas flow limitation discussed in this paper could decrease lung compliance "seen" by unidirectional flow during a maximal exhalation. The mechanisms would cause $\dot{V}_{LE}$ increased resistance (and possibly inertance as well) which, in turn, would decrease change in volume for a given pressure change. Also, increased gas turbulence with increasing V_{LE} would increase flow resistance.

Since it seems unlikely that flow limitation at wave speed or dynamic airway compression is significantly present before peak maximal expiratory flow is reached, it seems possible that selective merging is partly responsible for the extra drop in compliance. An increase in P_L gradient and change in pressure distribution patterns through the alveoli and airway tissues might change effective compliance of the tissues without an increase in lung resistance. We recall also, however, that selective merging is postulated to increase in extent with increase in lateral P_L gradient.

After the onset of peak expiratory flow, it is likely that all 3 major mechanisms of flow limitation are present in the lung air passageways in varying degrees. It is also probable that the relative extent of these 3 mechanisms varies both with lung volume (% of TLC) and between subjects. It is noted, for instance, that for subject 1 "dynamic compliance" for both slow and maximal $\dot{V}_{LE}$ reaches a peak at 50% of exhaled volume and then falls rapidly for both flows thereafter. For subject 3, peak compliance for maximal $\dot{V}_{LE}$ is reached at 60% of exhaled gas lung volume. For this subject, during slow exhalation, compliance continues to rise gradually and retains high magnitude to the last 90% of exhaled volume. Compliance during maximal exhalation, however, falls markedly at the 90% level of exhaled volume. These differences between subjects in dynamic patterns of lung compliance, during maximal and slow flow, suggest differences in physical states of the lungs between subjects as well as differences in relative dominance of flow limiting effects. For instance, for subject 1, the compliance drop for slow flow in the last half volume of exhalation suggests a direct measurement of the mechanical state of lung tissue. Lung resistance to both slow $\dot{V}_{LE}$ and $\dot{V}_{L\sim}$ remains low and nearly identical until about the 80% level of exhaled volume for single exhalations for subject 1. Therefore, it appears that measured compliance is not reduced by any of the 3 major flow limiting mechanisms throughout slow exhalations up to the 80% level of exhaled volume. This statement could also be made for subjects 2 and 3 for slow exhalations. These observations suggest that the methods of this research applied to slow exhalations might be substituted for the standard static compliance test, eliminating shutter use.

If lung compliance measured throughout slow exhalations is directly related to the physical state of lung tissue, what additional information is gained by measuring effective lung compliance during a maximal exhalation? One answer is that the slow flow test measures only a kind of spatial average of lung compliance at a given lung volume. The relative presence of the 3 flow-limiting mechanisms in the lungs during maximal exhalation might be used to suggest local variations in lung compliance (and resistance). When compared with the results from slow exhalations, flow-limiting sites at specific lung volumes might be estimated.

Parallel Compliance of Throat-Mouth-Nose Cavities

Compensation was not made in the methods for the parallel compliance of the throat, mouth and nose cavities or the series resistance and inertance of the throat and mouth. The series resistance and inertance and the parallel compliance are common to both $\dot{V}_{LE}$ and $\dot{V}_{L\sim}$. However, the uncompensated parallel compliance of the throat, mouth and nose cavities introduces the possibility of frequency dependence of measured resistance.

Michaelson, et al. measured the respiratory impedance of the throat-mouth-nasal cavities at FRC during Valsalva using forced random noise (22). They measured a parallel compliance of .0008 ℓ/cm H_2O. Their measurements were made with a closed glottis, resulting in a higher compliance than would be found for our experiments. The glottis was at least partially open during all exhalations, reducing its compliance effect.

Our calculations showed that, even using Michaelson's parallel compliance value, the frequency effects were not sufficient to account for the results of our experiments (19, p.172). Neglecting this calculation would not significantly affect the results.

It should be possible to compensate for the frequency effect of the parallel compliance of throat-mouth-nose cavities in the same way that compensation was made for the compressibility of the gas between the mouth and expiratory pneumotachograph. If C_2 is the compliance to be compensated, this would be done by making a gas flow correction equal to $C_2(d\,P_{ao}/dt)$ in the digital data processing program. Here P_{ao} = pressure at the subject's mouth.

Suggestions for Future Research

A much more complete picture of the dynamic behavior of lung mechanics during slow and maximal single exhalations can be constructed if values for "sinusoidal" compliance and inertance are available. If the sinusoidal data from two like exhalations (e.g., 2 slow flow exhalations) of close, but different, frequencies is combined in a matrix of separate data points, the problem of multicollinearity could be solved. This technique assumes that lung mechanical parameters for a given subject are not significantly different for the two breaths. Other methods may be possible for this calculation.

Drazen, et al. used gases of different density and viscosity with dogs as subjects (23). They applied a 4 Hz sinusoidal pressure signal to the trachea of dogs at FRC reference lung volume and varied flow rate between .25 ℓ/sec. and 1.0 ℓ/sec. They concluded that their methods allowed them to predict the predominant site of airway constriction without an airway catheter. Gases of different densities and viscosities could be used in lieu of or in addition to the 100% O_2 used in the research of this paper to make an even more accurate spatial description of lung mechanics possible.

The methods of this research should be applied to diseased lungs with "normal" lungs as a control. These methods would provide a new perspective on the relationships between lung mechanics and lung disease states. The eventual results could be much more accurate and sensitive new methods for the early detection of lung disease.

Random noise inputs could be superimposed on the $\dot{V}_{LE}$ instead of sinusoids. The apparatus could apply this noise throughout fast and slow exhalations. Michaelson used forced random noise to estimate repiratory mechanical parameters at FRC and slow inspiration (24). Lándser, et al. used random noise during 16-second intervals of quiet breathing (25).

Lastly, it may be possible to minimize the use of the esophageal balloon in measuring lung mechanical parameters by using mathematical methods of nonlinear programming (system identification) to estimate $\hat{P}_L$ as a fourth unknown (in addition to R_L, $(\hat{1/C_L})$ and $\hat{I}_L$).

These suggestions are representative but, of course, not exhaustive.

ACKNOWLEDGEMENTS

The author wishes to particularly thank three of the many persons who contributed to the successful completion of this research. Dr. George A. Bekey was both sponsor for a Special Research Fellowship and dissertation chairman. His skillful direction is much appreciated. Dr. Fred S. Grodins provided laboratory space and the use of equipment. Additionally, his perceptive instruction in all aspects of research was especially valuable. Dr. Jack D. Hackney and his staff at Rancho Los Amigos Hospital shared their extensive knowledge of pulmonary function testing methods.

REFERENCES

1. Hyatt, R.E., D.P. Schilder, and D.L. Fry: "Relationship Between Maximum Expiratory Flow and Degree of Lung Inflation." J. Appl. Physiol. 13:331-336, 1958.

2. Van de Woestijne and A. Zopletal: "The Maximum Expiratory Flow-Volume Curve: Peak Flow and Effort-Independent Portion." In Airway Dynamics, ed. Arend Bouhuys, pp. 61-72, Charles C. Thomas, 1970.

3. Einthoven, N.: "Ueber die Wirkung der Bronchialmuskeln, Noch Einer Neuen Methode Untersucht und Ueber Asthma Nervosum." Arch. Ges. Physiol. 51:367-445, 1892.

4. Fry, D.L. and R.E. Hyatt: "Pulmonary Mechanics. A Unified Analysis of the Relationship Between Pressure, Volume and Gasflow in the Lungs of Normal and Diseased Human Subjects." Am. J. Med. 29:672-689, 1960.

5. Mead, J., J.M. Turner, P.T. Macklem, and J.B. Little: "Significance of the Relationship Between Lung Recoil and Maximum Expiratory Flow." J. Appl. Physiol. 22:25-108, 1967.

6. Pride, N.B., S. Permutt, R.L. Riley, and B. Bromberger-Barnea: "Determinants of Maximal Expiratory Flow from the Lungs." J. Appl. Physiol. 23(5):646-662, 1967.

7. Elliott, E.A.: "A Model of Maximal Expiratory Flow in Lungs." M.S. Thesis, Massachusetts Institute of Technology, April, 1974.

8. Dawson, S.V. and E.A. Elliott: "Wave-Speed Limitation on Expiratory Flow - A Unifying Concept." J. Appl. Physiol.: Respirat. Environ. Exercise Physiol. 43(3):498-515, 1977.

9. Bramwell, J.C. and A.V. Hill: "The Velocity of the Pulse Wave in Man." Proc. of the Royal Society of London, Series B, 93:298-306, 1922.

10. Jones, R.T.: "Blood Flow." Ann. Rev. Fluid Mech. 7: 223-244, 1969.

11. MacDonald, D.A.: Blood Flow in Arteries, 2nd Ed. Baltimore, Md.: Williams & Williams, 1974, pp. 283-308.

12. Elliott, E.A. and S.V. Dawson: "Test of Wave-Speed Theory of Flow Limitation in Elastic Tubes." J. Appl. Physiol.: Respirat. Environ. Exercise Physiol. 43(3): 516-522, 1977.

13. Winter, D.C., R.L. Pimmel, and J.M. Fulton: "Wave Speed in the Dog Trachea." Proc. 30th ACEMB, Los Angeles, CA, Nov. 1977, p. 68.

14. Downing, W.G., Jr. and G.A. Bekey: "Lung Compliance Decrease with Expiratory Flow Rate." Fed. Proc. 37(3): 870, March, 1978.

15. Jacobs, C.A.: "A New Theory on the Mechanisms Determining Maximum Expiratory Flow Velocity from the Lungs." Ph.D. Dissertation, University of Southern California, Los Angeles, California, August, 1971.

16. Coanda, H.: "Device for Deflecting a Stream of Elastic Fluid Projected Into an Elastic Fluid." U.S. Patent No. 2,052,869, September 1936.

17. Angrist, Stanley W.: "Fluid Control Devices." Scientific American. 211:80-88, 1964.

18. Streeter, V.L.: "Fluid Amplifiers." Fluid Mechanics, Fifth Ed., McGraw-Hill Book Co., Inc., 1971, pp. 488-492.

19. Downing, W.G., Jr.: "A Unified Theory of the Causes of Maximum Expiratory Flow." Ph.D. Dissertation, University of Southern California, Los Angeles, 1977.

20. Clement, J., D.C. Stanescu, and K.P. van de Woestijne: "Glottis Opening and Effort-Dependent Part of the Isovolume Pressure-Flow Curves." J. Appl. Physiol. 34:18, 1973.

21. Frank, Charles R., Jr.: Statistics and Econometrics. New York: Holt, Rinehart and Winston, Inc., 1971.

22. Michaelson, E.D., E.D. Grossman, and W. R. Peters: "Measurement of Respiratory Impedance with Forced Random Noise.: Fed. Proc. 3(1):558, April, 1974.

23. Drazen, Jeffrey M., S.H. Loring, and R.H. Ingram, Jr.: "Localization of Airway Constriction Using Gases of Varying Density and Viscosity." J. Appl. Physiol. 41(3):396-399, 1976.

24. Michaelson, E.D., E.D. Grossman, and W.R. Peters: "Pulmonary Mechanics by Spectral Analysis of Forced Random Noise." J. Clin. Invest. 56:1210-1230, 1975.

25. Lándser, F.J. Nagels, M. Demedts, L. Billiet, and K.P. Van de Woestijne: "A New Method to Determine Frequency Characteristics of the Respiratory System." J. Appl. Physiol. 41(1):101-106, 1976.

THE PRESSURE/FLOW RELATION IN BRONCHIAL AIRWAYS ON EXPIRATION

Jay C. Hardin and James C. Yu
NASA Langley Research Center, Hampton, Va.

John L. Patterson and Waring Trible, Jr.
Medical College of Virginia, Richmond, Va.

ABSTRACT

This paper reports an experimental effort to determine the work of breathing caused by resistance to flow through the branching bronchial tree of the human lung. The earliest attempts to quantify this resistance employed the assumption of laminar Poiseuille flow through the bronchi which led to values much lower than expected from clinical tests. More recently, the observation of vortices in models of bronchial flow has resulted in the suggestion that the pressure drop relation for turbulent flow might be more appropriate even though the flow remains laminar. However, no definitive tests of this hypothesis exist in the literature. In this study, a model, which was made of glass and scaled in dimensions to represent four orders of the bronchial tree, was placed in a nominally quiescent plenum with the largest order exhausting into the atmosphere. A sensitive pressure transducer was utilized to measure the plenum pressure while a hot wire aneometer determined the velocity at the exit of the model for Reynolds numbers corresponding to those found in the human lung. From these data, the pressure/flow relation for the model was determined. The model data are then scaled to derive an improved pressure/flow relation for the human lung and to estimate the resistance of each order of bronchi.

INTRODUCTION

In recent years, there has been much interest in the determination of the resistance to flow through the human lung[1] in the hope that a better technique might be developed for the early diagnosis of lung disease. A change in this resistance is known to accompany many lung disorders, such as emphysema, and is one of the primary ways in which the lung responds to irritating stimuli.

Attempts to understand the pressure/flow relation in the branching bronchial tree of the human lung date at least to the work of Rohrer[2] in 1915. However, these early studies utilized essentially the laminar flow relation for pressure drop on the basis that the Reynolds number ($Re = Ud/\nu$ where U is the mean axial velocity, d is the diameter and ν is the kinematic viscosity) in all bronchi is less than the critical Reynolds number (Re_c = 2000) for flow in a straight circular pipe during normal breathing maneuvers in a healthy lung. Such estimates tend to underpredict the pressure drop measured in clinical tests[3].

More recently, Schroter and Sudlow[4] made flow visualization studies in large scale glass models of portions of the bronchial tree. The results of these studies indicate that, although the branching causes the flow to remain laminar up to a Reynolds number of 4500, each bronchus contains violent secondary motions consisting of axial vortices. Two vortices are formed in each bronchus during inspiration and four vortices are present during expiration. A visualization of the expiratory flow pattern obtained by the present authors is shown in Figure 1. Because of these strong secondary motions, one would no longer expect the laminar pressure drop relation to be valid.

On the basis of these observations, at least two attempts have been made to develop more accurate pressure/flow relations for the inspiratory case. Olsen, Dart and Filley[5] used an empirical relation to attempt to account for the pressure drop required for branching while Pedley, Schroter and Sudlow[6,7] calculated the dissipation from measured velocity profiles in their glass model. Both of

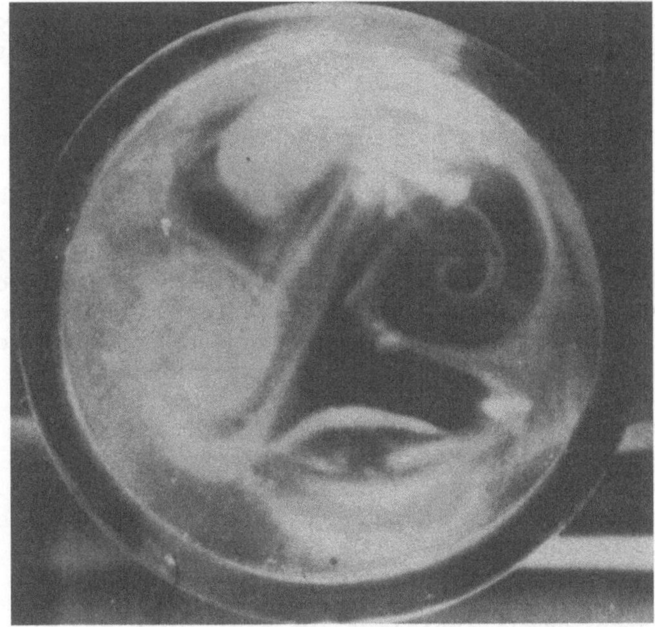

Figure 1: Expiratory Flow Pattern

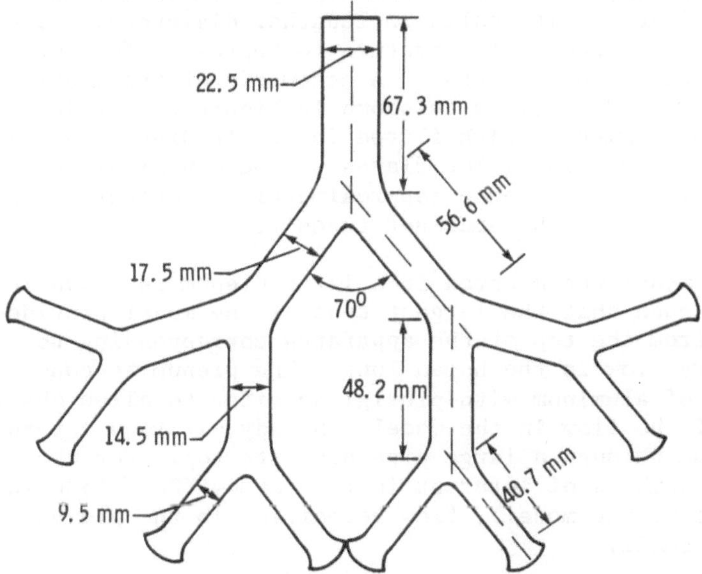

Figure 2: Schematic of the Experimental Model

these techniques produced better agreement with clinical measurements when applied to the human lung than did the laminar flow assumption.

However, as noted by Pedley[8], very little work on the pressure/flow relation during expiratory flow has been attempted. Although Schroter and Sudlow[4] did make velocity measurements corresponding to the expiratory case, they were insufficient to estimate the dissipation. Thus, Pedley was forced to suggest that the pressure drop in each bronchi be approximated by that appropriate for turbulent flow at the same Reynolds number. This hypothesis has, however, not been tested experimentally.

This paper reports an experimental investigation to determine the pressure/flow relation corresponding to expiration in a glass model of four orders of bronchi. The relation developed is then utilized to estimate the resistance to expiration in the human lung.

EXPERIMENTAL APPARATUS AND DATA ANALYSIS

The airway model which was employed in this study is made of glass and is scaled in lengths, diameters, branching angles and radii of curvature to represent four orders of the human bronchial tree. A schematic of the model with relevant dimensions is shown in Figure 2. It is planar and symmetric with flared inlets leading into the smallest tubes. The inlet flares produced negligible pressure loss on entrance (approximately 3 percent of the kinetic energy in the smallest tubes[9]).

The model was mounted in a large plenum as shown in Figure 3 such that the largest tube of the model provided an exit from the top of the apparatus corresponding to expiratory flow in the human lung. The plenum is constructed of aluminum with plexiglass sides to allow observation of the flow in the model. Steady air flow enters the plenum through a large pipe near its top. For the Reynolds numbers of interest in this study ($Re < 4500$ in all tubes of the model), flow velocities in the plenum are very small.

Figure 3: Experimental Apparatus

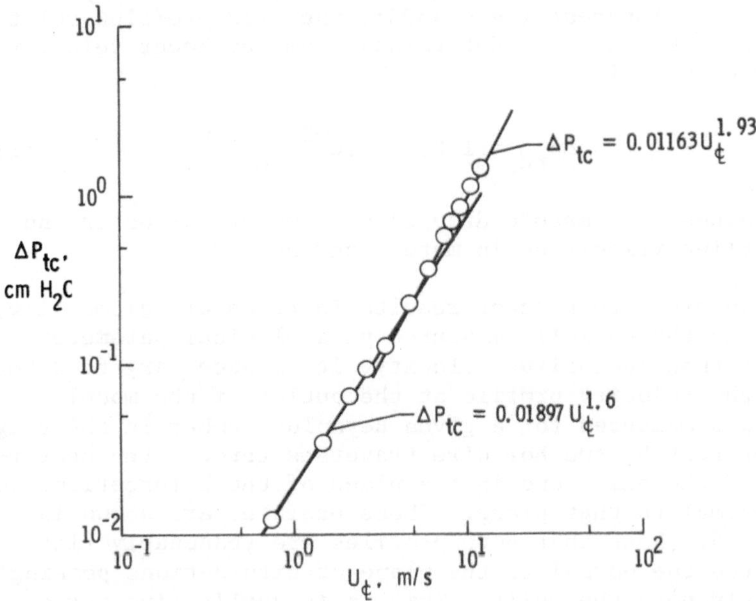

Figure 4: Pressure/Flow Relation in Glass Model

In the experiment, the difference between total pressure in the plenum and the ambient condition, $\Delta p_{tc}=p_{tc}-p_a$, where p_{tc} is the total plenum pressure and p_a is the ambient pressure in the surrounding atmosphere, was measured by a sensitive pressure transducer. In addition, the velocity at the outlet of the model was determined by a hot wire aneometer positioned at the center of the model exit approximately 1.5 mm downstream of the model exit plane. The relationship of these two variables is shown in Figure 4. Data were taken both above and below the critical Reynolds number of 4500 where Schroter and Sudlow[4] found the flow to become turbulent. Flow velocities were limited on the low end by the difficulty of measuring such small pressure drops. As can be seen, the data might have been acceptably fit by a single straight line on the log-log plot. However, the data are much better fit by two straight lines which cross in the neighborhood of $U_{c\ell}=3$m/s. As this value is near the velocity at which the flow in the largest tube exceeds the critical Reynolds number, it is the authors' contention that this change in the pressure/flow relation is real and is caused by the physical difference of the two flow regimes. Thus, for the Reynolds numbers of interest ($Re < 4500$), the pressure/flow relation will be taken as that determined from the lower velocity measurements, i.e.

$$\Delta p_{tc} = 1.897 \times 10^{-2}\, U_{c\ell}^{\;1.6} \qquad (1)$$

for values of pressure drop in centimeters of water and centerline velocities in meters per second.

In order to present results in terms of volume flow, which is the normally measured physiological parameter, rather than centerline velocity, it is necessary to determine the velocity profile at the outlet of the model. This was measured for a given Reynolds number in the range of interest by two hot wire traverses through the centerline of the exit, one in the plane of the bifurcations and one normal to that plane. These profiles are shown in Figure 5. Note that both profiles are reasonably flat with the one normal to the plane of bifurcations peaking slightly near the walls. This is in qualitative agreement with the profiles measured by Schroter and Sudlow[4]. There is a slight difference in the centerline velocity of

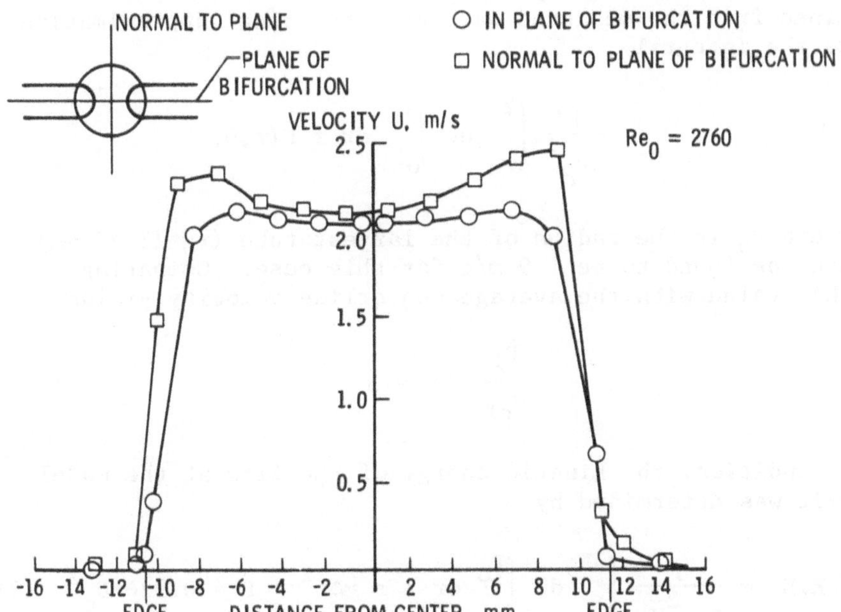

Figure 5: Velocity Profiles at Model Exit

the two profiles, for which there are two possible reasons. One is the difficulty of making hot wire measurements at such low velocities (nominally 2.1 m/s) in which case the difference might be attributed to experimental error. The second, and most likely, is the fact the hot wire was always aligned with the direction of traverse. Since a hot wire responds only to the magnitude of flow normal to its length and not its direction, it is possible that contamination by the secondary flows caused the difference. In any case, the difference is negligible for the purpose of this study.

The mean velocity, $\overline{U}_o$, at the model exit was determined from the measured data by a numerical approximation to the integral

$$\overline{U}_o = \frac{1}{\pi r_o^2} \int_o^{2\pi} d\theta \int_o^{r_o} dr\ r\ U(r,\theta)$$

where r_o is the radius of the largest tube (r_o=11.25 mm) and was found to be 1.9 m/s for this case. Comparing this value with the average centerline velocity yields

$$\frac{\overline{U}_o}{U_{c\ell}} = 0.913 \qquad (2)$$

In addition, the kinetic energy of the flow at the model exit was determined by

$$K.E. = \frac{1}{\pi r_o^2 \overline{U}_o} \int_o^{2\pi} d\theta \int_o^{r_o} dr\ r \cdot \tfrac{1}{2}\rho U^2 \cdot U = 0.526\rho\overline{U}_o^{\,2} \qquad (3)$$

which may be compared with the value $\frac{1}{2}\rho\overline{U}_o^{\,2}$ which would be valid for a uniform velocity profile. Pedley, Schroter and Sudlow[7] found such integrals of the velocity profile to be reasonably independent of Reynolds number for the inspiratory case. Similar independence for the expiratory case will be assumed in this work. With this assumption, the application of Eq. (2) in Eq. (1) yields

$$\Delta p_{tc} = 2.195 \times 10^{-2}\, \overline{U}_o^{1.6} \tag{4}$$

The total pressure drop may be further decomposed into two parts,

$$\Delta p_{tc} = \Delta p_V + \Delta p_{KE} \tag{5}$$

where Δp_{KE} is that portion of the pressure drop utilized to impart axial kinetic energy to the fluid and Δp_V is the remaining portion which has come to be called the viscous pressure drop. In actuality, this portion contains not only the energy lost as heat but also the kinetic energy of nonaxial motions such as the velocities induced by the axial vortices found in the bifurcating tubes. Nevertheless, the present authors will adhere to the standard terminology.

The kinetic energy of the exit flow is given by Eq. (3). For the model, this is a nonnegligible portion of the pressure-drop, approaching 50 percent at the higher velocities. Utilizing Eqs. (3) and (4) in Eq. (5), the viscous pressure drop is given by

$$\Delta p_V = 2.195 \times 10^{-2}\, \overline{U}_o^{\,1.6} - 6.443 \times 10^{-3}\, \overline{U}_o^{\,2} \tag{6}$$

in centimeters of water for the mean axial velocity in meters per second. Note that this pressure drop does not depend upon velocity to the first power as is true for laminar flow in a straight pipe[10] and, in fact, more nearly follows the 1.75 dependence upon axial velocity found in turbulent flow[10] in partial support of Pedley's suggestion[8]. However, the pressure drop is lower than would be found for fully turbulent flow.

In order to apply this result to the lung, it is of interest to determine the viscous pressure drop across a single tube in such a tree arrangement. The model employed in this study consisted of four orders of branches. Suppose they are labeled 0 through 3 beginning with the largest. Then, if Δp_n is the viscous pressure drop across a single tube of the nth order,

$$\Delta p_V = \sum_{n=0}^{3} \Delta p_n$$

This equation may be nondimensionalized by the dynamic pressure

$$\frac{\Delta p_V}{\frac{1}{2}\rho \overline{U}_o^{\,2}} = \sum_{n=0}^{3} \frac{\Delta p_n}{\frac{1}{2}\rho \overline{U}_o^{\,2}} \tag{7}$$

Now, if the total area of the branches of order n, A_n, is introduced, conservation of mass requires that

$$\dot{V} = A_o \overline{U}_o = A_n \overline{U}_n \tag{8}$$

where $\dot{V}$ is the volume flow through the model and $\overline{U}_n$ is the mean axial velocity in a tube of order n. Using Eq. (8) in Eq. (7) yields

$$\frac{\Delta p_V}{\frac{1}{2}\rho \overline{U}_o^{\,2}} = \sum_{n=0}^{3} \left(\frac{A_o}{A_n}\right)^2 \frac{\Delta p_n}{\frac{1}{2}\rho \overline{U}_n^{\,2}} \tag{9}$$

However, on the basis of previous work on straight tubes[9], one would expect

$$\frac{\Delta p_n}{\frac{1}{2}\rho \overline{U}_n^{\,2}} = \left(\frac{\ell_n}{r_n}\right) \lambda(\mathrm{Re}_n) \tag{10}$$

where ℓ_n and r_n are the length and radius of tubes of the nth order and λ is a function depending only upon the Reynolds number of flow in the nth order. For laminar flow[10], $\lambda = 16(\mathrm{Re})^{-1}$, while for turbulent flow, $\lambda = 0.133(\mathrm{Re})^{-0.25}$.

Now, using Eqs. (6) and (10) in Eq. (9) yields

$$C_1 \overline{U}_o^{\,-0.40} - C_2 = \sum_{n=0}^{3} \left(\frac{A_o}{A_n}\right)^2 \left(\frac{\ell_n}{r_n}\right) \lambda(\mathrm{Re}_n)$$

where $C_1 = 3.584$ and $C_2 = 1.052$ for velocity in meters per second. This relation is satisfied if $\lambda(Re_n)$ takes the form

$$\lambda(Re_n) = K_1 (Re_n)^{-0.40} - K_2 \tag{11}$$

Thus,

$$K_1 = \frac{C_1 \bar{U}_o^{-0.40}}{\sum_{n=0}^{3} \left(\frac{A_o}{A_n}\right)^2 \left(\frac{\ell_n}{r_n}\right) (Re_n)^{-0.40}} = \frac{C_1}{\sum_{n=0}^{3} \left(\frac{A_o}{A_n}\right)^2 \left(\frac{\ell_n}{r_n}\right) \left(\frac{r_o^2}{2^{n-1} r_n \nu}\right)^{-0.40}}$$

and

$$K_2 = \frac{C_2}{\sum_{n=0}^{3} \left(\frac{A_o}{A_n}\right)^2 \left(\frac{\ell_n}{r_n}\right)}$$

Note that these relations depend only upon the geometry of the model and the known pressure drop relation.

Performing these calculations for the known geometry of the model yields the nondimensional constants

$$K_1 = 3.10 \text{ and } K_2 = 0.062$$

Thus, from Eq. (10), the viscous pressure drop for a single tube of order n is given by

$$\Delta p_n = \left(\tfrac{1}{2}\rho \bar{U}_n^2\right)\left(\frac{\ell_n}{r_n}\right)\left[3.10(Re_n)^{-0.40} - 0.062\right] \tag{12}$$

and its resistance[1] is

$$R_n = \frac{\Delta p_n}{\pi r_n^2 \bar{U}_n} = \frac{\rho \bar{U}_n \ell_n}{2\pi r_n^3}\left[3.10(Re_n)^{-0.40} - 0.062\right] \tag{13}$$

Note that the resistance varies linearly with length, is approximately proportional to the square root of the axial velocity but is inversely proportional to approximately the 3.5 power of radius. This relation will be utilized in application of the model data to the human lung.

APPLICATION TO THE HUMAN LUNG

In this section, the pressure drop relation developed by the previous analysis will be utilized to estimate the pressure drop on expiration in the human lung. Although the authors recognize the complexity of flow in the lung (the facts that the bronchial tree is nonplanar and non-symmetric, the bronchi are not precisely circular and have elastic walls, the flow is nonsteady and the lung varies somewhat from person to person, etc.), the present data should provide a more realistic analysis than anything heretofore available. In addition, there do exist measurements of the expiratory pressure drop in human lungs for comparison.

As a model for the human lung, the symmetric model proposed by Weibel[11] will be utilized. This model presents average dimensions of bronchi corresponding to three quarters maximal inflation of the lung and consists of 24 orders of bronchi numbered from the trachea ($n = 0$) to the aveoli ($n = 23$). Again letting Δp_n be the viscous pressure drop in the nth order of bronchi

$$\Delta p_{TA} = \sum_{n=0}^{23} \Delta p_n \tag{14}$$

where Δp_{TA} is the total viscous pressure drop between the aveoli and the trachea. Assuming that the Reynolds number in all of these orders is in the range ($50 < Re_n < 4500$) for which Eq. (12) was developed yields

$$\Delta p_{TA} = \sum_{n=0}^{23} \left(\tfrac{1}{2}\rho \bar{U}_n^{\,2}\right)\left(\frac{\ell_n}{r_n}\right)\left[3.10(Re_n)^{-0.4} - 0.062\right] \tag{15}$$

Of course, the Reynolds numbers in the smallest bronchi are much lower than 50. However, as will be shown later, these orders make a negligible contribution to the total pressure drop. Again utilizing the relation,

$$\dot{V} = A_n \overline{U}_n$$

yields

$$\Delta p_{TA} = \frac{\rho \dot{V}^2}{2} \sum_{n=0}^{23} \frac{\ell_n}{r_n A_n^2} \left[3.10(Re_n)^{-0.4} - 0.062\right] \quad (16)$$

from which the viscous pressure drop may be calculated for a given flow rate. This relation is shown in Figure 6 and can be seen to yield values much greater than would be predicted by a Poiseuille flow model. At the higher flow rates, the Reynolds number in some of the upper orders exceeds the critical value of 4500. Thus, these bronchi will actually contain turbulent flow and Eq. (16) will slightly underestimate the total pressure drop. Also shown on the figure are measured data on human subjects by Hyatt and Wilcox[13]. In obtaining such data, it is impossible to actually measure the aveoli pressure and thus esophageal pressure measured by a balloon inserted into the esophagus is utilized. This tends to overestimate the aveoli pressure due to the viscous tissue resistance and thus to overestimate the total pressure drop. However, such a comparison does tend to validate the relation developed herein.

From Eq. (16), it can be seen that the total resistance of the lung is given by

$$R_T = \frac{\Delta p_{TA}}{\dot{V}} = \frac{\rho \dot{V}}{2} \sum_{n=0}^{23} \frac{\ell_n}{r_n A_n^2} \left[3.10(Re_n)^{-0.4} - 0.062\right] \quad (17)$$

Note that this relation may be rewritten as

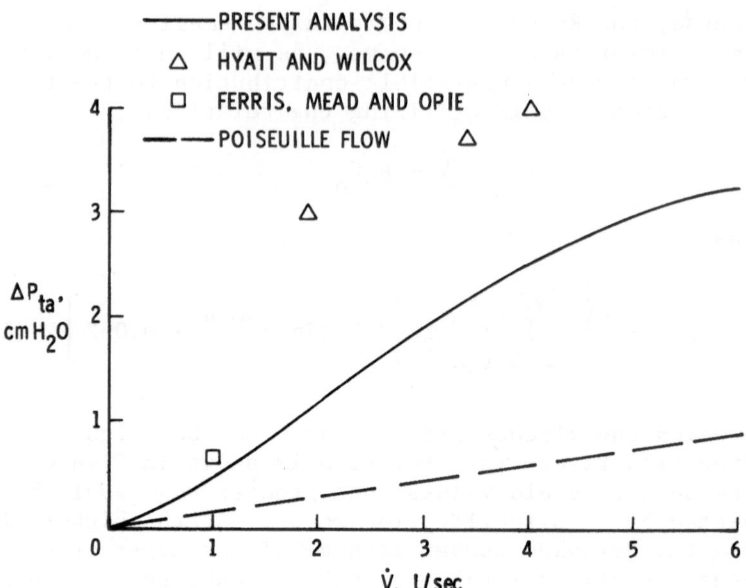

Figure 6: Viscous Pressure Drop on Expiration

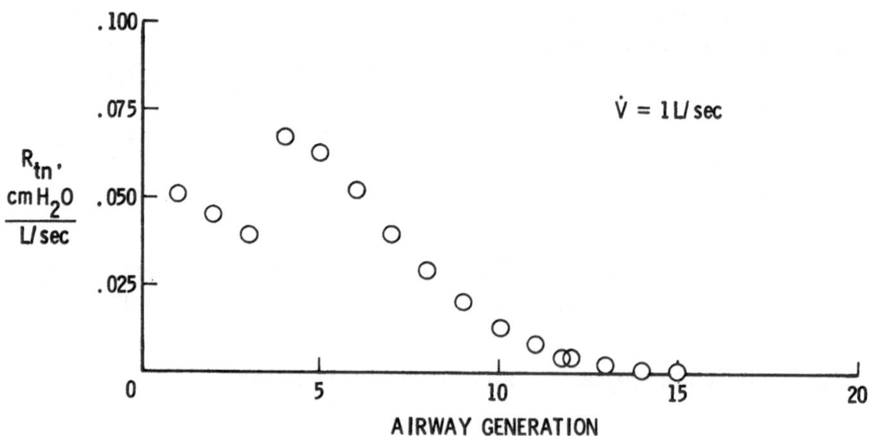

Figure 7: Total Resistance of Airway Orders

$$R_T = \sum_{n=0}^{23} \frac{\rho \overline{U}_n \ell_n}{2\pi r_n^3 (2)^n} \left[3.10(Re_n)^{-0.4} - 0.062 \right] = \sum_{n=0}^{23} \frac{R_n}{(2)^n}$$

utilizing Eq. (13). This relation shows that the total resistance to flow of the nth order, R_{Tn}, is given by

$$R_{Tn} = \frac{R_n}{2^n} \tag{18}$$

ie., the resistance of a single tube divided by their number. This is in agreement with the classical relation for resistors in parallel and is the reason why the smallest bronchi are of little importance to the overall pressure drop. The total resistance of the various orders for a volume flow rate of 1L/sec is shown in Figure 7. At this flow rate, the Reynolds number in all orders of bronchi is below critical. Curves for other volume flow values are similar. Note that the major resistance is found in the upper orders of airway with a peak occuring near the fourth order in agreement with experimental results[14]. The resistance values and the location of the peak resistance agree quite well with those obtained by Pedley, Schroter and Sudlow[3] for the case of inspiratory flow. However, their curve rises from order zero up to the peak while the present analysis yields a curve which falls before the peak even though both used the Weibel model[11] for the lung. The apparent discontinuity of the present analysis is caused by a complex interaction between the total area of the bronchial order, which reaches its minimum in the third order, and the length of the bronchi which exhibits a discontinuity similar to that shown by the resistance between the third and fourth orders.

CONCLUSION

This paper presents an analysis of experimental data on the pressure/flow relation in a large scale glass model of four orders of the human bronchial tree during expiratory flow. The data are utilized to develop an empirical expression for the pressure drop across a single branch in such an arrangement which is then applied to the human lung. Expressions for the total viscous pressure drop across the lung and viscous resistance of each order of bronchi are developed which agree reasonably well with the small amount of human data available. The pressure drop is shown to be greater than would be predicted by a laminar flow model but less than that for fully turbulent flow.

REFERENCES

1. Comroe, H. H.: Physiology of Respiration, Chap. 10, Year Book Medical Publishers Inc., Chicago, 1974.

2. Rohrer, F.: Der Strömungswiderstand in den menschlichen Atemwegen, Pfugers Arch. ges. Physiol., Vol. 162, pp. 225-259, 1915.

3. Pedley, T. J.; Schroter, R. C. and Sudlow, M. F.: The Prediction of Pressure Drop and Variation of Resistance within the Human Bronchial Airways. Respiration Physiology, Vol. 9, pp. 387-405, 1970.

4. Schroter, R. C. and Sudlow, M. F.: Flow Patterns in Models of the Human Bronchial Airways. Respiration Physiology, Vol. 7, pp. 341-355, 1969.

5. Olsen, D. E.; Dart, G. A. and Filley, G. F.: Pressure Drop and Fluid Flow Regime of Air Inspired into the Human Lung. Journal App. Physiol., Vol. 28, No. 4, pp. 482-494, 1970.

6. Pedley, T. J.; Schroter, R. C. and Sudlow, M. F.: Energy Losses and Pressure Drop in Models of Human Airways. Respiration Physiology, Vol. 9, pp. 371-386, 1970.

7. Pedley, T. J.; Schroter, R. C. and Sudlow, M. F.: Flow and Pressure Drop in Systems of Repeatedly Branching Tubes, J. Fluid Mech., Vol. 46, Pt. 2, pp. 365-383, 1971.

8. Pedley, T. J.: Pulmonary Fluid Dynamics. Annual Review of Fluid Mechanics, Vol. 9, pp. 229-274, 1977.

9. Marks, L. S.: Mechanical Engineers Handbook. McGraw-Hill Book Company, 5th Edition, p. 1624, New York, 1951.

10. Prandtl, L. and Tietjens, O. G.: Applied Hydro- and Aeromechanics. Dover Publ. Chap. 3, New York, 1934.

11. Weibel, E. R.: Morphometry of the Human Lung. Academic Press, Inc., Chap. 11, New York, 1963.

12. Hyatt, R. E. and Wilcox, R. E.: The Pressure-Flow Relationships of the Intrathoracic Airway in Man. Jour. Clin. Investigation, Vol. 42. No. 1, pp. 29-39, 1963.

13. Ferris, B. G., Jr.; Mead, J. and Opie, L. H.: Partitioning of Respiratory Flow Resistance in Man. Jour. App. Physiol., Vol. 19, pp. 653-658, 1964.

14. Macklem, P. T. and Mead, J.: Resistance of Central and Peripheral Airways Measured by Retrograde Catheter Technique. Jour. App. Physiol., Vol. 22, pp. 395-401, 1966.

MODELING THE PRESSURE-FLOW RELATION OF BIFURCATING NETWORKS

David B. Reynolds*

University of Virginia, Division of Biomedical Engineering
Charlottesville, Virginia 22908

INTRODUCTION

Despite a considerable number of measurements of the variation of pressure drop with flow rate through the bronchial tree (or physical models of it), reducing the data to a single relation is difficult. Much of the difficulty arises because of differences in geometry and experimental method. Integration of data obtained *in vivo* or in excised lungs depends strongly on controlling the experiment under isodimensional conditions. For example, during graded expirations the increasing pressure gradient with flow and the resulting decrease in bronchial caliber suggest that isovolume conditions may not be isodimensional. Consequently, experiments investigating pressure gradient in rigid models may be helpful in modeling the more complex situation *in vivo*.

Inspiratory pressure gradient was modeled by Pedley et al. (1) based on previous experiments (2) measuring velocity profiles in a two generation model of the bronchial tree. The tracheal Reynolds number investigated (275-1100) is on the lower end of the physiologically important range and some investigators have questioned the accuracy of determining energy dissipation from velocity profiles (3).

*Present address: Mayo Clinic
Thoracic Disease Research Unit
Plummer Bldg., S-3
Rochester, Minnesota 55901

However, their modeling approach allows extrapolation to higher Reynolds number and is an important aspect of their paper.

Douglass and Munson (4) report measurements of pressure drop and velocity profiles obtained with a model geometrically similar to Pedley's but for tracheal Reynolds number between 10^4 and 10^5. In this range, they found that dissipation increases more rapidly with Reynolds number than found by Pedley et al.

The purpose of this work is to investigate pressure drop for Reynolds number between 10^3 and 10^4, a range which is lacking experimental data. Expiratory as well as inspiratory flow is studied because data for the former are also meager. Our approach to modeling the pressure gradient in a branch is different from others in the literature in that it is determined from pressure drop across entire models rather than from velocity and pressure distribution in several branches. The small diameter of the branches and viscous fluids used increased pressure drop and measurement accuracy. Converting to dimensionless parameters enables application to gas flow. Due to geometrical differences between the models used and even idealized representations of the bronchial tree, the approach to modeling is probably as important as the actual results obtained. We believe it is a logical step toward a realistic model of the pressure gradient in the bronchial tree.

EXPERIMENTAL METHOD

Symmetric bifurcating networks were built using laboratory Y-connectors of constant internal diameter d. To help insure a smooth wall between adjacent connectors, each connector leg was cut with a diamond wafering blade to give a desired length ℓ between bifurcations. The connectors were force-fit together with rigid plastic or stainless steel tubing to form the airway model, shown schematically in the insert of Fig. 1.

Each connector had a parent tube from which two offspring tubes branched. An offspring in turn became the parent of two more peripheral branches. Thus, the network can be considered a sequence of generations of branching

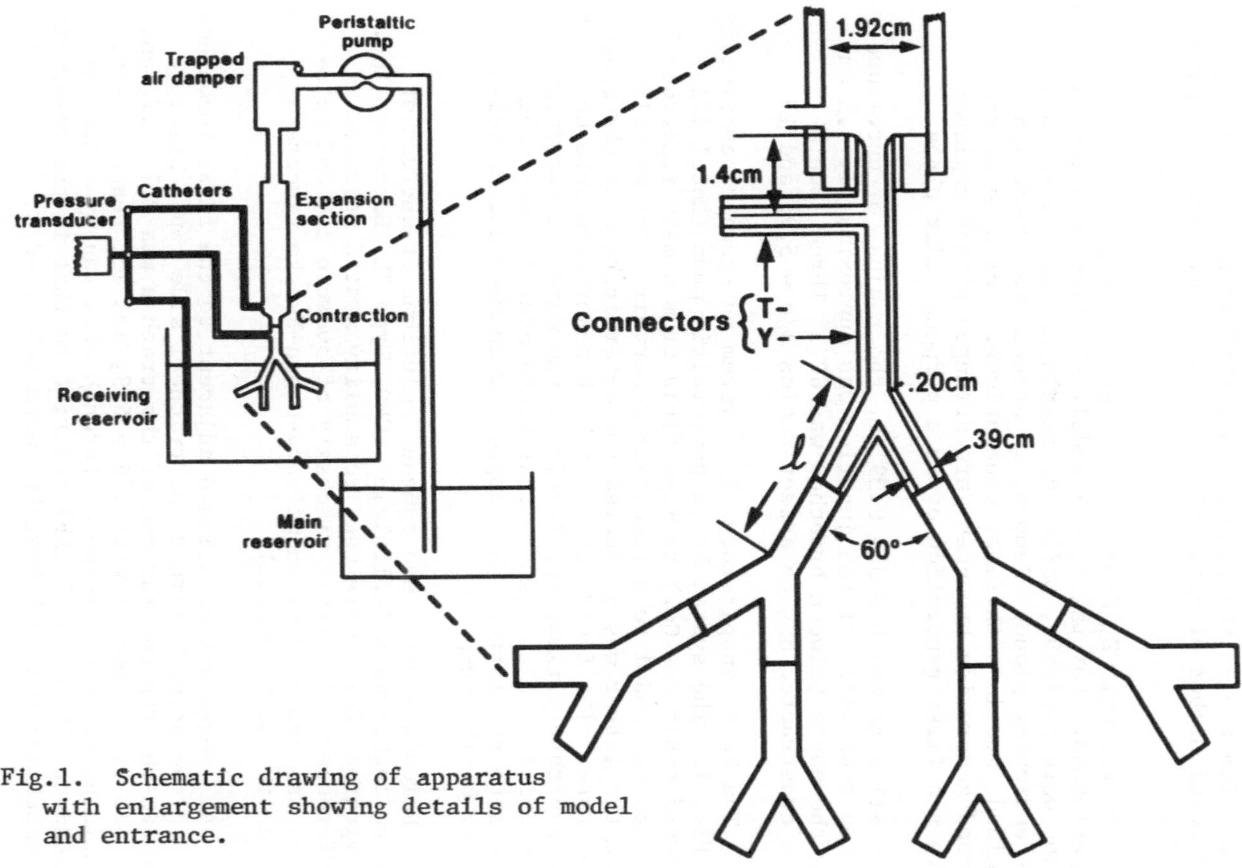

Fig.1. Schematic drawing of apparatus with enlargement showing details of model and entrance.

with order denoted by z. The "trachea" corresponds to generation z = 0, the primary "bronchi" to z = 1, and so on. The total number of generations below the trachea is denoted by Z_T.

Models with Z_T from one through five were used in experiments. For the Z_T = 5 model, the first two generations were coplanar, while the remaining three were in a perpendicular plane. Removal of successive generations yielded models with fewer generations. For Z_T = 2, the second generation branches were arranged either coplanar with the first generation or in a perpendicular plane.

All branches had d = 0.39 cm, the radius of curvature of the tube walls at the junction was approximately 0.2 cm, and the angle between branches was 60°. Three models were constructed having aspect ratios ℓ/d = 5,9, and 13.

The basic inspiratory flow system is shown schematically in Fig. 1. The system had a peristaltic pump (Model 1215, Harvard Apparatus Co.) to move fluid from a main reservoir through the model to a receiving reservoir. The overflow from the latter was collected for determination of the steady flow rate. The fluid levels of both reservoirs were constant. For expiratory flow, the levels of the main and receiving reservoirs were changed so that fluid overflow was from the former to the latter. A trapped air chamber damped pulsatility of the pump.

Deionized water and aqueous solutions of sucrose or sodium carboxymethylcellulose (CMC, Type 7L, Hercules, Inc., Wilmington, Del.) were used to amplify static pressure differences. Kinematic viscosity of sucrose and CMC solutions was measured by an Ostwald-Canon-Fenske viscometer (size 150, 0.78 mm bore).

A T-connector positioned adjacent to the first Y-connector in the network formed the tracheal side tap. Distance between the tap and the end of the trachea was 2.7, 3.7, and and 4.6 cm for models with ℓ/d = 5,9, and 13, respectively. Static pressure was measured through this tap by a Statham transducer (P23Db or P23BB) relative to that in the reservoir in which all terminal branches were submerged.

Although total pressure in the trachea was not measured, the static pressure drop across a smoothly rounded contraction to tracheal diameter was measured at several inspiratory flow rates in models with $\ell/d = 5$ and 9. Distance from the contraction entrance to the T-connector was 1.4 cm. Eliminating much of the contraction loss by rounding permitted use of the Bernoulli equation to estimate tracheal kinetic energy during inspiratory flow. In experiments with $\ell/d = 13$, a straight tube of tracheal diameter and 146 cm long replaced the expansion-contraction section shown in Fig. 1.

ANALYSIS AND MODELING

Because effects of gravity were avoided by using a single transducer to measure static pressure difference, the head loss across the model is given by the total pressure difference. The total pressure at a given location is the sum of the static pressure P and kinetic energy KE. Assuming uniform static pressure distribution across the tracheal cross-section, the inspiratory loss ΔPt_i between the trachea and reservoir (r) can be written

$$(Po + KEo) - (Pr + KEr) = \Delta Pt_i \qquad (1)$$

Because the fluid velocity in the reservoir was minimal, $KEr = 0$. Since pressure was measured relative to reservoir pressure (i.e. $\Delta Ps = Po - Pr$), for an average tracheal velocity $\bar{U}o$ and fluid density ρ, Eq. (1) becomes

$$\Delta Ps + \tfrac{1}{2}\, \beta \rho Uo^2 = \Delta Pt_i \qquad (2)$$

where β is the kinetic energy correction factor.

For expiratory flow, the drop in total pressure ΔPt_e is from the reservoir to the trachea, thus

$$\Delta Ps - \tfrac{1}{2}\, \beta \rho \bar{U}o^2 = \Delta Pt_e \qquad (3)$$

To integrate the results obtained with several fluids and make them more readily applicable to gas flow, the data were transformed into dimensionless parameters. A common method for pipe flow is to normalize frictional loss by KE. The dimensionless pressure is then plotted against Reynolds number (Re). White (5) describes this as "rather an outrageous idea, since it introduces a Reynolds number to a regime that is in fact independent of Reynolds number." Alternatively, pressure drop across the model is normalized by the Poiseuille pressure drop ΔPd_o over a length of one tracheal diameter d_o. The latter is also the Poiseuille pressure gradient in terms of a normalized length variable x/d. For generation z, this becomes

$$\Delta Pd_z = \frac{dP}{d(x/d_z)} = \frac{32\ \mu \bar{U}z}{d_z} \tag{4}$$

where μ is the fluid viscosity. The dimensionless pressure is then $\Delta P/\Delta Pd_o$ where ΔP is either static or total pressure drop. The Reynolds number in generation z is given by

$$Re_z = \frac{\rho \bar{U}z d_z}{\mu} \tag{5}$$

Plotting $\Delta P/\Delta Pd_o$ against Re_o integrates the data from a particular model on a single curve, which was fitted by linear regression:

$$\Delta P/\Delta Pd_o = A + B\ Re_o \tag{6}$$

This equation is the basis of our modeling of the pressure gradient in a branch of the network, as will now be shown.

The overall loss is considered as the sum of that in each generation of the network. The pressure drop in each generation ΔPz can be separated into a pressure drop ΔPp_z as if the flow were Poiseuille plus an excess $\Delta Pz+$ resulting from the complex flow field. The pressure gradient in the zth generation is assumed to vary with x, and from the above arguments becomes

$$\frac{dPz}{d(x/d_z)} = (1 + f(Re_z, x/d_z))\ \Delta Pd_z \tag{7}$$

in which the limiting case f = 0 is Poiseuille flow. Because the overall pressure-flow data were well fit by a dimensionless Rohrer equation, letting pressure gradient vary as $(x/d)^{\kappa}$ gives

$$f(Re_z, x/d_z) = (\gamma + \delta \cdot Re_z)(x/d_z)^{\kappa} \tag{8}$$

where the constants γ, δ, and κ are assumed to be independent of generation. The pressure drop in generation z is the integral of Eq. (7) from $x = 0$ to $x = \ell$.

For inspiratory flow, the tracheal pressure gradient of the entrance type flow is assumed to be on the same order of magnitude as that downstream of a bifurcation. Assuming also that flow divides equally at a bifucation, with constant tube diameter Eq. (4) and (5) become

$$Re_z = \frac{Re_o}{2^z} \tag{9}$$

and

$$\Delta Pd_z = \frac{\Delta Pd_o}{2^z} \tag{10}$$

Setting the equations fitting the overall loss equal to the sum of the Poiseuille and excess loss in each generation yields the modeling equations for inspiratory flow:

$$\begin{aligned} \Delta Pt_i \Big|_{\ell/d=5,9} = \Delta Pd_o [x/d_o \Big|_{3.6}^{10.5,13.1} & + (\gamma/(\kappa+1))(x/d_o)^{\kappa+1}\Big|_{3.6}^{10.5,13.1} \\ & + \sum_{z=1}^{5} 2^{-z}(\ell/d+(\gamma/(\kappa+1))(\ell/d)^{\kappa+1}) \\ & + Re_o\delta/(\kappa+1)((x/d_o)^{\kappa+1}\Big|_{3.6}^{10.5,13.1} \\ & + \sum_{z=1}^{5} 4^{-z}(\ell/d)^{\kappa+1})] \end{aligned} \tag{11}$$

The notation indicates an equation can be written for each model. Since each empirical equation for ΔPt_i has two constants, Eq. (11) yields four equations and three unknowns γ, δ, and κ. The lower limit of the x/d_o terms arises from the constant distance between contraction entrance and tracheal pressure tap while the upper limit reflects differences in tracheal length between the models.

Similarly, the following modeling equations for expiratory flow are obtained:

$$\Delta Pt_e \Big|_{\ell/d=5,9} = \Delta Pd_o [x/d_o \Big|_0^{6.9,9.5}$$

$$+(\gamma/(\kappa+1))(x/d_o)^{\kappa+1} \Big|_0^{6.9,9.5}$$

$$+\sum_{z=1}^{5} 2^{-z}(\ell/d+(\gamma/(\kappa+1))(\ell/d)^{\kappa+1})$$

$$+ Re_o\delta/(\kappa+1)((x/d_o)^{\kappa+1} \Big|_0^{6.9,9.5}$$

$$+\sum_{z=1}^{5} 4^{-z}(\ell/d)^{\kappa+1})] \tag{12}$$

Losses at the terminal branches due to sudden expansion with inspiration or contraction with expiration are negligible for five generation models but may become significant for models with fewer generations.

RESULTS

Inspiratory Flow

Dimensionless static pressure drop across the five generation models with ℓ/d = 5,9, and 13 is plotted against tracheal Reynolds number in Fig. 2. In addition to results obtained with water, results for ℓ/d = 13 using a sucrose solution (μ = 1.73 cp) and for ℓ/d = 9 with a CMC solution (μ = 3.0 cp) are also shown. Density of the solutions was not significantly different from water.

The data plotted this way show a marked increase in slope with increasing aspect ratio. Regression coefficients fitting the inspiratory data are in Table 1A. Because zero slope on this plot corresponds to a linear relation between pressure and flow, the small negative slope for ℓ/d = 5 indicates that its static pressure-flow relation is slightly concave with the flow axis. On the other hand, the positive slope for ℓ/d = 9 and 13 indicates that the static pressure-flow relation is curvilinear and convex with the flow axis.

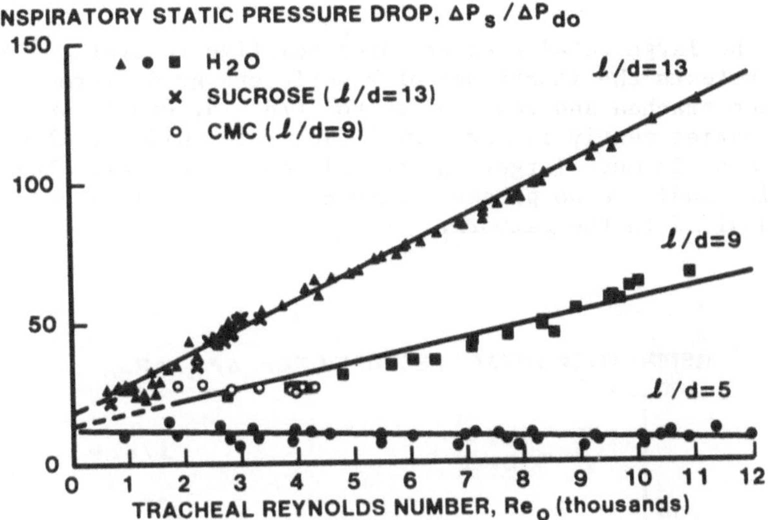

Fig. 2. Dimensionless static pressure drop as a function of tracheal Reynolds number for inspiratory flow through models with ℓ/d = 5,9, and 13.

The sucrose and CMC data appear to lie on the same curve as the water data. However, the former lie on a portion of the curve where some departure from the linear relation of Eq. (6) may be occurring. This is most evident in the ℓ/d = 13 experiment, which differed from the others in that more data were obtained for $Re_o < 2500$ and tracheal flow was probably fully developed due to the entrance length provided. The curve appears to dip starting at $Re_o = 2500$ and tends to become horizontal around $Re_o = 1300$.

Preliminary measurements of flow and static pressure drop across the contraction (6) indicated that the tracheal velocity profile was approximately uniform, as expected in the initial region of an entrance flow. Thus we can take $\beta = 1$ to convert static pressure drop to total pressure drop for experiments with ℓ/d = 5 and 9. The resulting dimensionless total pressure drop is plotted against Re_o in Fig. 3, with fitting coefficients in Table 1A. In the latter figure, note the scale change of the total pressure ordinate from that in Fig. 2.

The large total pressure drop relative to static pressure drop attests the importance of kinetic energy difference between trachea and reservoir. For $\ell/d = 5$, static pressure drop varies nearly linearly with flow rate while total pressure drop is much larger and curvilinear with flow. These results indicate no pressure recovery and thus loss of tracheal KE in the network.

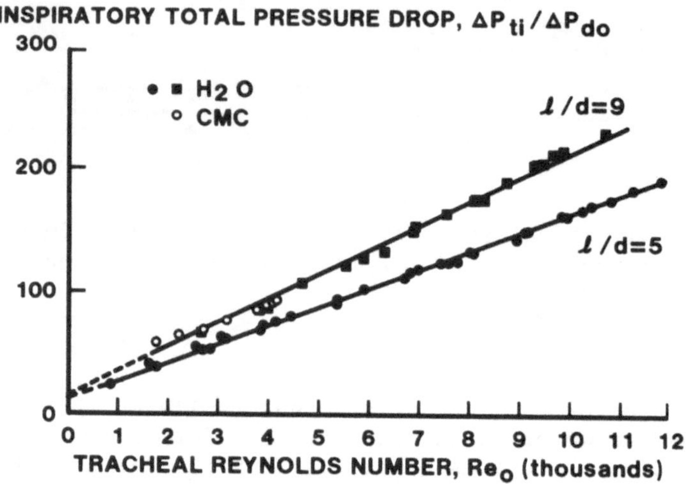

Fig. 3. Dimensionless total (frictional) pressure drop as a function of tracheal Reynolds number for inspiratory flow as computed from the data in Fig. 2.

Solving the modeling Eq. (11) yielded the values for γ, δ, and κ given in Table 1A. These constants quantify the excess pressure gradient in a branch of the model during inspiratory flow given by Eq. (7) and (8). At very low Re_z (i.e. creeping flow), $\gamma = 0$ indicates that flow is Poiseuillean. As Re_z increases, δRe_z contributes an increasing amount so that the pressure gradient can become many times that for Poiseuille flow. In the range of aspect ratio tested, the pressure gradient decreases approximately as the square root of the distance downstream of the bifurcation.

Experiments were also done in which successive generations of the $\ell/d = 9$ model were removed, yielding models with $Z_T = 4$ through 1. The total pressure drop across the latter models was predicted using γ,δ, and κ determined from the five generation models. With the exception of the $Z_T = 1$ model, the predictions are within 5% of the experimentally determined values. Assuming that expansion loss at the terminal branches of the $Z_T = 1$ model is the exit kinetic energy ($\simeq$ KEo/4), its loss is also predicted within 5%. These results support our assumption that γ,δ, and κ are independent of generation. Also, for $Z_T = 2$ models with second generations in different configurations, the total pressure drop of the one with the second generation coplanar with the first was less than 3% greater than that with second generation perpendicular to the plane of the first.

Expiratory Flow

Dimensionless static pressure drop across the five generation $\ell/d = 5$ and 9 models is plotted against tracheal Reynolds number in Fig. 4. Results for $\ell/d = 9$ include those obtained with water and a CMC solution ($\mu = 3.7$ cp).

The data so plotted have a positive slope which increases with aspect ratio. This corresponds to a curvilinear static pressure-flow relation which is convex to the flow axis. The CMC data appear to lie on the same curve as that obtained with water. Regression coefficients fitting the expiratory data are in Table 1B.

Recent measurements of expiratory velocity profiles downstream of a junction (7) have shown that after flow has passed through several generations of branching, the profile becomes virtually uniform within one diameter of the junction. Thus, we took $\beta = 1$ to compute the drop in total pressure, which is made dimensionless and plotted against Re_o in Fig. 5. Again note the scale change of the ordinate from that in Fig. 4.

Table 1. Regression coefficients of $\Delta P/\Delta Pd_o = A + BRe_o$ for $Z_T = 5$ and computed modeling constants γ, δ, and κ.

A. Inspiratory Flow

ℓ/d	ΔPs		ΔPt		Correlation	γ	$\delta(x10^3)$	κ
	A	$B(x10^3)$	A	$B(x10^3)$	r			
5	11.7	-0.40	11.7	15.2	0.99			
9	13.4	4.5	13.4	20.1	0.98			
						0	3.7	-0.5
13	17.7	10.1	17.7	25.7	0.99			

B. Expiratory Flow

ℓ/d	ΔPs A	ΔPs $B(x10^3)$	ΔPt A	ΔPt $B(x10^3)$	Correlation r	γ	$\delta(x10^3)$	κ
5	27.7	20.5	27.7	4.9	0.99			
9	31.2	23.1	31.2	7.5	0.99			
						0.6	0.5	0.15

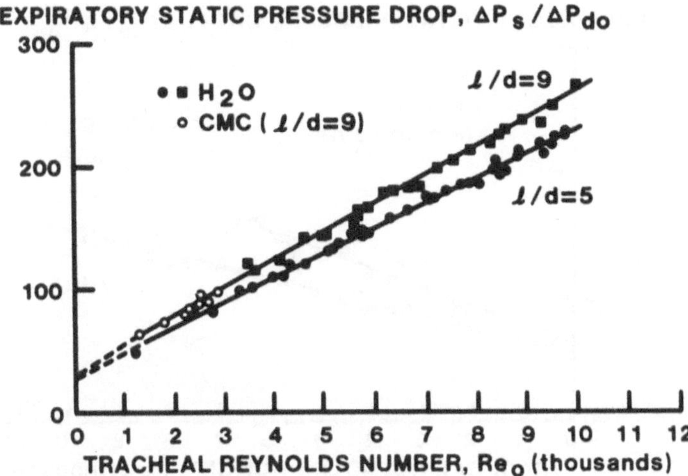

Fig. 4. Dimensionless static pressure drop as a function of tracheal Reynolds number for expiratory flow as computed from the data in Fig. 4.

The considerable difference between static pressure and total pressure drop is due to significant generation of tracheal kinetic energy. A legitimate question would be below what Re_o does the generation of KEo contribute less than 10% of the static pressure drop? Assuming a uniform velocity profile, we computed $Re_o = 200$ for the $\ell/d = 5$ model. Of course, the Re_o below which KEo can be neglected increases with frictional loss. With $\ell/d = 9$, for example, the value is 235. A velocity profile with $\beta > 1$ would decrease the lower limit of Re_o. The point is that in many physiological flow geometries, the Reynolds number is high enough so that kinetic energy changes may be **important** in determining frictional loss.

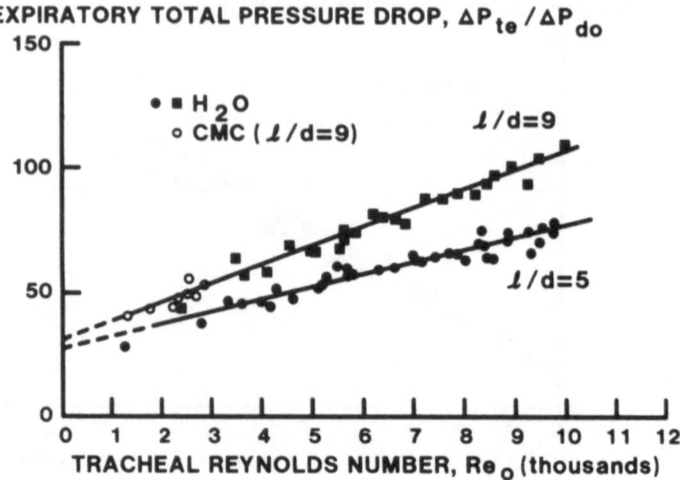

Fig. 5. Dimensionless total (frictional) pressure drop as a function of tracheal Reynolds number for expiratory flow as computed from the data in Fig. 4.

The modeling equations for expiratory flow were solved and gave the values of γ, δ, and κ listed in Table 1B. Unlike the finding for inspiratory flow, $\gamma = 0.6$ indicates that even at very low Re_z, the pressure drop is greater than that for Poiseuille flow in the series of tube segments. The much lower δ and positive but small κ are also striking differences from inspiratory flow. These results indicate that in the range of aspect ratio tested, the pressure gradient increases slightly with distance downstream of a bifurcation.

The frictional pressure drop across $\ell/d = 9$ models with Z_T from 1 through 4 was predicted using γ, δ, and κ in Table IB. The predictions are within 5% of the experimental results except for $Z_T = 1$, which had a larger drop than predicted. Assuming contraction loss at the terminal branches of the latter is about half the local kinetic energy (8), i.e., $KE_o/8$, its loss was also predicted within 5%, indicating that γ, δ, and κ are reasonably independent of generation.

DISCUSSION

Accurate static pressure differences across symmetric bifurcating networks were measured at several rates of steady inspiratory or expiratory flow. The networks were built with common laboratory Y-connectors having constant internal diameter and branching angle. Distance between junctions was varied in order to model the pressure gradient. A significant kinetic energy difference between trachea and reservoir was reasonably estimated and total pressure drop computed. Dimensionless pressure drop plotted against tracheal Reynolds number was well fit by linear regression. Our modeling of the pressure gradient downstream of a bifurcation yields a simple expression which accounts for a significant excess gradient over that for Poiseuille flow and is applicable to Newtonian fluid flow in similar geometries for Re in a physiologically important range.

Differences in geometry, experimental method and modeling make it difficult to compare our results with others in the literature. The physical model used by Pedley et al.(1) differed from ours in branching angle (70° and 60°, respectively) and area ratio of successive generations ($Az+1/Az$ = 1.22 and 2, respectively). Their model was constructed to simulate bronchial tree geometry while ours were inexpensively built to investigate an intermediate range of Re_o and demonstrate a modeling technique which, to our knowledge, have not been reported previously. Despite differences in geometry, method and modeling from Pedley et al., similar inspiratory pressure gradients are predicted for Re between 40 and 1500. Above 1500, our expression predicts an increasingly larger gradient. Jaeger and Matthys (9) measured pressure drop across an eight generation, $\ell/d = 5$ network with $Az+1/Az = 2$ during steady inspiratory flow of several gases. Their results are well correlated using a dimensionless approach, but they apparently neglected to consider kinetic energy changes which we have shown to be important in determining the overall loss. Douglass and Munson (4) measured inspiratory viscous dissipation in a two generation bronchial tree model which was similar to ours in branching angle (60°), but differed in that $Az+1/Az = 1.17$. They investigated $10^4 < Re_o < 10^5$ and found that dissipation is proportional to Re_o^3, a finding which we would predict at high Re_o. Using our finding that $\gamma = 0$ and $\kappa = -0.5$ for inspiratory flow, we computed that $\delta = 1.5 \times 10^{-3}$ closely fit their data for dissipation across both the entire model and just the first generation, supporting our assumption that the pressure gradient expression is independent of generation. Comparing δ for our model with theirs indicates that dissipation increases when A_{z+1}/A_z is increased from 1.17 to 2. Viscous dissipation measured by Round et al. (10) in models with A_{z+1}/A_z = 0.73, 1.07, and 1.33 indicate that dissipation decreases with increasing area ratio. This evidence suggests an optimal area ratio for a given branching angle. Interestingly, Wilson (11) predicts that by minimizing entropy production for a given alveolar ventilation, A_{z+1}/A_z in the bronchial tree should be $2^{1/3}$ or 1.26, which is close to the average of values reported in man for bronchi larger than 3 mm (12). For bronchi with diameter 1-3 mm, area ratio mesured by the latter group varies between 1 and 1.2. In this range, Weibel (13) shows higher area ratios, increasing from 1.35 in 3 mm bronchi to 1.8 in alveolar ducts. Branching angle also increases dramatically for bronchi smaller than 3 mm (12) thus two geometric changes may tend to increase losses in these airways. Because of low Re in smaller airways, loss proportional to Re may become insignificant and only the laminar flow loss important. That this loss in the smaller airways may be a small portion of the total loss in the normal bronchial tree led Jaffrin and Kesic (14) to conjecture that branching angles are distributed to minimize losses while still achieving the necessary changes in direction to reach alveoli. Area ratio appears to be distributed to minimize losses in the larger bronchi but increase surface area of respiratory bronchioles and alveolar ducts.

Expiratory flow appears to have a larger loss independent of Re than inspiration. This may further accentuate the role of small airways in resistance to gas flow. Very few model studies have been reported for expiratory flow, but as noted by Pedley et al. (7), the velocity profiles undergo rapid changes within one diameter of the junction and appear flat further downstream. These studies suggest that wall shear increases with distance downstream and may help explain our finding a positive κ for the expiratory pressure gradient.

Although our modeling approach has assumptions some of which have been only indirectly supported (eg., pressure gradient independent of generation) we believe it is a first step to reduce data to a form usable in modeling more complex phenomena such as flow in asymmetric, deformable tubes like bronchi. We have already attempted to model pressure gradient in a cast of the bronchial tree by substituting a symmetric analog which preserves area ratio between generations and average aspect ratio within a generation (6). Modeling of maximum expiratory flow depends strongly on the manner in which the bronchi deform with distending pressure. In the lung, questions remain not only about bronchial pressure-diameter behavior but also about the distending pressure of bronchi imbedded in parenchyma. A useful model of maximum expiration would have to include reasonable models of bronchial deformation and pressure gradient and be subject to the constraint that fluid velocity cannot exceed the propagation speed of pressure waves in elastic tubes, described in the lung by Dawson and Elliott (15). Although the latter theory qualitatively predicts the observed plateau of expiratory isovolume pressure-flow curves, a more complete model would describe the details, such as the maximum flow, location and pressure of the choke point, and changes that would occur when expiring gases of different properties than atmospheric air.

Based on the present study, we recommend additional studies to investigate: (1) the effects of entrance conditions, i.e., the effects of the upper airways on flow in the lower airways; (2) the pressure gradient in symmetric and asymmetric branching having more realistic geometry; (3) static pressure and velocity distribution in the trachea and daughter branches; (4) differences between inspiratory and expiratory pressure gradient for $1000 < Re_o < 10000$; and (5) application to the modeling of maximum expiratory flow. We feel that our approach will help unify and better quantify the considerable number of measurements of the pressure-flow relation in bifurcating networks.

REFERENCES

1. Pedley, T.J., R.C. Schroter and M.F. Sudlow. Energy losses and pressure drop in models of human airways. Resp. Physiol. 9:371-386, 1970.

2. Schroter, R.C. and M.F. Sudlow. Flow patterns in models of the human bronchial airways. Resp. Physiol. 7:341-355, 1969.

3. Brech, R. and B.J. Bellhouse. Flow in branching vessels. Cardiovasc. Res. 7:593-600, 1973.

4. Douglass, R.W. and B.R. Munson. Viscous energy dissipation in a model of the human bronchial tree. J. Biomech. 7:551-557, 1974.

5. White, F.M. Viscous Fluid Flow. McGraw-Hill, New York, 1974, p. 123.

6. Reynolds, D.B. Modeling studies of the pressure-flow relationship of the central airways. Ph.D. Dissertation, Charlottesville, University of Virginia, 1978.

7. Pedley, T.J., R.C. Schroter and M.F. Sudlow. Gas flow and mixing in the airways. Bioengineering Aspects of the Lung. J.B. West, ed. Marcel Dekker, Inc. New York, 1977.

8. Streeter, V.L. Fluid Mechanics. McGraw-Hill, New York, 1966, p.266.

9. Jaeger, M.J. and H. Matthys. The pressure-flow characteristics of the human airways. Airway Dynamics. A Bouhuys, ed. Charles C. Thomas, Springfield, Ill. 1970, p. 21-32.

10. Round, G.F., T.G. Pal and I.A. Feuerstein. Viscous energy dissipation for steady flow in models of arterial bifurcations. J. Biomech. 10:725-734, 1977.

11. Wilson, T.A. Design of the bronchial tree. Nature 213: 668-669, 1967.

12. Phalen, R.F., H.C. Yeh, G.M. Schum, and O.G. Raabe. Application of an idealized model to morphometry of the mammalian tracheobronchial tree. Anat. Rec. 190:167-176, 1970.

13. Weibel, E.R. Morphology of the Human Lung.Academic Press, New York, 1963.

14. Jaffrin, M.Y. and P. Kesic. Airway resistance: a fluid mechanical approach. J. Appl. Physiol. 36:354-361,1974.

15. Dawson, S.V. and E.A. Elliott. Wave-speed limitation on expiratory flow - a unifying concept. J. Appl. Physiol.: Respirat. Environ. Exercise Physiol. 43: 498-515, 1977.

MODERN MANAGEMENT OF CHRONIC RENAL FAILURE BY DIALYSIS AND TRANSPLANTATION

GEORGE R. HARVEY

THE QUEEN'S UNIVERSITY OF BELFAST

SUMMARY

The past ten years has witnessed changes in the treatment of patients with progressive renal disease. Presently there are about 50 Renal Units in Great Britain, but during the last fifteen years the change in treatment has been so marked that most patients are now treated by self-dialysis.

Dialysis treatment poses many therapeutic problems with patients at the extremes of life.

Prospective studies have shown that the interface between home dialysis and the emphasis towards transplantation can be clearly defined. The combined treatments of dialysis and transplantation have improved over the years and resulted in a higher survival rate for patients. The long-term life expectancy from combined treatment of patients with chronic renal failure is currently about 70%.

INTRODUCTION

Patients suffering from progressive renal disease and receiving conservative treatment can have the symptons of uraemia minimized or postponed with long term treatment by dialysis. Both maintenance haemodialysis and renal transplantation have passed through the experimental stage and are now acceptable treatments for terminal chronic renal failure. At first they were sometimes regarded as rival forms of treatment. It is now appreciated that both must be available for optimum treatment. Maintenance haemodialysis is essential for the adequate preparation of patients for renal transplantation and must be available for those transplanted patients in whom irreversible graft rejection occurs.

What is the extent and what are the results of these treatments?

In Britain the exclusive operation of an open market in health care is precluded. Free from economic constraints the demand for nephrological and associated medical services exceeds the supply. A natural and humane tendency has been to try and treat those with the most lethal prognosis. In a democracy the correctness of using public money in this manner may be criticised, questioned, contested and changed if necessary. But such debates require accurate information.

The European experience as of December 1975 was as follows: over 900 centres were performing dialysis or transplantation, or both. Out of a total estimated population of 491.4 millions, more than 18,000 patients were undergoing hospital haemodialysis, more than 4,000 were on home dialysis, and just over 5,000 patients had functioning renal transplants. The results of treatment varied from centre to centre and from country to country, but the overall European experience was that the percentage patient survival at three years from hospital dialysis, home dialysis, and after the first cadaveric transplant was 65%, 81%, and 61% respectively. Graft survival after the first cadaveric transplant was 39% at three years.

During the past 10 years dialysis units have proliferated throughout the world. At present there are about 50 in Britain. A recent report[1] indicates that while nephrological services are unevenly distributed throughout the country the incidence of patients with renal diseases is probably evenly distributed.

In 1965 the fact became obvious to many consultant nephrologists that hospital-based haemodialysis would provide for only a few deserving patients. Consequently the idea of home dialysis was developed and is now so successful that in Britain most patients are treated by self-dialysis in their own homes.

Dialysis treatment poses many therapeutic problems with patients at the extremes of life. In young children the main problems relate to retardation of growth and sexual maturation whereas with geriatric patients the prognosis is very poor. However, with continuing research into the metabolic problems of renal failure these age barriers are being gradually pushed back especially in the case of therapeutic studies with children.

The presence of other diseases in which chronic renal failure is but one symptom, e.g. diabetes mellitus, systemic lupus erythematosus, etc. were previously considered a contraindication to long-term treatment, but with increasing experience such patients are nowadays often accepted for haemodialysis, with encouraging results.

Prospective studies have shown that emphasis must be towards transplantation. Not only is it less expensive than dialysis but it also results in a better degree of rehabilitation. Patients suffering from terminal renal failure who can benefit from renal replacement are evenly distributed throughout the country[2, 3, 4].

Haemodialysis and renal transplantation should aim at a complete interchange with each other; if a transplant fails the patient can return home without delay.

Continuous management of chronic renal failure is essential. Careful diet control undoubtedly contributes to the continued wellbeing of many patients with slowly progressing renal failure but it is generally less effective when function declines rapidly. This decline may be a feature of the specific underlying disease, but it can also be due to inadequate control of blood pressure, infection, and electrolyte balance or even the effects of drug treatment, e.g. tetracyclines.

Nephrologists usually define a patient who is no longer able to lead a normal life because of 'uraemic' symptoms as ready for dialysis. Biochemical abnormalities consistent with the symptoms lend weight to this decision. Preparation for dialysis is made when endogenous creatinine clearance approaches 5 ml/min but because of the extreme shortage of dialysis places those patients who remain relatively well may have to continue on beyond this stage, or be temporarily treated by peritoneal dialysis. When complications such as uraemic pericarditis and peripheral neuropathy arise these are absolute indications for commencing dialysis. Any further delay may well result in a permanently crippled patient or death.

Long term dialysis treatment was made possible by the introduction of the arteriovenous shunt by Quinton et al[5] in 1960. The modified Kiil artificial kidney was adopted by the Seattle group because of its advantages over existing coil dialysers. These advantages include low priming volume and low internal resistance to fluid flow. Twenty years later the same basic principles of dialysis are still being used, but of course equipment has become much more sophisticated[6]. Safety[7] and reliability[8] have increased so that patients may now be trained to dialyse themselves with confidence. Arteriovenous shunts have a limited life span. Manufactured from Teflon and Silastic rubber they are easy to handle but because of their tendency to clot and become infected they are not used to the same extent as previously. The arteriovenous fistula was a welcome innovation. First described by Brescia et al[9] in 1966 it can function for many years without any undesirable complications. The shunt, however, is still used in the treatment of children and those adult patients in need of prompt haemodialysis. Ideally the arteriovenous fistula requires to be left to develop for several weeks before being cannulated.

HAEMODIALYSIS

The most effective form of dialysis treatment is a process of selective diffusion through an artificial kidney membrane of toxic molecules from the blood to a specifically prepared dialysate solution. The concentrations of solutes in this solution are much the same as normal plasma with the exception of a low potassium concentration and the absence of phosphate ions. Specially designed proportioning machines are now available that mix a concentrated solution of these salts with softened or deionised tap water to produce the dialysate solution. The concentration of potassium is purposely kept to a minimum and acetate is substituted for bicarbonate so as to avoid precipitation with calcium and magnesium. The resulting fluid is warmed to 39°C, ideally deaerated, and checked for conductivity before being pumped through the artificial kidney. Inside the artificial kidney or dialyser the patient's blood is separated from the dialysing fluid by a thin synthetic semi-permeable membrane, which allows exchange of molecules across it. In this way toxic metabolites from the blood are removed and washed away. Acid-base balance is corrected by the passage across the membrance into the blood of basic anions. The pH value of the dialysing fluid is maintained within the range 7.25 to 7.30.

Patients suffering from kidney failure can excrete only small quantities of urine, or even none at all, and it is therefore necessary to remove water during the treatment with the artificial kidney. The membranes used for dialysis are freely permeable to water and it is easily possible to adjust the pressure gradient between the blood and the dialysate sides of the membrane to remove the amount of water required from the individual patient.

A new approach in dialysate supply has been developed recently. A small volume of prepared dialysate (5.5l) is continuously recirculated through a sorbent cartridge which contains layers of active chemicals designed to remove dialysed toxins thus providing a continuous purified supply of dialysate. This system, however, is expensive and running costs are high but it has the advantages of portability, economy of water usage, and needs minimal rearrangement of the room should it be installed in the patient's home.

Duration of a Dialysis Period

The duration of a dialysis period depends on a number of factors. Most people agree that dialysis should be performed in such a way and at times that it will not interfere with the patient's life style. It should be sufficiently effective as to maintain reasonable good health, a suitable haematocrit, and prevent long term side effects such as peripheral neuropathy and renal bone disease. Short period dialyses of three hours per day keep the concentration of small molecular mass toxins under control. Middle molecular mass substances, however, are dialysed only very slowly and require longer, though less frequent dialysis. Regulated patients practise a regimen of six to eight hour dialyses every other night or three times per week. Long-term studies have shown periods such as these to be adequate but duration and frequency of treatment depend on the type of dialyser used.

Choice of Dialyser

The optimum dialyser should have a low priming volume, minimum internal resistance, good clearance values for small and middle molecules, good ultrafiltration and washback characteristics, low blood leak rate and low cost. A disposable dialyser has added advantages.

Many suitable models are now available designed on a coil principle, parallel plate, or made of hollow capillary fibres. Disposable kidneys have a higher cost which precludes their universal use, so that, at least in Great Britain, many haemodialysis units still use the Meltec Multipoint dialyser, the disadvantages of which are only their size and the routine maintenance necessary to strip and rebuild them at intervals.

General Management of Dialysis Patients

When a patient has been established on dialysis dietary restrictions should be withdrawn. Normal protein intake is then permitted to offset amino-acid losses by dialysis and maintain positive nitrogen balance. An adequate calories intake is required so as to make the most efficient use of protein, but overstepping of due limits for carbohydrate and lipid intake will induce hyperlipoproteinaemias. Crude balance studies are used to judge electrolyte and fluid allowances. A hypertensive patient showing evidence of fluid retention will need a low sodium intake, and fluid allowance should be judged on the patient's total output. Restriction of potassium intake is usually necessary. Folic acid and vitamin B supplements are required to replace dialysis losses. Phosphate restriction is required and is a safe method of controlling plasma phosphate - safer than aluminium hydroxide, but the latter may be needed intermittently.

About 2g of iron/year are necessary to replace blood losses. When a patient is on dialysis, blood transfusions are rarely necessary and then only to replace losses due to bleeding.

Regular monitoring of plasma urea, electrolytes, creatinine, calcium, phosphate, alkaline phosphatase and proteins concentrations provide information on the patient's metabolic state and the quality of dialysis. However when a patient is stabilised excessive investigation can be dispensed with, in order to conserve blood.

Complications of Dialysis

The clinical course of long-term dialysis patients is usually free from impediments but nevertheless some dialysis disequilibrium may occur from over-efficient treatment when uraemic patients are first introduced to dialysis. Early treatment to lower uraemic toxin levels by a course of short dialyses or by peritoneal dialysis will avoid many of the early problems.

Insufficient or inadequate dialysis may lead to peripheral neuropathy, but this symptom can be halted or even improved by increasing the frequency at which dialysis is performed. Renal osteodystrophy sometimes worsens and may even develop during dialysis. To improve dietary calcium absorption and bone mineralisation, vitamin D analogues may be used, but to prevent metastatic calcification careful control of plasma phosphate is necessary.

Dialysis osteomalacia has been recorded in hypophosphataemic patients and a number of units have described a progressive dementia in some patients. These complications may be related to aluminium hydroxide treatment.

Generally, hypertension patients are controlled with dialysis but some may need additional hypotensive agents. Even with careful dialysis about 5% of all patients have severe uncontrollable hypertension due to high plasma renin activity and this can lead to a bilateral nephrectomy.

Infection can result from cannulation of arteriovenous fistulae or from shunts and prompt treatment of septicaemia with appropriate antibiotics is necessary together with revision of the affected fistulae or shunts.

Districts having a supply of hard water to the dialyser will need to have the water treated. Failure to soften the dialyser supply will give rise to acute hypercalcaemia and hypermagnesaemia, causing acute hypertension, tachycardia, vasodilation, headaches, and vomiting. The plasma amylase concentration may also be increased due to the hypercalcaemia.

Modern dialysis equipment has greatly reduced the risk of death due to an air embolus arising from failure of the extra-corporeal blood circuit[7, 8]. In such an emergency immediate lowering of the bed head, turning the patient to left lateral position and administering oxygen may be sufficient to resuscitate the patient.

Hypertension due to over-ultrafiltration may be corrected by fluid replacement. Patients with uraemic pericarditis should be regionally heparinised to protect them against haemorrhagic effusion with the risk of tamponade. Febrile reactions on dialysis are unpleasant and usually no bacterial infection can be found in shunts or in the blood stream, consequently they are attributed to pyrogens. Often organisms can be isolated from water supplies and it is possible that their exotoxins may be responsible. The problem can be eliminated with regular and complete sterilisation of the dialysate and water circuits.

Serum hepatitis is a potential hazard in any dialysis unit but careful screening, limitation of movement between units, screening of essential blood transfusions and isolation of carriers has reduced the incidence almost to nil.

RENAL TRANSPLANTATION

Renal transplantation is no longer regarded as an experimental procedure. When performed in close conjunction with haemodialyses units it will give satisfactory treatment. A successful live donor graft will restore patients virtually to normal health, but unfortunately only about 40% of all cadaver kidneys are capable of this[10]. Fig. 1 shows the 4-year graft survival rate.

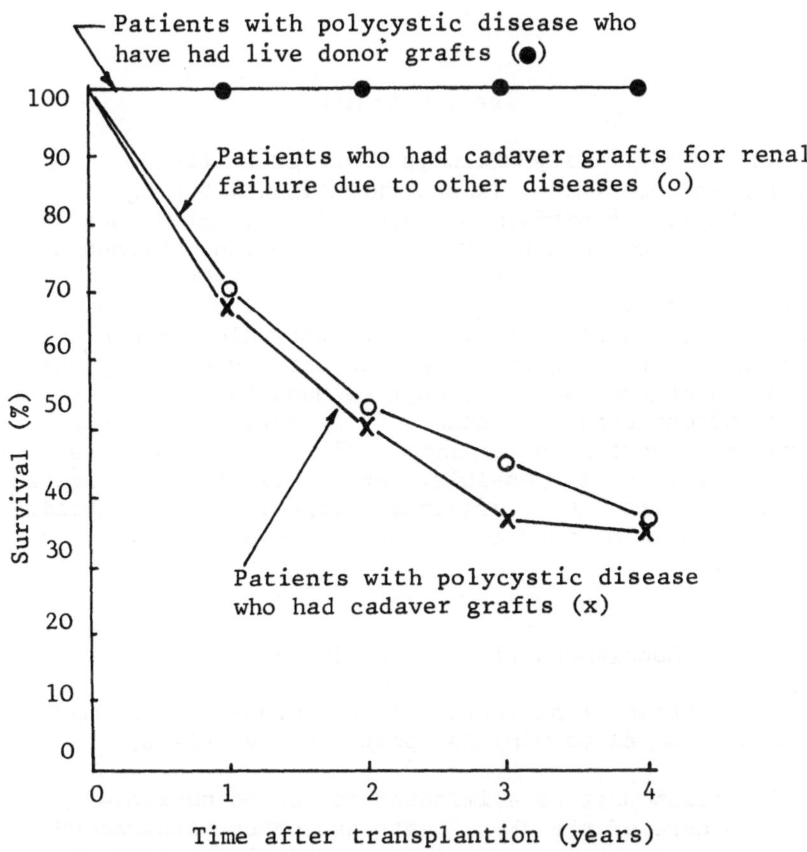

Figure 1 SURVIVAL RATE FOR TRANSPLANTATION

In Great Britain relatives rarely offer to donate kidneys, so the National Health Service depends almost entirely upon cadavers.

Patients who have had suitable dialysis treatment can withstand transplant surgery satisfactorily and with good anaesthetic administration and careful fluid and electrolyte control mortality is low. It is depressing that after 12 years' experience, 30-40% of all grafts are still lost during the first six postoperative weeks from irreversible rejection, infection, or technical problems.

Tissue Matching

ABO compatibility is always observed in tissue matching and HLA (homol. leucocytic antibods.) typing of recipients may be performed prospectively, but donors have to be typed just before transplantation. Direct crossmatching of donor cells and recipient serum is usually performed to detect cytotoxic antibodies and to avoid second-set rejection. Though the value of HLA matching is not altogether clear in primary renal transplants, an attempt should always be made to select the closest matching pairs, because if rejection does occur the range of antibodies induced will be confined to as few specificities as possible. When patients are selected for second grafts their previous antigenic incompatibilies will have to be avoided and closer matching will be necessary.

Management of Preoperative Patients

Preparation of patients for transplant surgery must be arranged so as to minimise postoperative risks,

(1) Infection must be eliminated as far as possible with care of the skin, teeth and upper respiratory and urinary tracts receiving particular attention. Patients undergoing treatment for chronic infection such as bronchiectasis or tuberculosis may cause considerable problems in the face of immunosuppressive drugs and are best excluded from transplantation.

(2) To detect peptic ulcers or diverticular disease, barium meal and enema examinations should be performed.

(3) Skeletal surveys and studies of calcium metabolism should be done to assess the parathyroid state.

(4) It is essential to control hypertension.

(5) Bladder neck function and detection of ureteric reflux are assessed during investigation of the lower urinary tract.

(6) Cervical smears should be examined.

Appropriate surgical correction of gastrointestinal tract abnormalities, the removal of infected refluxing kidneys and ureters, and gynaecological problems should all be dealt with before transplantation.

CADAVER DONORS

An ideal kidney donor should have a normal blood pressure and renal function before death, be from 10 to 50 years old and should not have suffered a long period of premortem hypertension. Evidence of localised or generalised antemortem infection should be regarded suspiciously and treated energetically if time allows. Many donors are the result of auto-mobile accidents and hence time for assessment of function is not usually available so that much will have to depend on the appearances of the kidneys on removal and their perfusion characteristics assessed by the surgeon. Patients dying of cerebrovascular accidents or cerebral tumours may often be suitable donors, but those with extra-cranial tumours should never be considered. Evidence of pre-existing renal disease should naturally exclude the potential donor.

The time from circulatory arrest to removal and perfusion of the kidneys should be as short as possible and certainly not longer than one hour. In Great Britain permission for kidney removal from next of kin or coroner, or both, is essential, and this may take considerable time in unexpected death. Many suitable organs are inevitably lost because permission is unobtainable in this limited time. Cold perfusion will retard metabolic deterioration of the organs so that time is available to prepare suitable recipients once the kidneys have been cooled.

If living related donors are considered normal, renal function, blood pressure, intravenous urography, and renal arteriograms are mandatory. HLA identity and a negative mixed lymphocyte reaction are desirable for the best results.

The Operation

The actual transplant operation is now a relatively routine procedure, the kidney being placed extraperitoneally in one or other iliac fossa. The vascular connections are made between the renal vessels and the ureter implanted into the bladder through a submucosal tunnel.

Immunosuppressive Treatment

Corticosteroids and azathioprine are usually started at the time of transplant. Dose regimens vary but moderately large doses of steroids are usually given initially (100-200 mg prednisone) are gradually reduced to a maintenance dose of 10-20 mg per day. Azathioprine is given in the highest possible dose tolerated by the individual as judged by the leucocyte and platelet counts. Doses are curtailed when renal functions is poor. Antilymphocyte globulin is used by some departments.

Rejection episodes are treated by increasing the steroid dose temporarily with or without the addition of actinomycin D, cyclophosphamide, and sometimes local irradiation.

Complications

The complications encountered with transplantation are related to four main factors: (a) the quality of the donor kidney; (b) technical problems at operation; (c) graft rejection; and (d) immunosuppressive drugs.

A degree of ischaemic renal failure is not uncommon in the early postoperative period. Diuresis normally follows within one to three weeks unless other complications supervene. Unfortunately, acute rejection often occurs during this oliguric period and may be difficult to diagnose with certainty. Other complications such as ureteric obstruction or leakage and vascular thrombosis may be masked and should always be considered. Dialysis treatment with regional heparinisation must be continued during this period and constitutes a further potential hazard. When immediate diuresis occurs, complications are more easily diagnosed since any deterioration in urinary output and renal function can be immediately investigated.

A relatively safe and reliable investigation in cases of suspected rejection, is the use of a percutaneous needle biopsy. The histological changes of acute rejection are well known. They include, (a) interstitial haemorrhage; (b) mononuclear cell inflitration; (c) arterial and arteriolar necrosis; and (d) glomerular thrombosis and necrosis.

When the immunological reaction is agressive and severe vascular changes are seen, there is little hope of success with present antirejection treatment, so that the graft should be removed and the patient re-established on dialysis.

Doses of immunosuppressive drugs, given at regular intervals, may lead to dangerous complications. These are principally infective in nature and may be due to bacterial, fungal, viral or protozoal organisms, often complicated by bone marrow depression. Treatment is difficult for specific chemotherapeutic agents some of which can be toxic. Immunosuppressive drugs may have to be withdrawn.

Gastrointestinal complications may be extremely serious because of exsanguination and infection and need prompt surgical intervention. Ureteric obstruction or leakage also demand prompt and expert surgical correction. Localised or systemic infections sometimes lead to rupture of the vascular anastomosis with consequent loss of the graft.

Complications can develop months or years after a successful graft. These may include alopecia, thinning of the skin, recurrent infections, avascular necrosis of weight-bearing joints, cataracts, glaucoma, and neoplastic diseases.

RESULTS AND COMMENTS

Despite the many complications which arise the haemodialysis machine/kidney transplant interface can be defined and many patients benefit from successful renal transplantation. Although 40-50% of kidney transplants fail during the first 6 months, the rate of loss of function after the end of the first year is low and in some cases 20-25% of the transplants are still functioning at 10 years. The combined treatments of dialysis and transplantation have improved over the years and resulted in a higher survival rate for patients. The long-term life expectancy from combined treatment of patients with chronic renal failure is currently about 70%.

The number of transplants carried out in Great Britian is steadily rising, but the waiting list for a transplant gets longer.

Last year 729 transplants were performed - 121 up on the year before and the first time that more than 700 had been exceeded. There were 17 transplants in Northern Ireland.

The waiting list has now increased from 1160 to 1250 and according to Health Department figures, 1500 transplants per year are needed if demand is to be met.

To treat more patients economically in the future, home dialysis and transplantation units must continue working closely together with greater emphasis on transplantation. More donor kidneys both from living relatives and cadavers are urgently needed. A change in the law regarding cadaver donation may well help to solve the problem of donor shortage. Immunosuppressive treatment has not improved over the last decade and furthermore attempts must be made to improve immunological methods of selection since HLA matching as such has not contributed as much to graft acceptance as was originally hoped. Multicentre sharing of kidneys, e.g. the National Organ Matching and Distribution Service, Eurotransplant, etc. theoretically should help to ensure better HLA matching, but this has to be weighed carefully against the possible disadvantages of increasing ischaemic time intervals.

REFERENCES

(1) Distribution of nephrological services for adults in Great Britain. Report of the Executive Committee of the Renal Association. British Medical Journal, Vol. 2, 1976, pp. 903-906.

(2) Branch, R.A., Clark, G.W., Cochrane, A.L., Jones, J.H. and Scarborough, H., "Incidence of Uraemia and Requirements for Maintenance Haemodialysis, British Medical Journal, Vol. 1, 1971, pp. 249.

(3) Prendreigh, D.M., Heasman, M.A., Howitt, L.F., Kennedy, A.C., MacDougall, A.I., MacLeod, M., Robson, J.S., Stewart, W.K., "Survey of Chronic Renal Failure in Scotland", Lancet, Vol. 1, 1972, pp. 304.

(4) McGeown, M.G., "Chronic Renal Failure in Northern Ireland", Lancet, Vol. 1, 1972, pp. 307.

(5) Quinton, W., Dillard, D., and Schribner, B.H. Transactions American Society of Artificial Internal Organs, Vol. 6, 1960, pp. 104.

(6) Harston, G., Beattie, M., and Ivison P., "Development of a portable dialysing system", Engineering in Medicine, Vol. 7, No. 4, 1978, pp. 233-234.

(7) Harvey, G.R., Lyness, J., McGeown, M.G., "Monitor for arterial line during haemodialysis", Lancet, Vol. 2, 1970, pp. 247.

(8) Harvey, G.R., "Increasing the reliability of machines for regular haemodialysis", Proceedings 2nd International Bio-engineering Conference, Milan, Italy, 1973.

(9) Brescia, M.J., Cimino, J.E., Appel, K., and Hurwitch, B.J., "Chronic Haemodialysis using venipuncture and a surgically created arteriovenous fistula", New England Journal of Medicine, Vol. 275, 1966, pp.1089.

(10) Mitcheson, H.D., Williams, G., and Castro, J.E., "Clinical aspects of polycystic disease of the kidneys", British Medical Journal, Vol. 1, 1977, pp. 1196-1199.

HUMAN REPRODUCTION: BIOENGINEERING ASPECTS OF CONTRACEPTION APPLIED TO THE DEVELOPMENT OF A NEW FEMALE CONTRACEPTIVE

Bruce W. Vorhauer, Ph.D., President

Vorhauer Laboratories, Inc.
130 McCormick Avenue, Building 104
Costa Mesa, California 92626

FOREWORD

"There is no such thing as a problem without a gift for you in its hands. You seek problems because you need their gifts."

Richard Bach, ILLUSIONS

ABSTRACT

This paper has two purposes: the first is to describe the basic physiology of human reproduction, with particular emphasis on intervention in the reproductive process for the purpose of contraception; the second is to present a new barrier method of contraception for females which has been developed by Vorhauer Laboratories, Inc.

During the design and development of this non-implantable, vaginal contraceptive sponge (VCS) device, the bioengineering aspects of human sexual intercourse and the esthetics of this activity were basic considerations. As a background for the rationale involved, the physiology of the human male and female reproductive systems, the response of human anatomy during coitus, and the fundamentals of current contraceptive techniques are reviewed. Details of the development process are given to provide insight into the medical-bioengineering interfaces involved, and the clinical evaluation programs establishing the safety and effectiveness of the VCS are also outlined.

The COLLATEX™ Contraceptive Sponge developed by VLI is a diaphragm-like device, molded from a new biomaterial, a hydrophilic polyurethane. The VCS is impregnated with a conventional spermicide, Nonoxynol-9. Because of its compatibility with the vaginal environment, the device is intended for a single 2-day use period, with multiple coital episodes possible during this time. Sexual spontaneity is thus inherent in the method since no preparations are required for contraception, other than insertion of the sponge (which can be done up to two days prior to intercourse). The VCS can be inserted and removed as desired by the user. In certain lesser developed countries where price is an extreme consideration, the product will undoubtedly be washed and reused, although this is not recommended since the spermicide can be depleted by repeated washings. Because of its material compliance, the VCS adapts to anatomical variations which are a function of the user and her daily activities. Currently, one size of the sponge is intended for all users.

Appropriate regulatory compliance filings have been made with the United States FDA, and the VCS is currently undergoing extensive clinical effectiveness evaluations. Once marketing approval has been obtained, the product will be sold as an over-the-counter (OTC) contraceptive. Initial marketing of the product outside the United States is scheduled for late 1980.

INTRODUCTION

Worldwide, the concern about population growth and methods for voluntary control of births is growing steadily. More and more governments are recognizing that population size and its rate of increase directly affect the quality of life of their citizens. For India, as an example, with a net increase in population of over one million people per month, the problem is in fact a crisis. In the U.S. with a population growth essentially at zero (ZPG), the concern has been directed more to specific problems such as the epidemic of one million teenage pregnancies annually. Among the developed countries, only East Germany, Bulgaria, New Zealand, and Romania have higher teenage fertility rates [1]. Also, both pro- and anti-abortion factions are beginning to realize that of the three alternatives (abstinence, childbirth, and contraception) to the over one million abortions annually in the U.S., only the last presents an opportunity to deal with this problem. A thorough analysis of fertility and contraception in the U.S. is presented in the recent U.S. House of Representatives Select Committee on Population Report [2].

Obviously, there are great opportunities in this area of great problems. Safe, effective, and simple methods of contraception are direly needed to increase the choices available to those who wish to limit the size of their families. Bioengineering approaches to intervention in the reproductive process are numerous, particularly in the medical devices field. This paper will concentrate on reversible contraception for females; reversible since sterilization must be considered permanent (although this is not absolute) and female since ultimately it is the female who must bear the physiological consequences of conception.

PHYSIOLOGY OF REPRODUCTION

Figure 1 shows the female and male reproductive systems simplified for the purposes of this presentation. Upon ejaculation by the male in the female's vagina, approximately 500 million sperm are deposited in the immediate vicinity of the cervix. These sperm are suspended in a biochemically complex seminal fluid (semen) which contains, among other substances, fructose, proteins, enzymes, citric acid, and prostaglandins. The function of the seminal fluid has not yet been determined, other than as a fluid vehicle for the sperm.

In the upper end of the vagina there is a small opening through the cervix called the os, which provides the path for the sperm to reach the uterus. This pathway, called the endocervix, also acts as a reservoir for the sperm and is filled with a viscous mucus. The ability of the sperm to move through the endocervix is affected greatly by the viscosity of this cervical mucus [3]. The physical and chemical characteristics of this mucus change during the menstrual cycle, with the viscosity decreasing at the time of ovulation; this decrease in viscosity allows the sperm to more readily penetrate the mucus, thus enhancing their ability to reach the ovum.

Travelling through the endocervix, up the uterus, then into the oviducts (or Fallopian tubes), the sperm finally reach the ovum, assuming that ovulation has recently occurred.

Incidentally, sperm can enter the endocervix within three minutes of ejaculation in the vagina, and can typically reach the oviducts in less than a half hour. Since the sperm swim at a rate of only 100 microns/second or 1 centimeter/100 seconds, it is thought that contractive motive forces in the female act on the sperm to move them so rapidly to the oviducts.

The egg is fertilized (conception) in the oviduct and then is known as a zygote; it reverses the path of the sperm until it reaches the uterus where it implants and becomes an embryo, nourished by the placenta. About 35% of the zygotes will not implant and about 25% of those that do spontaneously abort. If the egg is not fertilized during the 24 hours after ovulation, no conception will occur and the egg will disintegrate and be expulsed with the menstrual flow. Living sperm have been observed in the uterus more than five days after intercourse. However, sperm lose their fertility before their motility, and currently it is believed that sperm can remain viable for fertilization for about 48 hours once inside the uterus. As a point of reference, about 90 women of every 100 sexually active not using contraception will become pregnant during a one year period. References [4-9] provide more detailed explanations of the reproductive process.

ANATOMY DURING INTERCOURSE

Considerable anatomical changes occur in the female during sexual stimulation. Masters and Johnson [7] documented these as related to four phases of sexual response: excitement, plateau, orgasmic, and resolution. Among the most pertinent of these changes relative to contraception are the ballooning and lengthening of the upper two-thirds of the vagina during the excitement phase, the elevation of the cervix during the plateau phase, the expulsive contractions of the uterus during orgasm, the post-orgasm dilation of the cervical os, plus the vaginal shrinking and cervical lowering of the resolution phase. All of these dynamic effects are obviously altering the physical configuration of the female sexual anatomy; clearly they must be considered in the design of any contraceptive device intended to reside in the vagina or uterus. The effects of anatomical factors on one particular contraceptive method, the vaginal diaphragm, have been reported by Masters and Johnson [10]; for example, they found the female-superior coital position to considerably enhance the likelihood of diaphragm displacement in the sexually stimulated female. This is due to the combination of the ballooning of the upper portion of the vagina and gravitational force.

Further temporary anatomical changes occur because of the presence of and movement of the male's penis during intercourse; besides the extraordinary compliance and elasticity of the vagina in response to the penis, the pistoning effect and changes in pressure within the vagina and uterus are factors to consider. Using again the impermeable diaphragm contraceptive device as an example, differential pressures between the vagina and the uterus caused by such a piston effect may induce displacement of this device.

CURRENT METHODS OF CONTRACEPTION

Recognized efficacious reversible methods of birth control currently available for female use and involving the use of drugs and/or devices are limited to the following (in decreasing frequency of use): Oral Contraceptives (OC's); Intrauterine Devices (IUD's); Intravaginal Spermicidal Drugs; and Diaphragms/Cervical Caps. The only device for male use is the condom. In addition, two surgical procedures are now commonly used to induce sterility: vasectomy (male) and tubal ligation (female). A schematic illustrating these various techniques is shown in Figure 2. Finally, the rhythm method of avoiding coitus during the period of ovulation, and withdrawal (coitus interruptus) are two methods requiring no drugs, devices, or surgery, but with significantly lower contraceptive effectiveness. The book by Shapiro [11] contains excellent coverage of the current methods of birth control. In addition, there are numerous comprehensive references discussing these and various still experimental techniques in detail [12-18]; therefore they will not be covered here except to summarize their effectiveness [see Figure 3], and some of the problems and inconveniences currently associated with their use. Finally, Himes gives a thorough presentation of the long history of contraception in his book [19].

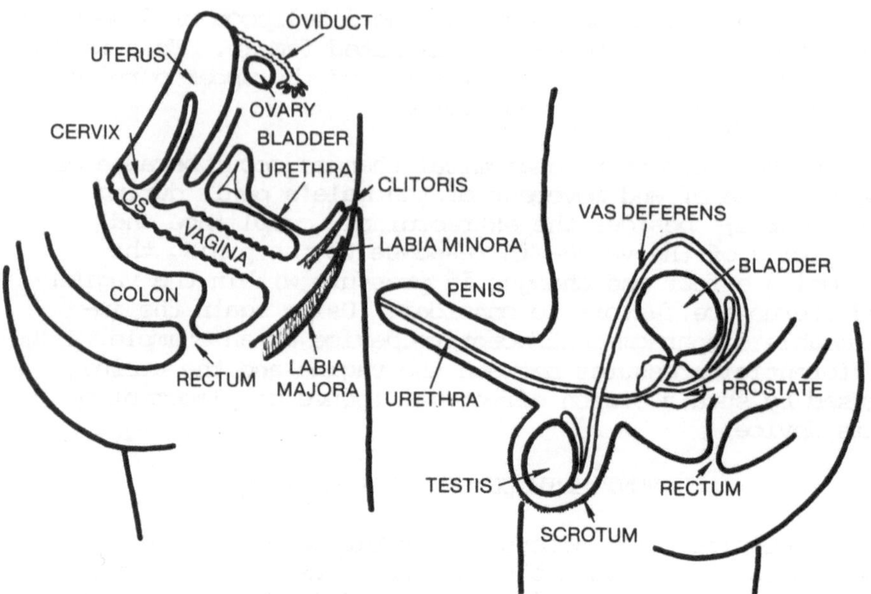

Figure 1. Diagram of female and male reproductive systems.

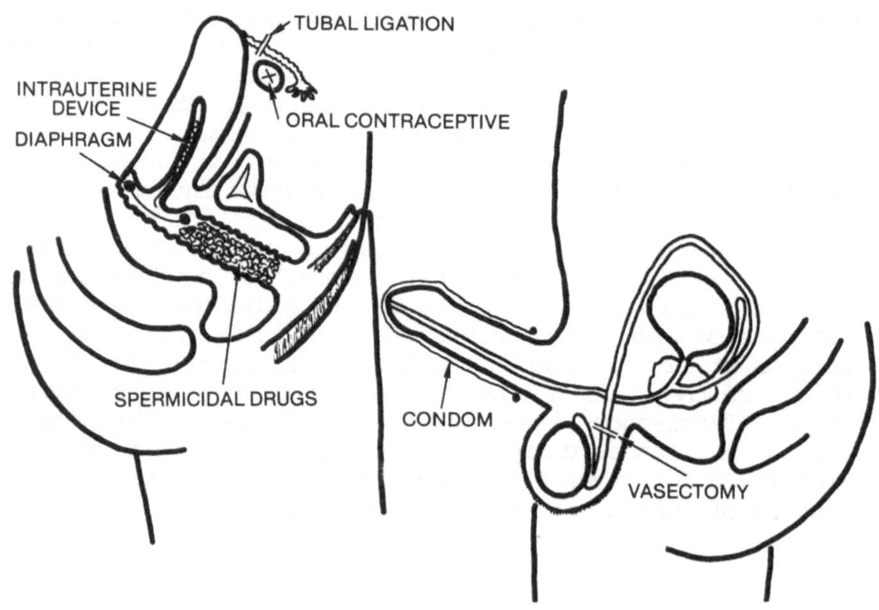

Figure 2. Methods of contraception for females and males.

Method	Used Correctly & Consistently (Theoretical Effectiveness)	Average of 100 U.S. Women Who Want No More Children (Use Effectiveness)
Abortion	0	0+
Abstinence	0	?
Tubal Ligation	.04	.04
Vasectomy	.15	.15+
Oral Contraceptive	.34	4-10
Condom + Spermicide	less than 1	5
IUD	1-3	5
Condom	3	10
Diaphragm + Spermicide	3	17
Spermicidal Foam	3	22
Coitus Interruptus	9	20-25
Rhythm (calendar)	13	21
Douche	?	40
Chance (sexually active)	90	90

Figure 3. Approximate number of pregnancies during the first year of use per 100 non-sterile women initiating method. [Data from Reference 12, Table 1, page 20]

PROBLEMS WITH THE CURRENT METHODS

OC's

The most obvious drawback to such drugs for birth control is that they act systemically, thus affecting the neurohumoral balance of the whole body. OC's provide higher levels of synthetic estrogen and progesterone which act on the hypothalmus and the pituitary gland preventing ovulation. The side effects associated with this systemic action have been well publicized recently, with some problems becoming apparent only after long term use of certain formulations [12, 13, 20, 21].

IUD's

The IUD's action in contraception is not yet fully understood except that it is known that the presence of a foreign body in the uterus prevents the implantation of the fertilized egg (zygote) in the uterine lining. Thus, the growth of the fertilized egg into an embryo is somehow prevented. Although it was thought that the IUD accelerates the journey of the egg through the oviduct thus preventing fertilization, this now seems unlikely. This device must be inserted and removed by trained medical personnel using sterile techniques, and is considered to be an implanted device. Problems with IUD's have also been well publicized and involve pain and cramps, bleeding, expulsion, infection, ectopic pregnancy (zygote growth in the oviduct), and uterine perforation [12, 14, 22].

Diaphragms/Cervical Caps

The diaphragm is prescribed by a physician since it must be individually sized for each woman. Although it is easily thought of as a mechanical barrier preventing sperm from passing through the cervical os into the uterus, it is in fact primarily a holder for spermicidal gel or cream when positioned over the cervix. Since the diaphragm can easily move about during intercourse (particularly with the female superior) and become displaced from its proper position, the reliability of this device is lower than those methods above. Also, the requirements to manually fill with spermicide and insert before intercourse, and to remove and wash six hours postcoitally, are bothersome and distasteful to many women [11, 12, 15, 24].

The cervical cap is a small thimble-shaped cup which covers the cervix; it is deeper, more rigid, and smaller in diameter than the diaphragm. Usually no spermicide is used inside the cap and it can be left in place for longer periods than the diaphragm. Caps are available in three main sizes and must be individually fitted. The major disadvantage to the cap is the considerable difficultly involved in inserting and removing the device since it is held in place over the cervix by suction [11, 15]. There are no caps approved by the FDA for marketing in the U.S.

Spermicidal Drugs

These chemicals (primarily Nonoxynol-9) are principally non-ionic surfactants in a foam carrier; they act to immobilize and kill the sperm after the male ejaculates in the female's vagina. They must be placed in the vagina usually within one hour before intercourse, and must be applied again each time there is intercourse. Side effects reported from these preparations have been minimal and limited to infrequent minor irritations of the vaginal mucosa and penile tissue. The main concerns with products in this area are lower effectiveness, messiness, and the need to prepare for intercourse/lack of spontaneity [11, 12, 15, 23, 24].

Summary

Thus, in the reproductive process, OC's control the time of ovulation, IUD's prevent implantation of the fertilized egg, spermicidal foams entrap and kill the sperm before they traverse the cervical os, diaphragms block the cervical os with spermicidal gel/cream, and caps act as a simple mechanical barrier over the cervix.

DESIGN CONSIDERATIONS

The most important consideration in any contraceptive technique is user safety. With the increasing awareness of patients about their body functions and the risks associated with various contraceptive techniques, safety has become the prime concern of contraceptive users. As developers of such techniques, our primary responsibility should be for the safety of our products and methods. But besides the safety of the users, consideration must also be given to the safety of the fetus should a given technique fail and pregnancy termination not be a desired or possible option.

Clearly, the contraceptives must also have a high level of effectiveness, although this is becoming of lesser import with the wide spread availability of DES (diethylstilbestrol or morning-after pills), as an "emergency measure" only, and "menstrual regulation" (vacuum extraction for late menstrual periods). Parenthetically, induced abortions are now performed in about 25% of all U.S. pregnancies. Many women are switching to the lower effectiveness barrier methods, such as the diaphragm, and backing up any failure with menstrual regulation in the first week or two after a missed menstrual period.

In any contraceptive technique there is strong desire on the part of the users that it interfere minimally with spontaneity and naturalism. This, combined with their high effectiveness, is the major reason for the popularity of the OC's and IUD's - no precoital planning or preparations are required. The technique used should also be simple and "foolproof" for users with a wide spectrum of motivation, educational level, and intellect.

In summary, the basic desirable characteristics of a contraceptive are:

<u>Mandatory</u>:

- User Safety;
- High Effectiveness.

<u>Preferred</u>:

- Local rather than systemic effect;
- Administered by lay person or users themselves;
- Adaptable to or unaffected by anatomical changes during intercourse;
- No sizing or custom fitting required;
- Preparation before intercourse not required;
- Postcoital steps not necessary;
- Easily discontinued if pregnancy is desired (reversible);
- No side effects or characteristics noticeable by users;
- Reasonable cost (say, less than OC's).

<u>Bonus</u>:

- VD Prophylaxis;
- Ability to select sex of child.

These last two points concern possibilities of pharmacological methods which would be hostile to sexually transmitted microorganisms, and to sperm which result in a zygote of the opposite sex desired by the couple.

It seems that the earlier in the reproductive process that intervention can occur, the simplier will be the task of preventing pregnancy. For example, preventing the sperm from entering the vagina is the most straightforward solution. Two common methods for doing this are condoms [12, 15] and male vasectomies [11, 12, 16], both beyond the control of the female, thus being "male" methods. Given that sperm is deposited in the vagina, the first option is to immobilize and kill the sperm (such as by spermicidal foam). The next choice is to prevent the sperm from passing through the cervical os (diaphragm device). Next would be blocking the passage of the sperm through the oviducts (tubal ligation/blockage or female sterilization); this technique is generally not reversible. The fourth option would be to prevent the egg from being available for fertilization by the sperm (OC's). Fifth, implantation of the fertilized egg can be prevented (IUD's). If implantation does occur, inducement of abortion is the last option for preventing birth. But these last two methods are not "contraception" in the purest sense, since conception has already occurred.

Considering the above, and given that this discussion concerns female contraception and birth control techniques which are reversible, the simpliest and safest method appears to be simple mechanical blockage of the passage of sperm from the vagina through the cervix. Without this passage, conception is impossible. Coincidentally, this blockage method also potentially satisfies all of the design criteria cited above.

A NEW FEMALE BARRIER CONTRACEPTIVE

The human cervix is in effect a valve. This valve allows menstrual fluid to pass from the uterus to the vagina during menstruation, and sperm to pass from the vagina to the uterus. At the same time, the cervix inhibits the transmission of bacteria and organisms into the uterus; one complication which IUD wearers experience is a higher rate of uterine and pelvic infections, presumably due to the pathway provided by the IUD "string" through the endocervix.

In blocking sperm movement through the endocervix (female barrier contraception), two possibilities were considered: first, make the cervical mucus more viscous and essentially impenetrable by the sperm; second, temporarily block the cervical os with a spermicide while live sperm are present in the vagina. In developing our new female barrier contraceptive, both approaches have been studied using a compliant sponge drug reservoir/delivery system. The first involves pharmacological agents which will require extensive safety testing before human effectiveness studies can be justified. The second utilizes a spermicidal compound which has been used in contraceptive creams, jellies, tablets, etc., for over twenty years and is generally recognized as safe and effective by the FDA.

This new female barrier contraceptive, the COLLATEX™ Contraceptive Sponge, incorporates the latter approach of a mechanical blocking coupled with a spermicidal compound. The specific design parameters incorporated in this vaginal contraceptive sponge (VCS) include:

- No systemic pharmacological agents;
- Compliant, flexible device to allow for anatomical changes during sexual arousal and intercourse;
- Biocompatible to allow up to two-day residency in the vagina, thereby permitting spontaneity without need for preparation;
- Ability to resist displacement during normal daily activities and intercourse;
- Small mass and/or volume so it will not be noticed by user and male partner;
- Non-implantable (no penetration of the cervical os) to allow insertion by users with techniques similar to menstrual tampons;
- Easily inserted/removed;
- No systemic or local aftereffects.

The sponge concept was chosen for its potential mechanical compliance; open-celled, hydrophilic properties were incorporated to allow adsorption of the semen. This latter property is important since the number of sperm available directly affects the probability for conception; for example, if the semen contains less than 20 million sperm per cc, the male is generally considered infertile.

This proprietary contraceptive system [25] has become feasible because of the recent availability of a new biomaterial, a water-reacting polyurethane prepolymer [26]. Using this prepolymer as a component, the VCS is molded as cup-shaped sponge, about six centimeters in diameter and about 1.5 centimeters thick, incorporating a woven polyester removal loop and about one gram of the spermicide, Nonoxynol-9; it is currently packaged in a transparent, heat-sealed, laminated film bag [Figure 4].

The sponge is inserted by the user into her vagina before intercourse after having been moistened with warm tap water. It can remain in place for two days, with multiple coital episodes possible during the use period. It must not be removed sooner than six hours after the last intercourse since live sperm may still be present in the vagina hours after ejaculation. The VCS has the following modes of action:

- It covers the cervix, providing mechanical blockage of the cervical os;

- It acts like a sponge to soak up semen and thus make fewer sperm available for fertilization;
- It slowly releases a spermicide which renders the sperm inactive.

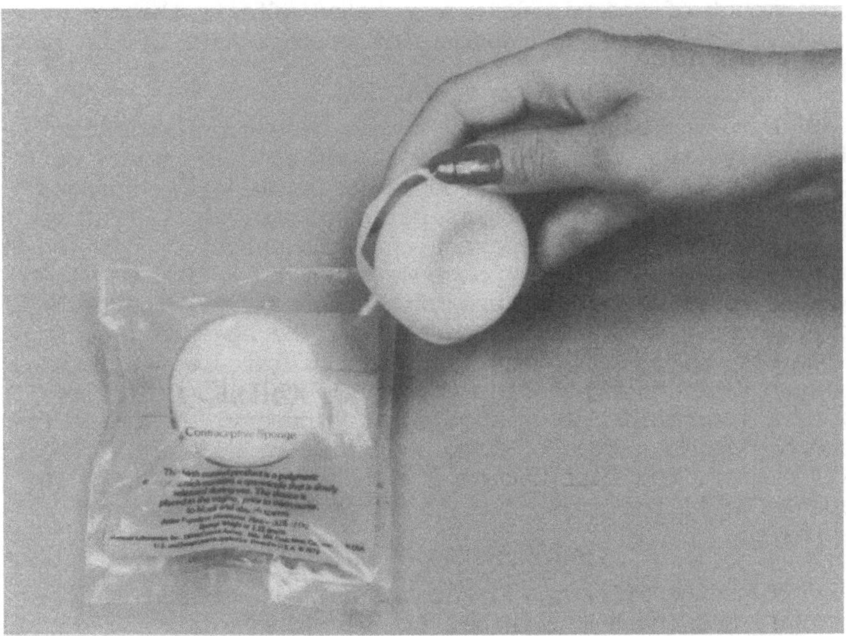

Figure 4. Current Version of the COLLATEX Contraceptive Sponge.

DEVELOPMENT HISTORY OF THE VLI CONTRACEPTIVE SPONGE

VLI began intravaginal sponge contraceptive research in early 1975 with a collagen sponge material composed of bovine protein fibers, without any spermicide. This material was initially attractive due to its potential biocompatibility. However, because of durability and economic considerations it was subsequently dropped from consideration and various synthetic polymers and co-polymers incorporating several types of spermicides were selected for evaluation. One particular polymer [26] continued to show significant advantages in the safety, drug delivery, and manufacturing areas during the early stages of testing, and eventually became the material of choice as it passed strenuous toxicological and mutagenic testing programs.

The major questions confronting us during the development process related to the size/shape of the sponge, the mechanical materials properties of the device, and the amount of spermicide (if any) required for contraceptive effectiveness. The last two factors are interrelated since the amount of spermicide incorporated affects the compressibility, cell size, and spermicide release rate of the sponge material.

Duration of use and reuse were also considerations. Initially, the VCS was first intended as a one-month contraceptive, with the user periodically removing the sponge, washing it, and either reinserting or storing for future use. It soon became apparent that the esthetics of removing and washing the sponge were not appealing to women; we now recommend single use of the sponge. Further, when spermicide was shown to be essential for acceptable effectiveness, single use was absolutely dictated since some users washed the sponge so vigorously that most of the active spermicide was washed out of the device; since the spermicide is a type of soap and foams or suds upon contact with water, the inclination of some women was to wash the sponge until the suds, i.e., the spermicide, no longer appeared.

Thus the VCS evolved through animal, laboratory (in vitro), and Phase I clinical (in vivo) acceptability trials from a one-month, multiple use collagen sponge with no spermicide, to a one-month polyurethane/collagen hybrid sponge with 10% by dry sponge weight of spermicide [#1 in Figure 5], to a ten-washing limit polyurethane sponge with 20% spermicide [#2 in Figure 5], to a three-washing limit rounded configuration with 30% spermicide [#3 in Figure 5], to the current dimpled polyurethane sponge with one gram of spermicide [#4 in Figure 5]. The Phase I clinical acceptability testing which generated this evolution of designs is discussed in the publication by Aznar [27]. Simultaneous with the clinical testing, in vitro techniques had to be developed to measure spermicidal release rates and spermicidal effectiveness of rinses from the sponges. While the in vivo studies were confirming the need for spermicide, the in vitro studies were demonstrating the performance effects of spermicide. These areas are discussed in more detail in the following sections.

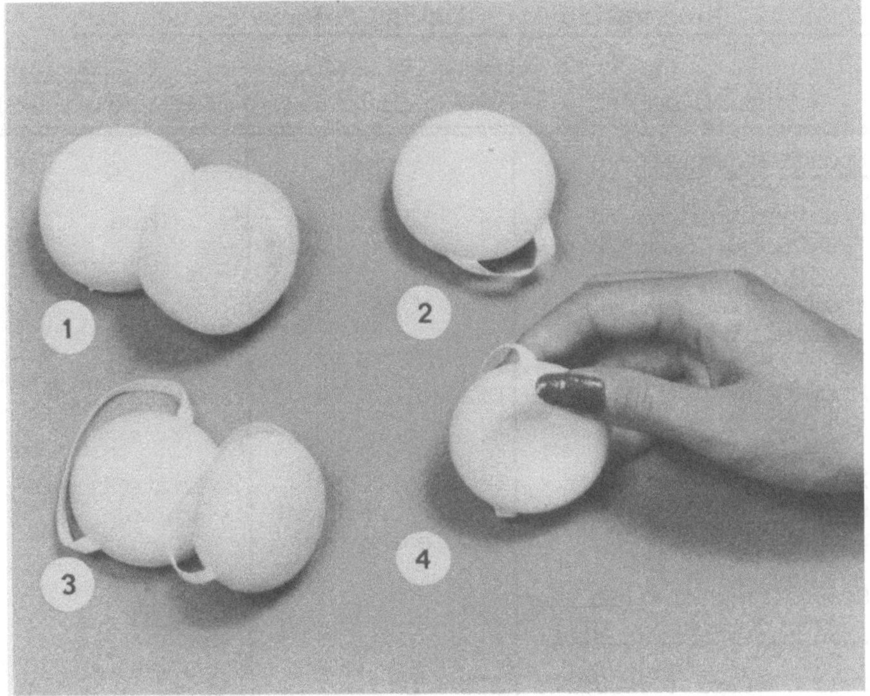

Figure 5. Evolution of the Vaginal Contraceptive Sponge to its Present Configuration (#4).

SAFETY STUDIES

The type of safety testing which a medical product should undergo before clinical (human) use depends on the type of contact with the intended user (e.g., external skin, mucous membrane, blood path, etc), and the criticality of use (e.g., a cardiac pacemaker vs. a menstrual tampon). An excellent current review of medical device safety testing has been prepared by Wallin [28].

For the VCS, short-term contact with the vaginal mucous membrane required thorough biocompatibility testing, short of long term implantation. The following test program was completed for the sponge, with and without the spermicide, to justify large scale clinical testing for effectiveness.

TABLE I. Biocompatibility Testing Program and Results

Test Methods	Urethane Sponge W/O Spermicide	Plus Spermicide (30% Nonoxynol-9)
Sponge Material:		
Free isocyanates by "Kubitz" & "Spot" tests	None	None
Aromatic amines by thin layer chromatography	None	None
Primary skin irritation-rabbits	Draize = 0.04	Draize = 0.25
Rabbit muscle implant seven days-USP with pathology	No significant tissue reaction; muscle ingrowth	Moderate tissue reaction
Sponge Extracts (Saline and/or Cottonseed Oil):		
Cell culture (WI-38) cytotoxicity MEM elution test-USP	CTE = 0	CTE = 4
Acute systemic injection mice-USP	Satisfactory	Unsatisfactory
Intracutaneous injection-USP	Satisfactory	Marked reactivity
Acute dermal toxicity-rabbits	No systemic tox. mild irritant	No systemic toxicity
48-hour dermal patch test-humans	May irritate	May irritate
Repeat insult dermal patch test-humans	Not a cumulative irritant; not a sensitizing agent	Not a cumulative irritant; not a sensitizing agent

TABLE I. continued

Ames salmonella mutagenicity	Negative	Negative
Mouse lymphoma assay	Negative	Negative

Interpretation of these test results is sometimes subjective since "acceptable" results for some of these have not yet been defined. Also, a particular test may not be entirely appropriate as was the case for the Acute Systemic Injection Test in mice; the extracts of the sponge without spermicide produced no effects on the mice, but the extracts of the sponges with spermicide resulted in the deaths of all the mice as would be expected with a "soap" injected into the cardiovascular system. Other tests, such as the mutagenicity assays, were made at two different test centers simultaneously because of their sensitivity and still-developing protocols. Satisfactory completion of this test battery justified the large-scale clinical trials described below.

ACCEPTABILITY/IN VITRO STUDIES

As we discussed earlier, two prime considerations in the design of the VCS were size and mechanical materials properties. Since there is no appropriate animal model for the human vagina, these aspects had to be defined using human volunteers, and by in vitro analyses of sponges worn by such women.

Some parameters were more qualitative than others; for example, the compressibility or "softness" of the sponge was determined to be acceptable by wearing, since comfort (of the user and her partner) was paramount. But, within the range of user comfort, too soft and the sponge would not maintain its position and respond to anatomical changes; on the other hand, high rigidity, low compressibility would possibly generate sufficient pressure against the vaginal wall to cause necrosis of the mucosal tissue, as has been reported with the rigid diaphragm. Density of the device was also important since it must not migrate down the vaginal tract due to gravitational force. And the desirability of having the sponge rapidly adsorb and bind semen dictated an open-celled foam, especially on the surface. Finally, the configuration and position of the woven polyester removal loop were determined purely by user feedback; some of the variations can be observed in Figure 5.

Size was a major decision. Many women commented that even a four centimeter diameter sponge seemed too large, seemingly overlooking the fact that a baby traverses the vagina as well. The size range studied was based on Masters and Johnson's research [7] involving in vivo measurements of vaginal distention during sexual arousal. We also performed similar studies using collagen sponges; the retention of sponges in women masturbating to orgasm was evaluated using a fiber optic duodenoscope remotely coupled to a motion picture camera. These studies confirmed both the lack of movement of the sponge during non-coital sexual excitation and the ability of sponges to adsorb a semen analog (water-glycerine) injected into the vagina through the duodenoscope. However, the major determinants of sponge size acceptability were postcoital observations of sponges by gynecologists, and comments of the users and their partners concerning ease of sponge insertion/removal and comfort during intercourse.

As the "softness" of the sponge material increases, its tear strength decreases. Tear strength is important since stress concentrations at the sponge/removal loop junction can cause tearing of the sponge upon removal. Thus, compressibility of the sponge for user comfort balances tear strength sufficient to enable removal of the sponge without tearing. Again, simple insertion/removal tests by human volunteers, plus feedback from subsequent clinical effectiveness testing, established experimentally the performance of the sponge. As a production quality control check for tear strength, sponge samples from each lot must survive a pull through a tapered polyethylene cone, 15 centimeters in length, 6 cm. in diameter at one end and 3 cm. at the other.

Initially we believed spermicide not to be an essential component of the VCS, since early postcoital testing [29] had indicated the adsorbing/blocking actions of a vaginal sponge to be sufficient for high theoretical contraceptive effectiveness. However, it soon became apparant, due to several pregnancies in the Phase I acceptability trials [27], that spermicide is crucial. With spermicide first incorporated at 10% of dry sponge weight, it was subsequently increased in two steps to its current concentration of 1 gram per sponge (30%). Since the spermicide is an active ingredient, it became necessary to develop quantitative tests showing how this component acted both in vitro and in vivo.

The in vitro spermicide tests basically involved measuring spermicidal activity and the rate of release from the sponge. This was done with these basic test protocols:

1) Sequential 5 minute soaks of sponge in 20 ml of physiologic saline; hand squeeze of sponge to collect the rinse eluate; and determination of amount of spermicide in each "rinse" (by UV spectroscopy) and spermicidal effectiveness of each rinse (by standard assay using fresh human ejaculate [30]). This test was basically an analog for user washing; recall that originally the VCS was intended for multiple use during a one month period. We thus had to demonstrate how the spermicide available was affected by washing. Sponges with four levels of spermicide (0, 10%, 20%, and 30% of dry sponge weight) were analyzed, along with the spermicidal effects of the preservatives used in the sponge and the buffered citric acid pH adjuster (pH is about 5.0 to correspond with the normal healthy vaginal acidity). These simple experiments showed that a transition occurs between 20% and 30%, wherein much more spermicide becomes available over a longer use period. Apparently a portion of the spermicide is bound into the urethane sponge as a surfactant; as the amount is increased, a level is reached where additional spermicide added is more readily released from its polymeric delivery system. This is one reason why we set our final product spermicide level at 30% of dry sponge weight. We also demonstrated that the preservatives and the citric acid have negligible spermicidal action relative to the Nonoxynol-9. A summary of the results from these tests is shown in Figure 6; see Ref. [31].
2) One, two, three, four, and five day "incubation" soaks of sponges in physiologic saline at body temperature. The eluates from these soaks were analyzed by UV for spermicide content and diluted sequentially to determine spermicidal potency. This test was intended as an analog to user wearing of the sponge; in this respect it was not satisfactory, although it did demonstrate little difference over time of spermicidal release. Results are summarized in Figure 7; see Ref. [31].

SPONGE CODE #	COMPOSITION						RINSE IN WHICH SPERMICIDAL EFFECTIVENESS LOST				
	URETHANE ONLY	10% N-9	20% N-9	30% N-9	PRESERVATIVES	CITRIC ACID	0	3	5	10	20
1	●						●				
2		●						●			
3			●						●		
4				●							●
5					●		●				
6				●	●						●
7						●	●				
8					●	●	●				
9				●	●	●					●

Figure 6. Spermicidal Activity of Washings From Various Types of Sponges.

INCUBATION TIME, DAYS	TOTAL SPERMICIDE CONTENT, mg (%)	AMOUNT OF TOTAL SPERMICIDE RELEASED, mg (%)	DILUTION AT WHICH SPERMICIDAL ACTIVITY LOST
1	1330 (10)	440 (33)	1:32
	3000 (20)	1120 (37)	1:64
	5140 (30)	1720 (34)	1:128
5	1330 (10)	550 (41)	1:32
	3000 (20)	1280 (43)	1:64
	5140 (30)	2330 (45)	1:128

Figure 7. Release of spermicide during one and five days of incubation in saline at 37 degrees C., and spermicidal potency of saline eluates.

3) Extract of spermicide from clinical sponges which had been a) washed and used for one month; b) worn one through five days continuously without removal or washing; and c) worn one day only, with and without coitus. These tests demonstrated that essentially all of the active spermicide is released during one month of use involving up to ten washings. When these results became available, the use period was limited to no more than three washings, with two days of wearing permitted between washings. This was subsequently modified to single use when the difficulties of defining a proper wash became apparent. The analyses of sponges worn for one through five days showed an exponential release of the spermicide, analogous to the soak tests described in 2) above. About 400 mg or 35 % of the spermicide was released during the first day of wearing. The analysis of the sponges with and without coitus is still underway, but indications are that the coital act increases this release to about 50% during one day of use. These test series will be presented in detail in Ref. [32].

Summarizing these test results, the current 30% spermicide level VCS maintains spermicidal effectiveness up to 20 simple rinses using 20 ml of physiologic saline; the preservatives and citric acid pH adjuster have essentially no spermicidal action; about 400 mg of spermicide are released in vivo during one day of wearing and about 600 mg are released during a one day period when intercourse occurs (note: these results are preliminary).

An additional human evaluation is currently underway to determine wear time tolerance of users under carefully controlled conditions. Users are instructed to wear the sponge for one week, changing to a new sponge every two days. The vaginal environment is being monitored using sophisticated gynecological diagnostic techniques (e.g., colposcopy, smears, cultures, etc.) for any pathological changes; no reports of such changes or of any problems associated with the sponge have been made from the field trials, other than two incidences of an apparent allergenic response to the spermicide (such allergenic response runs about 2-3% for users of other vaginal contraceptives containing Nonoxynol-9). Further, postcoital tests are being conducted wherein the cervix is examined for sperm about one and a half hours after intercourse at mid-cycle, both with and without use of the sponge; the women volunteers have all been sterilized [32]. These postcoital tests give an indication of the blocking and adsorbing actions in VCS effectiveness.

While these latter trials are still in progress and results cannot yet be given, indications are that the dimpled Sponge 4 [Figure 5] has a greater tendency to remain in place over the cervix than Sponge 3, as shown in vivo in Figure 8 and diagramatically in Figure 9. Therefore, the latest version of the VCS has the dimple incorporated [Figure 4]. Additional advantages of the dimple include a decrease in effective thickness of about one centimeter allowing deeper residence in the vagina and less chance of interference during intercourse, and easier removal due to a tendency to fold as shown in vitro in Figure 10.

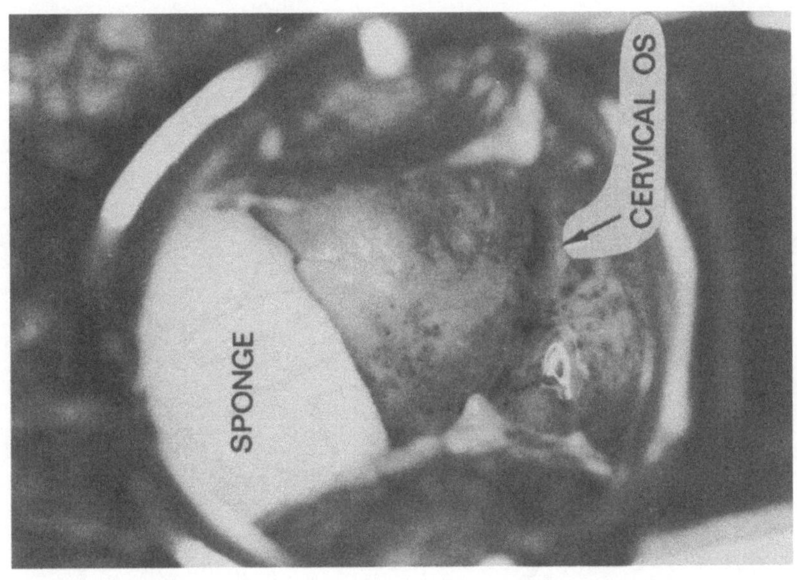

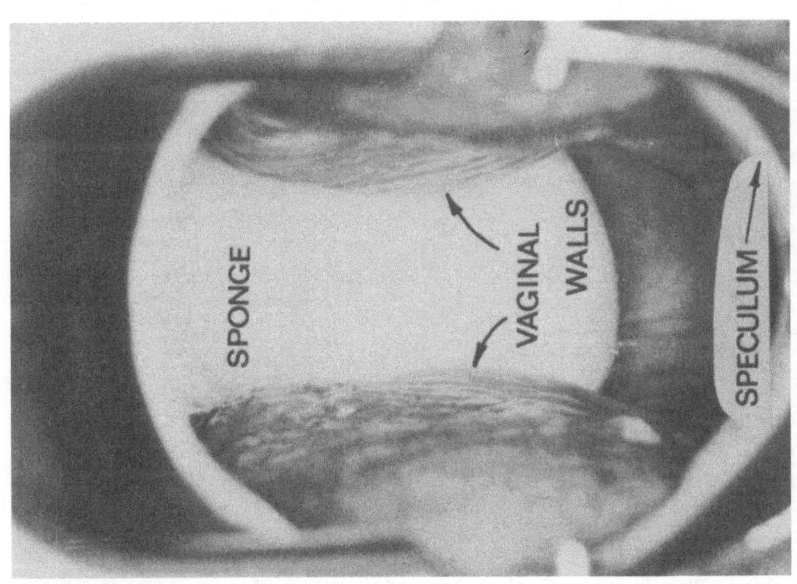

Figure 8. Intravaginal position of the COLLATEX Contraceptive Sponge, correctly over the cervix (left) and incorrectly displaced to the vaginal fornix exposing the cervical os (right); observed postcoitally using a gynecological speculum.

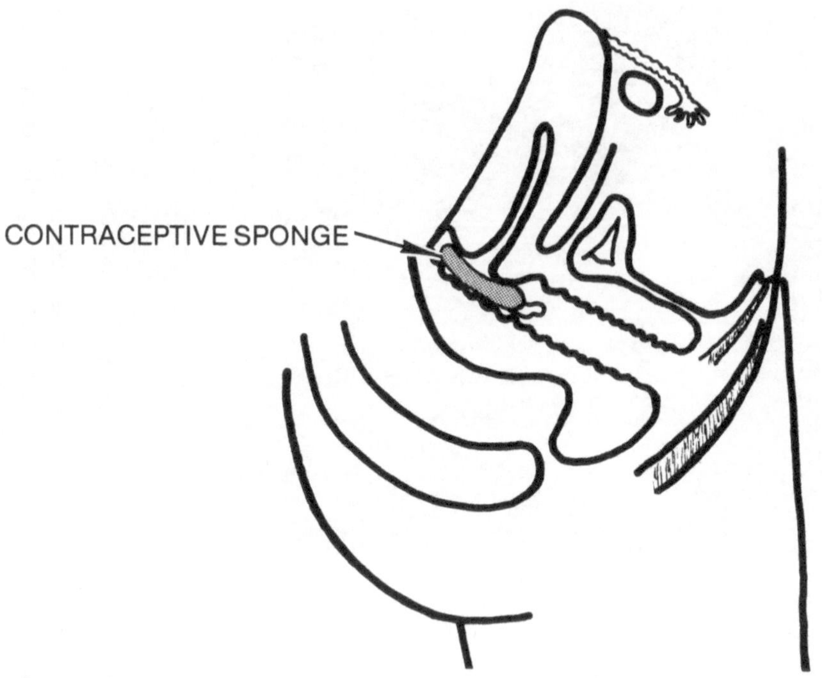

Figure 9. Diagram of correctly placed Contraceptive Sponge.

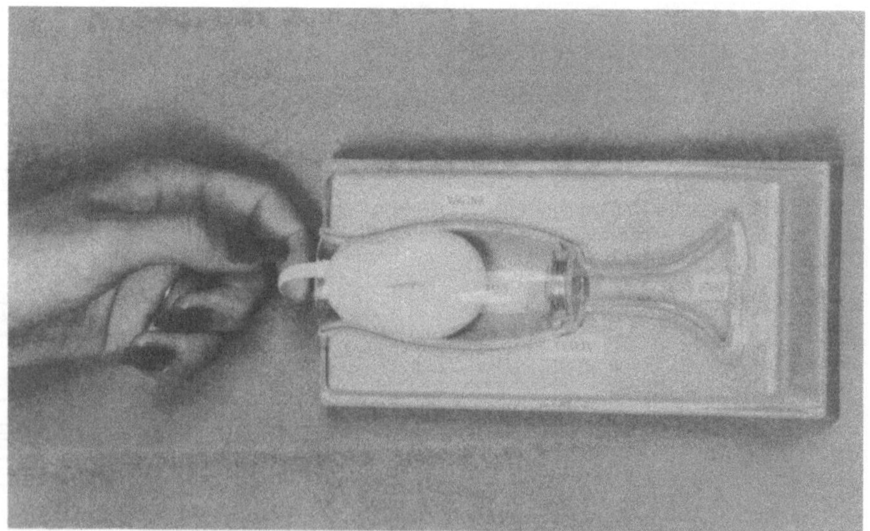

Figure 10. Tendency of the Contraceptive Sponge to fold upon removal, thereby decreasing pulling force required.

HUMAN EFFECTIVENESS STUDIES

The lack of a good animal model for the human reproductive system necessitates early clinical testing of contraceptive techniques once safety has been demonstrated. The first clinical trials of the VCS were intended to demonstrate acceptability of the method to the user, with effectiveness being a secondary consideration. These Phase I clinical trials were conducted under the auspices of the CIFE (Center for the Investigation of Fertility and Sterility) in Mexico City and PIACT (Program for the Introduction and Adaptation of Contraceptive Technology); they resulted in changes to the shape, size, removal loop, composition, packaging, and period of use of the VCS [27, 33]. In addition, because of unexpected pregnancies, the spermicide level was increased during this phase to the current 1 gram per sponge level. These trials involved 138 volunteers, were conducted from May, 1977, to January, 1979, and involved sponges 1 through 3 of Figure 5.

Confidence in the VCS method of birth control developed during the Phase I testing resulted in the IFRP (International Fertility Research Program, Research Triangle Park, N.C.) initiating Phase II trials with a limited clinic population at eight international centers in both developed and lesser developed countries. These tests gave preliminary effectiveness levels of 98.9% for developed countries, and about 96% over all; they are discussed in Ref. [34]. No comparative studies were made, although the IFRP has recently compared COLLATEX to a foaming tablet vaginal contraceptive using their computerized data banks for each product tested at similar centers [35]. Figure 11 shows the results of this recent comparison. The protocol for the Phase II trials called for the sponge to be used for no more than ten washings, and all data reflect usage of VCS on a wash and reuse basis. The protocol has been changed to single use for the Phase III trials discussed below. About 400 volunteers were enrolled in the Phase II trials which started during May, 1978, and were terminated in November, 1979. Sponge 3 of Figure 5 was used exclusively in these trials.

The Phase III effectiveness trials involve direct comparative studies against the currently used contraceptive methods of diaphragm/spermicide, spermicidal vaginal foam, and foaming vaginal tablets. These trials were begun in

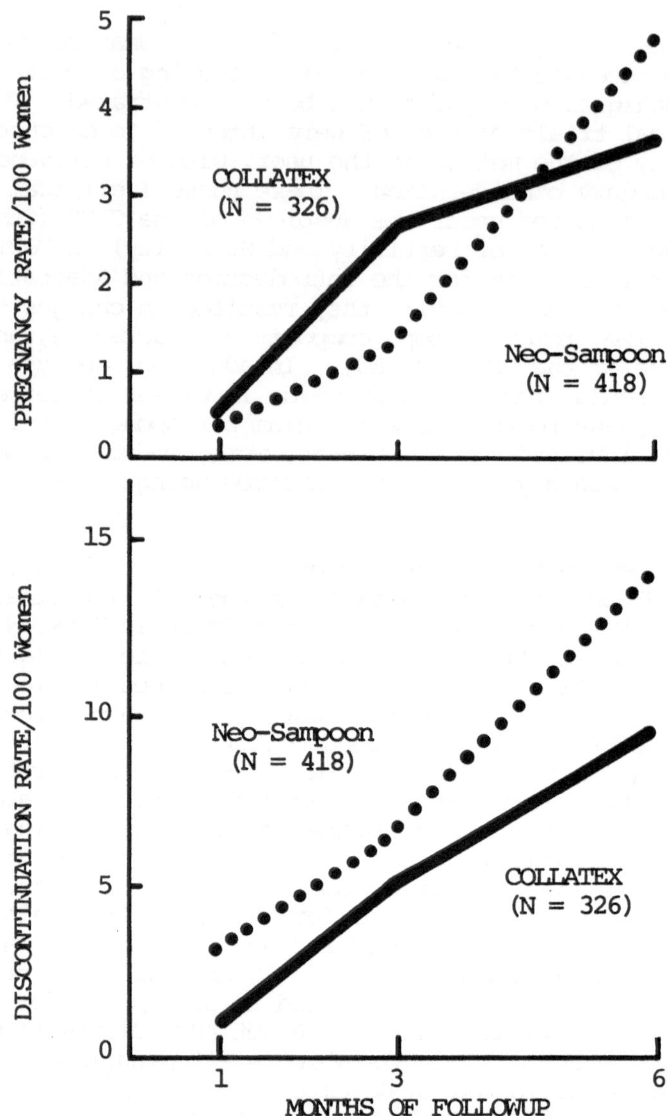

Figure 11. Six-month Cumulative Pregnancy and Discontinuation Rates for the COLLATEX Contraceptive Sponge and the Neo-Sampoon Foaming Vaginal Tablet (data from Ref. [35]).

October, 1979, and involve about 1400 women at eight international centers and 2000 women at twenty U.S. centers. After continuous one year data have been obtained, the trials will evolve into Phase IV followup trials for about 15% of the women to determine long-term effectiveness and continuation rates. The protocol for these trials calls for single use of the sponge (no washing) for a maximum two day wearing period; the sponge is to be left in place for at least six hours after intercourse, and multiple coital episodes are permitted during the wearing period. The shape of the sponge for the Phase III trials has been modified based on user and clinician feedback from Phase II, and as a result of postcoital observations of sponge position discussed previously. The final configuration of the sponge incorporates a dimple on the "inward" side facing the cervix (Figure 4 & Figure 5). The belief is that this will permit the sponge to behave somewhat like the cervical cap, positioning itself over the cervix thus enhancing the blocking action and further increasing effectiveness.

MANUFACTURING DEVELOPMENT

From the final selection of our material, manufacturing engineering has been a primary concern. Because we anticipated a high acceptance of the VCS in the lesser developed countries, we concentrated on reducing the manufacturing system to a reliable, modular, self-contained plant which could be readily containerized for shipping overseas. A major innovation permitting such a system was the development of a disposable mold for the sponge. This polypropylene mold snap-locks shut and has a ridged seal line at the mold junction to eliminate flashing (when shot size is properly in range). The surface finish on the inside of the mold was designed to obtain a skin-free surface on the sponge. These molds can in fact be the final package as well, although cost of the current molds makes this prohibitive. An additional feature of this mold system is that the shape and size of the sponge can be quickly modified by simply changing the disposable molds. Mold life is currently more than 5000 cycles.

The molds are locked in place on a continuous chain conveyor, the length of which is determined by the desired production rate and required cure time before mold opening.

The second generation production lines now being fabricated are about 4 meters long, weigh about two metric tons, and will produce 4 million sponges yearly on a three shift basis. Incidentally, based on an estimated typical use of 40 sponges yearly per user, one of the machines will supply about 100,000 users.

Summarizing the objectives obtained with this manufacturing system:

- Flexibility in sponge configuration/size through disposable molds;
- Low capital cost since molds are disposable;
- High production rate; molds are inexpensive, therefore line can readily be lengthened;
- Containerization of complete plant;
- Sponges can be manufactured close to market; local labor utilized; can be "nationalistic" (not imported) product;
- Packaging can be tailored to local requirements;
- With further packaging development, mold can become the package as well.

GOVERNMENTAL REGULATORY CONSIDERATIONS

The regulatory environment facing the small developer of medical products is a formidable one. This is the number one constraint on product innovation, and has considerably decreased the probability of successfully bringing a major new development to the marketplace. The caution and restraint forced by governmental regulations is certainly justified, for no one can argue against safety. The dilema arises in weighing risk against benefit, for this is necessarily subjective, especially when the product does not have a record of use.

The basic problem with regulatory schemes is that the only way the regulator can get into trouble is by saying "Yes", and then having something go awry. As long as he says "No", he takes no risks; of course, nothing happens either. In the drug and medical device field, the situation is further complicated because of consumer pressure. However, there are signs that this serious problem of hindered innovation is being recognized. For example, during 1979, a bill was introduced in Congress which will simplify the new drug approval process in certain cases. If this legislation is enacted, it probably will speed the marketing approval of products such as the COLLATEX Contraceptive Sponge.

As for our product, its regulatory status has changed during development. Originally the FDA classified the VCS as a medical device, regulated under the 1976 Medical Device Amendments to The Food, Drug, and Cosmetic Act. About one year later, the FDA determined that no opinion could be given concerning a specific product, but that the VCS would henceforth be regulated as a new drug because the spermicide, although not a new drug, was being delivered in a new manner. Therefore, we filed an Investigational New Drug application (IND) early in 1979 to obtain approval to continue our clinical trials. This is the first step in obtaining marketing approval, and simply permits the investigators to conduct human testing under approved protocols. The second step is to file a New Drug Approval application (NDA), and we currently plan this for late 1980.

SUMMARY

The development of sophisticated medical products involves a spectrum of activities peripheral to the engineering R&D. The COLLATEX Contraceptive Sponge development described in this paper is a typical example of this developmental process. The biomedical engineer not only faces the complexity of dealing with organisms [36], but also must deal with a myriad of regulatory constraints and guidelines, increasing consumer scrutiny and input, and a shrinking base of "risk" funds for such development work. The logic and process which generated the VCS reflects these factors.

And for the VCS the process is continuing: some areas where additional research will be required include reusability/washing, use as a menstrual tampon after contracepting, need for more than one size (function of number of childbirths, race, age, etc.?), inclusion of scents and flavorings, use of an insertion device to better facilitate placement, incorporation of drugs for disease prevention and/or treatment (sexually transmitted diseases are at epidemic levels worldwide), and potential for delivering other contraceptive drugs (e.g., prostaglandins).

REFERENCES

[1] "11 Million Teenagers", a pub. of the Alan Guttmacher Institute, 515 Madison Ave., New York, 1976.

[2] Fertility and Contraception in the United States, Report prepared by the Select Committee on Population, U.S. House of Representatives, 95th Congress, 2nd Session, 12/78.

[3] Blandau, R. & Moghissi, K., ed., The Biology of the Cervix, University of Chicago Press, Chicago, 1973.

[4] Epel, David, "The Program of Fertilization", Sci. Am., 11/77, pp. 129-138.

[5] Garcia, G. R. & Rosenfeld, D. L., Human Fertility: The Regulation of Reproduction, F. A. Davis Co., Phil., 1977.

[6] Hafez, E. S. F. & Evans, T. N., editors, Human Reproduction, Conception and Contraception, Harper & Row, Hagerstown, MD., 1973.

[7] Masters, W. H. & Johnson, V. E., Human Sexual Response, Little, Brown & Co., Boston, 1966.

[8] Segal, S. J., "The Physiology of Human Reproduction", Sci. Am., 9/74, pp. 53-62.

[9] Taylor, H., ed., Human Reproduction, Vol. 1, Physiology, MIT Press, Cambridge, 1976.

[10] Johnson, V. E. & Masters, W. H., "Intravaginal Contraceptives Study: Phase I. Anatomy", Western J. Surg. Obstet. Gynecol., 7-8/62, p. 202.

[11] Shapiro, H. I., The Birth Control Book, St. Martin's Press, New York, 1977.

[12] Hatcher, R. A., et al, Contraceptive Technology, 1978-1979, John Wiley, Irvington Publishers, N.Y., 1978.

[13] Population Report, Series A, #1-5, 9/75, Population Information Program, The Johns Hopkins Univ., 624 N. Broadway, Baltimore, 21205.

[14] Ibid, Series B, #2-3, 1/75.

[15] Ibid, Series H, #1-5, 1/76.

[16] Ibid, Series D, #1-3, 1/75.

[17] Bender, S. J. & Fellers, S., Contraception, by Choice or by Chance, Wm. C. Brown Co., Dubuque, Io., 1971

[18] Segal, S. J., "Advances and Opportunities in Fertility Research", Mt. Sinai J. of Med., Vol XLII, #4, 7-8/75, pp. 375-383.

[19] Himes, N. E., Medical History of Contraception, Schocken Books, N.Y., paperback, 1970.

[20] Bernstein, E., "Update on the Pill: Does Reliability Outweigh Risk?", Today's Health, 3/76, p. 15.

[21] FDA Drug Bulletin, "Risk of Myocardial Infarction in Users of Oral Contraceptives", 7-8/75, p. 10.

[22] Morrison, M., "Contraception with IUD's", FDA Consumer, 2/75, DHEW Pub. #75-4005.

[23] Bernstein, G. S., "Physiological Aspects of Vaginal Contraception, A Review", Contraception, Vol. 9, #4, 4/74, pp. 333-345.

[24] Bernstein, G. S., "Coventional Methods of Contraception: Condom, Diaphragm, and Vaginal Foam", Clin. Ob. Gyn., Vol. 17, #1, 3/74, pp. 21-33.

[25] Patent applications filed U.S. and foreign by VORHAUER LABORATORIES, INC.

[26] Hypol 2001 Prepolymer, W. R. Grace and Co., Organic Chemicals Div., Lexington, MA.

[27] Aznar, et. al., "Polyurethane Contraceptive Sponge: Product Modification Resulting from User Experience", submitted to Contraception, 1980.

[28] Wallin, R., "Safety Testing of Medical Devices, an Overview", Med. Dev. & Diag. Indust., Vol. 1, #6, 11/79, p. 25.

[29] Chvapil, M., Personal Communication to B. W. Vorhauer, 10/27/76.

[30] Moran, J., et. al., "Comparison of the Fractional Post-coital Test with the Sims-Huhner Post-coital Test", Int. J. Fertility, Vol. 19, 1974, p. 93.

[31] Bernstein, G. S., "Laboratory Studies of the Release of Nonoxynol-9 from Polymeric Intravaginal Contraceptive Sponges", submitted to Contraception, 1979.

[32] Bernstein, G. S., et. al., paper in preparation for submission to Contraception, 1980.

[33] Aznar, et. al., "A Clinical Appraisal of a Medicated Polyurethane Sponge Used for Contraception", in Vaginal Contraceptives: New Developments, Zatuchni, G., et. al., editors, Harper and Row, Hagerstown, in Press.

[34] Taylor, R., et. al., "Preliminary Results of a Multi-clinic Trial of a Polyurethane-Spermicide Contraceptive Sponge", ibid.

[35] Edelman, D. A., "Barrier Contraception - An Update", presented at the Assoc. of Family Planning Physicians Meeting in Philadelphia, 10/79, and to be published in Advances in Planned Parenthood, 1980.

[36] Biomedical Engineering Axiom No. 1 - Under carefully controlled experimental conditions, organisms behave pretty much as they please.

PERSISTENCE OF NON-NEWTONIAN SQUEEZE FILMS IN JOINTS

George Piotrowski, Ph.D

University of Florida

ABSTRACT

The fluid found between cartilage surfaces in joints is a dilute solution of an unbranched long chain polymer, which causes the synovial fluid to follow the power law model of a shear thinning fluid over a range of 5 orders of magnitude of the shear rate. An analysis was performed to determine the pressure distribution and load-velocity relationships for a squeeze bearing containing a shear thinning fluid, bounded by paraboloid surfaces, and subjected to a constant squeeze load.

In the simple case of parallel flat plates the pressure distribution was found to be independent of the film thickness, and varied slightly with the flow index; constriction of the edges of the fluid film caused the pressure distribution to flatten out. The squeeze time, defined as the time required for a pair of surfaces to be squeezed together at constant load from some initial film thickness to some final film thickness h_2, was found to vary as $h_2^{-(1/n+1)}$. Clearly, the squeeze times increase dramatically for shear-thinning fluids over isoviscous fluids, in thin films. For films with constricted edges, the increase in squeeze time due to the shear thinning behavior is even more pronounced.

These results suggest that the presence of hyaluronic acid in synovial fluid can explain the long times required for the extrusion of the fluid film from between cartilage surfaces. The formation of an edge constriction, which readily occurs with soft bearing surfaces, increases the squeeze times tremendously, indicating that the shear-thinning lubricant and the compliant surface, acting as a non-Newtonian elastohydrodynamic bearing, can explain the persistence of fluid films in synovial joints during high-load portions of the loading history.

INTRODUCTION

Synovial joints are places where bones connect, yet may move freely relative to each other. These joints are further distinguished from other joints in the body in that they contain a cavity filled with a clear liquid called synovial fluid. A typical synovial joint is shown in cross section in Figure 1, which has been adapted from Barnett _et al_. (1961). The opposing ends of the bone flare out into condyles, consisting of cancellous (spongy) bone covered with a thin layer of dense bone. The articulating surfaces are covered with hyaline cartilage which, together with the underlying bone, forms a resilient or compliant bearing surface.

The bone ends are connected by a tough, fibrous sheath or capsule, which encloses the joint like a sleeve and creates the joint space or cavity. Ligaments act with the capsule as elastic constraints on the relative motions of the joint surfaces. The interior surface of the capsule, except for the articulating surfaces, is lined with the synovial membrane which secretes the synovial fluid.

The fluid in synovial joints was described by Jones (1934) as being generally clear and straw colored with a specific gravity slightly higher than water and a viscosity that ranged from 2.5 to 10.0 times that of water. Normal synovial fluid from humans is difficult to obtain, due to the small volume present in any given joint. Most work on "normal" human synovial fluid has been done with fluid obtained from patients with traumatic effusions. By contrast, the joints of cattle contain copious amounts of synovial fluid. Thus, much of the data in the literature has been obtained from bovine fluid, which appears quite similar to human fluid in all respects.

Synovial fluid is a dialysate of blood plasma, with a protein-polysaccharide complex (sometimes referred to as

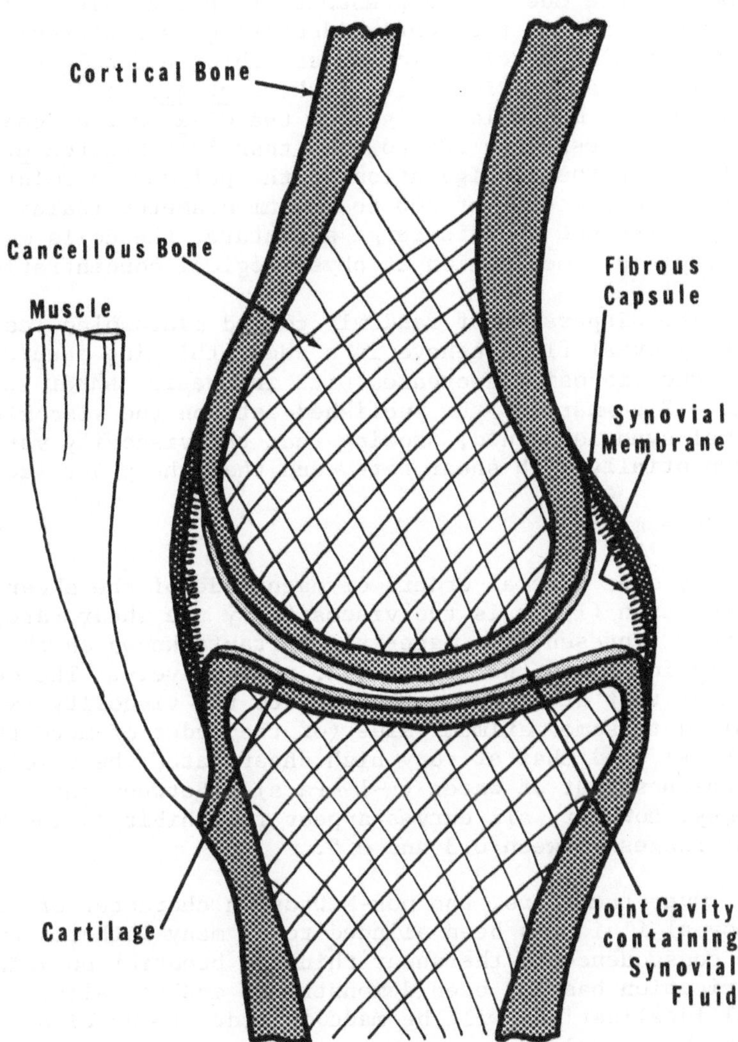

Figure 1. Schematic cross section through a typical synovial joint.

mucin or mucopolysaccharide) added. Most of the interesting properties of synovial fluid, particularly its viscous behavior, are due to the presence of this complex. The polysaccharide is frequently identified as hyaluronic acid (sodium hyaluronate), which is an unbranched polymer of repeating disaccharide units (Atkins et al., 1972). Estimates of the molecular weight of the disaccharide chain range from less than 10^6 to more than 10^7 (Preston et al., 1965), with the configuration of the polymer in solution being a random coil of 200 to 400 nm diameter (Balazs, 1966). Because of this large structure, the coils must interact in some fashion at physiological concentrations.

The dispersion of randomly coiled macromolecules makes bulk synovial fluid behave in a shear-thinning fashion, i.e. the viscosity decreases with increasing shear rate. Figure 2 summarizes the published data on the viscosity of normal synovial fluid, showing that the viscosity varies exponentially with shear rate, and that the power law model

$$\eta = m\dot{\gamma}^{n-1} \qquad (1)$$

applies over several orders of magnitude of the shear rate. In Equation (1), η is the viscosity, $\dot{\gamma}$ the shear rate, and m and n represent two material constants known as the viscosity index and the flow index, respectively. The curves suggest that at very low shear rates the viscosity is limited to some maximum value (of the order of more than 10 Pa·s), and that at very high shear rates the viscosity approaches that of water (η~1 mPa.s). Between those extremes, however, all curves appear to exhibit a flow index that ranges between 0.3 and 0.5.

Over the years, the non-Newtonian character of the synovial fluid has been alluded to by many authors, but the consequence of the shear thinning behavior on joint lubrication has not been demonstrated analytically. Several idealizations will be made in order to develop a tractable model.

THE ANALYTICAL MODEL

The cartilage surface will be modeled as a smooth, non-porous, elastic solid. Viscoelastic effects will be ignored since their time constants are much longer than a

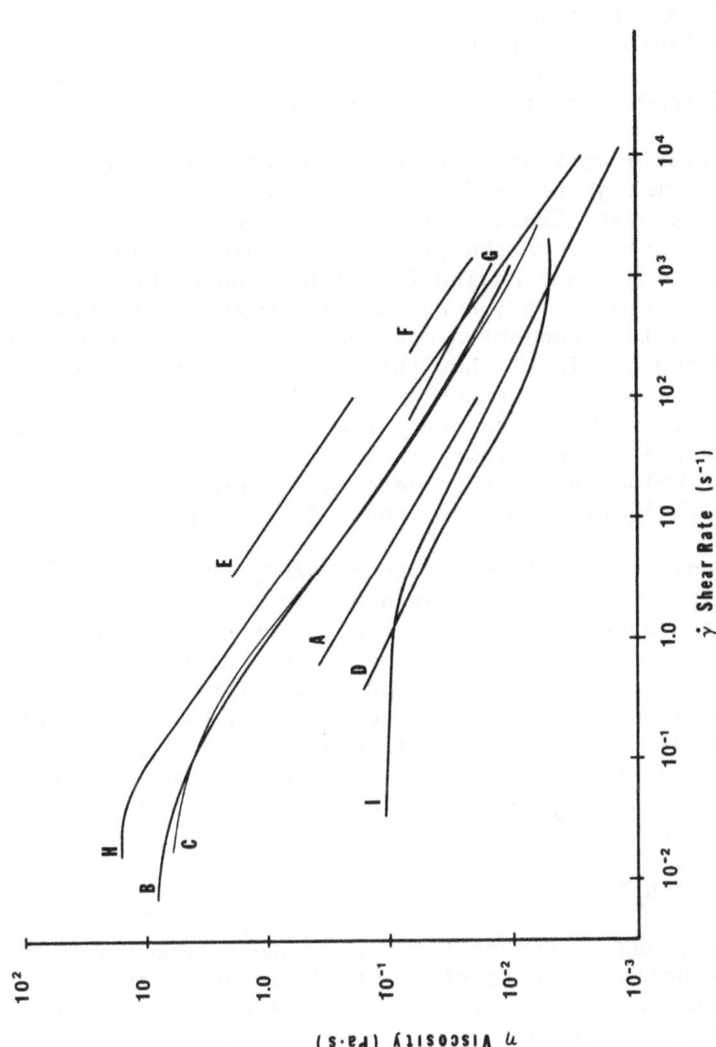

Figure 2. Effect of shear rate $\dot{\gamma}$ on synovial fluid viscosity η for normal synovial fluids. Sources of data: A, Ogston and Stanier (1953). B, Davies and Palfrey (1969). C and D, Davis (1966). E, Bloch and Dintenfass (1963). F, Vos and Theyse (1969). G, Caygill and West (1969). H and I, King (1966).

typical loading cycle, such as a gait cycle. The simultaneous solution of the fluid dynamics problem involving the flows in the synovial fluid film and the elasticity problem of the deformation of the bearing surfaces was not attempted, but the fluid flows between various geometries of articulating surfaces were analyzed to determine pressure distributions and load-velocity relationships.

Fein (1967) estimated the duration of some squeeze films in synovial joints and concluded that squeeze films, formed when synovial fluid is trapped between approaching cartilage surfaces, were the primary mechanism of joint lubrication. Gupta and Phelan (1964) had shown that the load capacity of a short journal bearing with oscillating speed and load is dependent almost entirely on the squeeze behavior of the bearing. Thus this analysis will involve an axisymmetric curved surface approaching a flat plate, with a power-law fluid separating the surfaces. A relationship between the force driving the surfaces together and the approach velocity shall be derived in terms of the geometry of the fluid film and the properties of the fluid.

The geometry of the model to be analyzed is illustrated in Figure 3. The two surfaces are assumed to be rigid, and the upper bearing surface is then described by $z = h(r,t)$, where $h(r,t)$ is the local, instantaneous gap height. This gap is filled by a power-law fluid which is being squeezed out from this "squeeze" bearing by a pair of axial forces W_o. Pressure distributions and load-velocity relationships will be developed for the simple case of two flat plates as well as for the more general case of a parabolic gap, i.e.

$$h(r,t) = h_o(t) + \beta r^2 \tag{2}$$

Let u,v, and w represent the r, θ, and z components of the fluid velocity respectively. Since the geometry is axisymmetric, the tangential fluid velocities and derivatives with respect to θ vanish

$$v = \frac{\partial}{\partial \theta} = 0 \tag{3}$$

For a gap height much less than the diameter of the bearing it can be shown that $\partial u/\partial z$ is much larger than any other velocity gradient. Likewise, the power-law viscosity reduces to

$$\eta = m\left|\frac{\partial u}{\partial z}\right|^{n-1} \tag{4}$$

The Navier-Stokes and continuity equations that describe the flow therefore reduce to

$$\frac{\partial p}{\partial z} = 0 \tag{5a}$$

$$\frac{\partial p}{\partial r} = m\frac{\partial}{\partial z}\left[\left|\frac{\partial u}{\partial z}\right|^{n-1}\frac{\partial u}{\partial z}\right] \tag{5b}$$

$$\frac{1}{r}\frac{\partial}{\partial r}(ru) + \frac{\partial w}{\partial z} = 0 \tag{5c}$$

with boundary conditions that

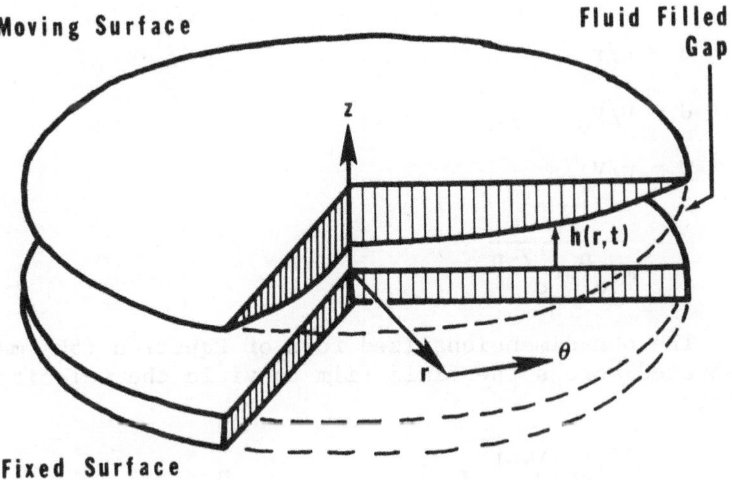

Figure 3. Geometry of the squeeze bearing.

$$u = 0 \quad \text{at} \quad z = 0 \tag{6a}$$

$$u = 0 \quad \text{at} \quad z = h \tag{6b}$$

$$w = 0 \quad \text{at} \quad z = 0 \tag{6c}$$

and

$$w = -V_o \quad \text{at} \quad z = h \tag{6d}$$

where V_o is the velocity of approach. The conditions that

$$\frac{\partial p}{\partial r} = 0 \quad \text{at} \quad r = 0 \tag{7a}$$

and

$$p = 0 \quad \text{at} \quad r = r_o \tag{7b}$$

where r_o is the outer radius of the bearing surfaces, must also be used in integrating Equation (5b). This completes the formulation of the problem.

The describing equations and boundary conditions can be non-dimensionalized by the following relations:

$$R = r/r_o \tag{8a}$$

$$Z = z/r_o \tag{8b}$$

$$P = p\pi r_o^{\,2}/W_o \tag{8c}$$

$$H = h/r_o \tag{8d}$$

$$U = u/V_o \tag{8e}$$

$$W = w/V_o \tag{8f}$$

and

$$M = \frac{W_o}{mV_o^{\,n}\pi r_o^{\,2-n}} \tag{8g}$$

The non-dimensionalized form of Equation (5b) may be integrated across the fluid film to yield the velocity profile

$$U = \frac{\left(M\left|\frac{\partial P}{\partial R}\right|\right)^{N-1}}{N}\left[\left(\frac{H}{2}\right)^{N} - \left(\frac{H}{2} - Z\right)^{N}\right], \quad Z \leq \frac{H}{2} \tag{9a}$$

$$U = \frac{\left(M\left|\frac{\partial P}{\partial R}\right|\right)^{N-1}}{N}\left[\left(\frac{H}{2}\right)^{N} - \left(Z - \frac{H}{2}\right)^{N}\right], \quad Z \geq \frac{H}{Z} \tag{9b}$$

where the substitution

$$N = \frac{1}{n} + 1 \tag{10}$$

has been made purely for the sake of convenience.

Now this velocity profile U(Z) can be substituted into the non-dimensionalized form of the continuity equation (5c), which is then integrated across the fluid film to yield an expression for the pressure gradient in the bearing of the form

$$\frac{1}{R}\frac{\partial}{\partial R}\left[\frac{2R}{N+1}\left(M\left|\frac{\partial P}{\partial R}\right|\right)^{N-1}\left(\frac{H}{2}\right)^{N+1}\right] = 1 \tag{11}$$

Solving for the pressure gradient and integrating to obtain the pressure profile leads to the relation

$$P(R) = \frac{2}{M}(N+1)^{n}\int_{R}^{1}\frac{R^{n}dR}{H^{2n+1}} \tag{12}$$

This integral,

$$\Phi = \int_{R}^{1}\frac{R^{n}dR}{H^{2n+1}} \tag{13}$$

cannot be solved in closed form except for the case of two flat plates, where H is independent of R. Since the integral of the pressure profile equals the applied load W_o,

$$2\int_{0}^{1} PR\, dR = 1 \tag{14}$$

and thus

$$P(R) = \frac{\Phi}{2\int_{0}^{1}\Phi R dR} \tag{15}$$

Equation (15) shows that the shape of the pressure distribution is determined solely by the geometry of the fluid film and the flow index of the fluid. The load-velocity relation is more complicated and can be shown to be of the form

$$V_o = \frac{n}{(2n+1)r_o^{\frac{2-n}{n}}} \left[\frac{W_o}{4\pi m \int_0^1 \Phi R dR} \right]^{\frac{1}{n}} \tag{16}$$

We can define a "non-dimensional" approach velocity, $\bar{V}$, which depends only on the bearing geometry and the fluid's flow index, as

$$\bar{V} = \left[\frac{1}{2 \int_0^1 \Phi R dR} \right]^{\frac{1}{n}} \tag{17}$$

in which case Equation (16) can be written as

$$V_o = \frac{n}{2n+1} \left[\frac{W_o}{2\pi m \; r_o^{2-n}} \right]^{\frac{1}{n}} \bar{V} \tag{18}$$

ANALYTICAL FINDINGS

For two flat plates, i.e.

$$H = H_o(t) \tag{19}$$

the pressure distribution can be derived in closed form to be

$$P = \frac{n+3}{n+1} \left(1-R^{\,n+1} \right) \tag{20}$$

It is interesting to note that the pressure profile is independent of the fluid film thickness. The non-dimensional approach velocity is

$$\bar{V} = (n+3)^{\frac{1}{n}} H_o^{\frac{1}{n}+2} \qquad (21)$$

and decreases with decreasing film thickness. Since the exponent of H_o is larger with shear thinning fluids (n<1) than with Newtonian fluids, this decrease of approach velocity with decreasing film thickness is more pronounced with shear-thinning fluids. The complete expression for the approach velocity becomes

$$V_o = \frac{n}{2n+1}\left[\frac{(n+3)h^{2n+1}W_o}{2\pi m\ r_o^{n+3}}\right]^{\frac{1}{n}} \qquad (22)$$

This agrees with the formulas derived by Scott (1931) for the case of the flat plate plastimeter.

Equation (22) can be integrated to find the time required to squeeze the fluid film down from some initial film thickness h_1 to some final value h_2. This time interval Δt is given by

$$\Delta t = \frac{(2n+1)(1+n)}{n^2}\left[\frac{2\pi m\ r_o^{n+3}}{(n+3)\ W_o}\right]^{\frac{1}{n}}\left[\frac{1}{h_2^{\frac{1}{n}+1}} - \frac{1}{h_1^{\frac{1}{n}+1}}\right] \qquad (23)$$

and is clearly seen to be quite sensitive to the value of the flow index. If h_1 is much greater than h_2, Δt essentially is inversely proportional to $h_2^{\frac{1}{n}+1}$; for a Newtonian fluid Δt is inversely proportional to h_2^2 but for a shear thinning fluid with n=0.5, Δt goes as $1/h_2^3$. Shear thinning fluids are much more difficult to extrude from the film than Newtonian fluids.

For the more general case of a parabolic gap, as described by Equation (2), the integral Φ of Equation (13) was obtained numerically for various values of the nondimensional curvature K, given by

$$K = \frac{2\beta r_o^2}{h_o} \tag{24}$$

As is shown in Figure 4, the pressure profiles for a gap with K=3.2 are the same for both Newtonian and shear thinning fluids. As K approaches 0, the upper bearing surface changes from convex to flat, and the pressure for the shear-thinning fluid is higher in the center of the bearing than it is for a Newtonian fluid, as shown by Equation (20). The discrepancy becomes more pronounced for a film with $K<0$, where h is maximum at the center and the fluid film is constricted at the edges of the film. Thus shear-thinning films with flat or concave bearing surfaces exhibit higher pressures in the center than Newtonian films.

A computer program was written to perform the integrals of Equations (13) and (15) for parabolic bearing surfaces numerically. These computer-calculated values for $\bar{V}$ were found to differ from the flat-plate values by a function of only K and n, such that

$$\bar{V} = f(K,n)\ \bar{V}_f \tag{25}$$

where $\bar{V}_f$ is the flat plate solution given by Equation (21). The function f(K,n) was found numerically to fit the relation

$$f(K,n) = \left(\frac{K+2}{2}\right)^{\frac{1}{n}+1} \tag{26}$$

to better than 0.5% for a wide range of values of h, n, and K. Thus the non-dimensional approach velocity for the case of a parabolic bearing surface is given by

$$\bar{V} = (n+3)^{\frac{1}{n}} \left(\frac{K}{2} + 1\right)^{\frac{1}{n}+1} H_o^{\frac{1}{n}+2} \tag{27}$$

where H_o is the non-dimensional fluid film thickness at the center of the bearing.

Figure 5 illustrates the relationship between $\bar{V}$, H_o n, and K [c.f. Eq. (24)]. Note that β is the actual curva-

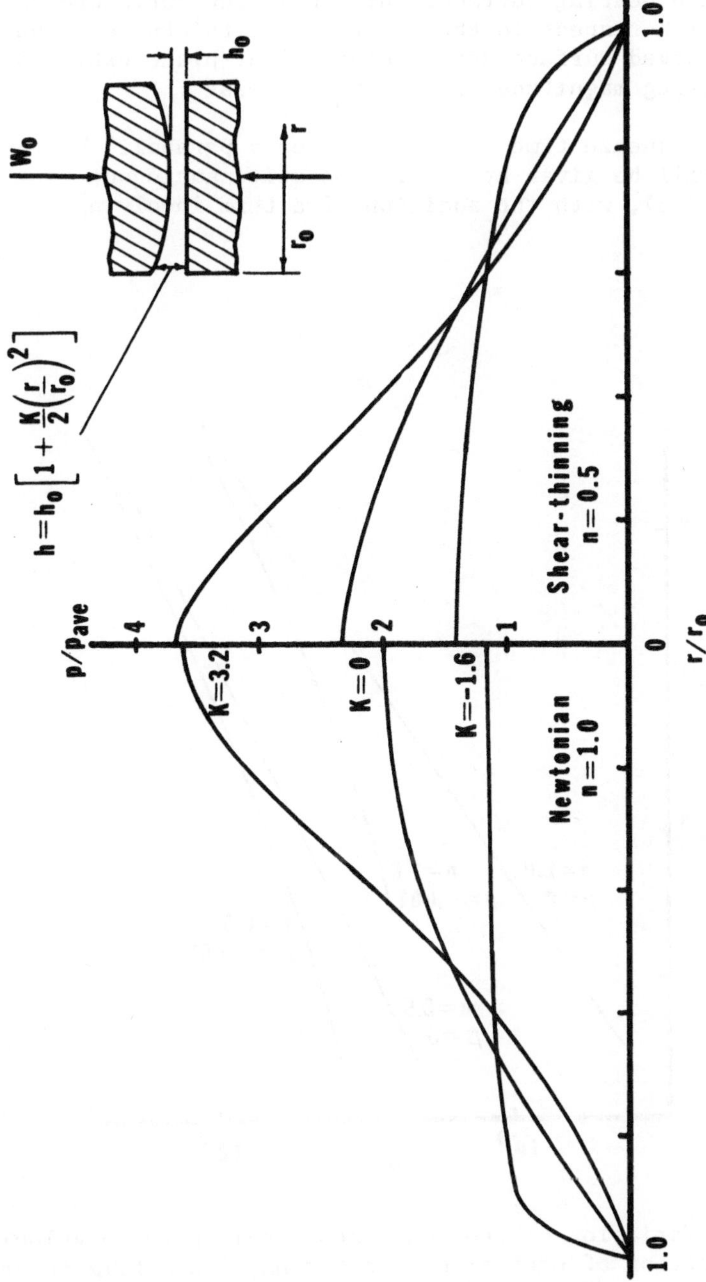

Figure 4. Pressure profiles as a function of gap geometry for Newtonian and shear-thinning fluids.

ture of the bearing surface, while K is the curvature normalized with respect to the central film thickness. Thus a rigid curved surface appraoching a flat plate exhibits an increasing magnitude of K as H_o decreases.

The squeeze time for the case of a parabolic bearing surface will be given by a relation quite analogous to Equation (23), with the addition of a term involving K:

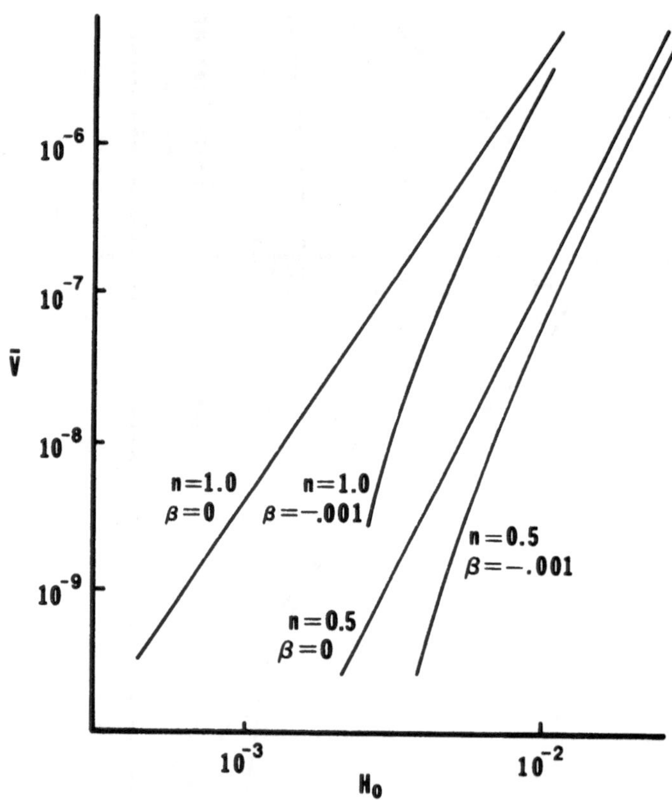

Figure 5. Variation of the non-dimensional approach velocity $\bar{V}$ as a function of central gap height H_o, fluid film curvature $K = 2\beta r_o/H_o$, and flow index n (for r_o = 20 mm).

$$\Delta t = \frac{(2n+1)(n+1)}{n^2} \left[\frac{2\pi m\, r_o^{3+n}}{(n+3)\, W_o} \right]^{\frac{1}{n}} \left(\frac{2}{K+2} \right)^{\frac{1}{n}+1}$$

$$\left[\frac{1}{h_2^{\frac{1}{n}+1}} - \frac{1}{h_1^{\frac{1}{n}+1}} \right] \tag{28}$$

As before, the squeeze time is inversely proportional to $h_2^{\frac{1}{n}+1}$, but in this case the squeeze time is magnified by the factor involving the curvature. When K is negative, meaning that the edges of the film are constricted, the squeeze times for shear-thinning fluids can become quite large. As Equations (27)-(28) show, squeeze films with shear thinning fluids and constricted edges are extremely persistent.

DISCUSSION AND CONCLUSIONS

Squeeze bearings containing a shear-thinning lubricant behave differently than bearings containing a Newtonian fluid. The shear-thinning lubricant generates higher pressures in the center of the bearing than a Newtonian lubricant, no matter what the viscosity index of the fluid. Shear-thinning fluid is harder to extrude out of a parallel-plate bearing with small film thickness. It is even harder to extrude when the edges of the film are constricted as was the case for the bearing surfaces with negative curvature.

A compliant spherical surface approaching a flat plate with a fluid in the gap will deform as the fluid pressure builds up. This was shown very clearly by Roberts and Tabor (1963). If a Hertzian contact is postulated, with a parallel gap, the contact stress will be distributed as shown in Figure 6. The fluid pressure, however, will be differently distributed, being higher in the center and lower at the periphery. Where the fluid pressure exceeds the required contact stress, further deformation of the surface will take place. Where the fluid pressure is less than the Hertzian stress, the surface will relax. Consequently, an elastohydrodynamic fluid film, i.e. a film with constricted edges, will form. This type of film will form more readily for a

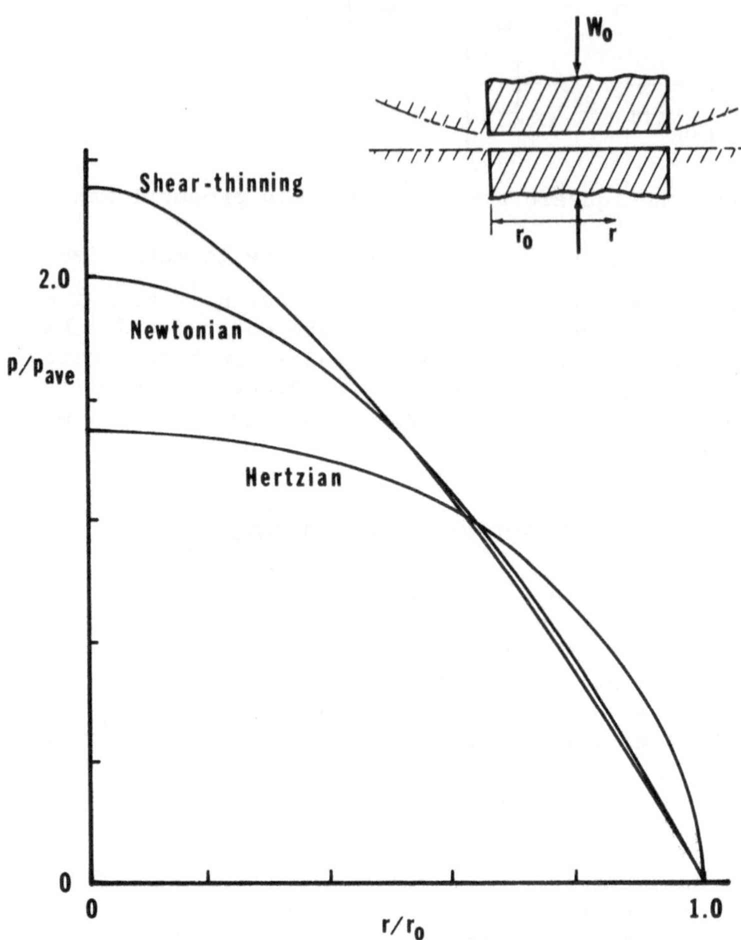

Figure 6. Comparison of the Hertzian contact stresses with the pressure distributions for Newtonian and shear-thinning fluids in a parallel squeeze film.

shear-thinning fluid since the pressure distribution is more peaked, and persist longer, as shown by the equations for the approach velocity (27) and for the squeeze time (28).

Joints, especially the heavily loaded joints in the lower extremities, are typically subjected to cyclic loads and motions, with periods of high loads coinciding with slow movements, and rapid motions occurring under relatively low loads. Consider the case of a circular contact area 30 mm in diameter loaded with a force of 750 N, with a fluid described by a flow index of n=0.5 and a viscosity index of m=1.0 Pa.$s^{0.5}$. When fluid film edges are assumed to be lightly constricted, and K is assumed to be constant at K=-0.5, a time of 0.5 seconds is required to squeeze the film from 6 μm to about 1 μm. If a more pronounced constriction is assumed, say K=-1.2, than a squeeze time of 9.5 seconds is required. Conversely, application of the load for 0.5 seconds only reduces the film thickness to 2.6 μm.

The above values are not intended to represent any one particular joint, but are representative of the conditions joints operate under. Clearly, a shear-thinning squeeze film can support substantial loads for a considerable time, as was hypothesized by Fein (1967). Walker et al. (1968) and Longfield et al. (1969) proposed the theory of "boosted lubrication" to explain the long squeeze times they observed when pieces of cartilage were pressed against glass with synovial fluid as lubricant. The equation describing a Newtonian squeeze film was used in the analysis, and yielded very high effective viscosities. This phenomenon can, however, be explained by a constriction of the edges of the squeeze film, or by the resistance to squeeze motion that a shear thinning fluid acquires at small film thicknesses, or both, without resorting to a hypothetical entrapment mechanism.

In contrast to the imbibition of fluid required for "boosted lubrication", McCutchen (1959, 1978) advocated the concept that fluid flow out of the cartilage, referred to as "weeping", is a significant factor. It can be shown (Piotrowski, 1975) that the presence of a weeping flow decreases the approach velocity by a factor $V_o/(V_o + 2w_n)$ where w_n is the fluid velocity out of the cartilage (assuming a uniform flow for all values of r). Conversely, fluid flow into the cartilage will increase the squeeze velocity. It must be kept in mind, however, that the pressure distribution is quite flat (c.f. Figure 4), and thus there is nothing to drive fluid radially within the cartilage to supply the weeping flow or to remove the imbibed flow. Since no estimates of the magnitude of w_n relative to V_o are available, the effect or likelihood of weeping or imbibition is not clear.

CLOSURE

The analysis presented in this paper demonstrates that extremely long squeeze times are possible in synovial joints. These squeeze films can readily be formed due to the cyclic loading of the joint and the compliance of the articulating surfaces. The analysis assumed rigid bearing surfaces, but *in vivo* the surfaces deform under pressure, and the geometry of the squeeze film undoubtedly changes appreciably with time. Analytical estimates of squeeze times for joints cannot be made until the combined fluid and elastic problem is solved. An iterative approach, using the computer program developed by Burstein (1968) in combination with the computer program developed for this analysis appears to be a feasible method for the simulation of *in vivo* squeeze film behavior.

REFERENCES

Atkins, E.D.T.; Phelps, C.F.; and Sheehan, J.K. 1972. The conformation of the mucopolysaccharides: Hyaluronates. *Biochem. J.* 128: 1255-1263.

Balazs, E.A. 1966. Sediment volume and viscoelastic behavior of hyaluronic acid solutions. *Fed. Proc*. 25; 1817-1822.

Barnett, C.H.; Davies, D.V.; and MacConaill, M.A. 1961. *Synovial Joints. Their Structure and Mechanics*. New York: Longmans, Green and Company.

Bloch, B., and Dintenfass, L. 1963. Rheological study of human synovial fluid. *Austral. N.Z.J. Surg*., 33: 108-113.

Burstein, A.H. 1968. Elastic analysis of condylar structures. Ph.D. Thesis, School of Engineering and Science, New York University.

Caygill, J.C. and West, G.H. (1969). The rheological behavior of synovial fluid and its possible relation to joint lubrication. Med. Biol. Eng. 7:507-516

Davies, D.V. 1966. Synovial fluid as a lubricant. Fed. Proc. 25: 1069-1076

Davis, D.V., and Palfrey, A.J. 1968. Some of the properties of normal and pathological synovial fluids. J. Biomech. 1: 79-88.

Fein, R.S. 1967. Are synovial joints squeeze-film lubricated? Proc. Instn. Mech. Engrs. 181 (Pt. 3J): 125-128.

Gupta, B.K., and Phelan, R.M. 1964. The load capacity of short journal bearings with oscillating effective speed. J. Basic Eng., Trans. ASME, Ser. D. 86: 348-354.

Jones, E.S. 1934. Joint lubrication. Lancet, 226: 1426-1427.

King, R.G. (1966). A rheological measurement of three synovial fluids, Rheol. Acta. 5: 41-44.

Longfield, M.D.; Dowson, D.; Walker, P.S.; and Wright, V. 1969. 'Boosted' lubrication of human joints by fluid enrichment and entrapment. Biomed. Eng. (London) 4: 517-522.

McCutchen, C.W. 1959. Sponge-hydrostatics and weeping bearings. Nature. 184: 1284-1285.

McCutchen, C.W. (1978). Lubrication of Joints. In The Joints and Synovial Fluid, pp.437-483. Edited by L. Sokoloff. New York: Academic Press.

Ogston, A.G., and Stanier, J.E. 1953. The physiological function of hyaluronic acid in synovial fluid: Viscous, elastic, and lubricant properties. J. Physiol. 119: 244-252.

Piotrowski, G. 1975. Non-Newtonian Lubrication of Synovial Joints. Ph.D. Thesis, Case Western Reserve University, Dept. of Biomedical Engineering.

Preston, B.N.; Davies, M.; and Ogston, A.G. 1965. The composition and physicochemical properties of hyaluronic acids prepared from ox synovial fluid and from a case of mesothelioma. Biochem. J. 96: 449-474.

Roberts. A.D., and Tabor, D. 1971. The extrusion of liquids between highly elastic solids. Proc. Roy. Soc. London. 325-A: 323-345.

Scott, J.R. 1931. Theory and application of the parallel plate plastimeter. I.R.I. Trans. 7: 169-186.

Vos, R. and Theyse, F. (1969). Lubricating properties of synovial fluid in human and animal joints. In Lubrication and Wear in Joints, pp. 29-38. Edited by V. Wright. Philadelphia: Lippincott.

Walker, P.S.; Dowson, D.; Longfield, M.D.; and Wright, V. 1968. Boosted lubrication in synovial joints by fluid entrapment and enrichment. Ann. rheum. Dis. 27: 512-520.

NOMENCLATURE

$f(K,n)$	ratio of non-dimensional approach velocity to flat plate value.
h	local fluid film thickness
h_o	film thickness at $r=0$
h_1	initial film thickness
h_2	final film thickness
H	non-dimensional local film thickness.
H_o	non-dimensional film thickness at $r=0$
K	non-dimensional curvature of bearing surface
m	viscosity index
M	load-velocity parameter [see Eq. (8g)]
n	flow index
N	$\frac{1}{n}+1$
p	fluid pressure
P	non-dimensional pressure
r	radius, radial coordinate

r_o	outer radius of bearing
R	non-dimensional radial coordinate
t	time
Δt	squeeze time
u	radial fluid velocity
U	non-dimensional radial fluid velocity
v	tangential fluid velocity
V_o	approach velocity
$\bar{V}$	non-dimensional approach velocity
w	axial fluid velocity
w_n	velocity of weeping fluid
W	non-dimensional axial fluid velocity
z	axial coordinate
Z	non-dimensional axial coordinate
β	curvature of bearing surface
$\dot{\gamma}$	shear strain rate
η	viscosity
θ	tangential coordinate
Φ	bearing form factor [see Eq. (13)]

A FLUID MECHANICS APPROACH TO THE PROBLEM OF SENSORY FEEDBACK IN PROSTHETIC DEVICES

Roy B. Davis, III and Daniel J. Schneck

Department of Engineering Science and Mechanics
Virginia Polytechnic Institute and
State University, Blacksburg, Virginia 24061

INTRODUCTION

Structural replacement of amputated limbs has been common for hundreds, even thousands, of years (1). However, the concept of restoring more than the lost structure and supporting musculature, for example, the lost sensory information, is relatively new. Even newer is the idea of using a fluid to transmit tactile information, rather than the more traditional transducers (e.g., strain gauges mounted on cables or various other parts of the terminal device). To use a fluid for this purpose increases the scope and potential for accessing sensory information. That is, the acquisition and transmission of such quantities as heat, prehensile slippage and/or prehensile forces (pressure and shear) is greatly simplified. On this basis, a new type of terminal device for an upper-extremity amputee has been designed and is described below. The device will allow fluids to act as the medium for the transmission of sensory feedback information to the prosthetic wearer. It also introduces a push/pull mode of operation which incorporates a versatile, previously unavailable, wrist mechanism.

BRIEF HISTORICAL PERSPECTIVE

During the past thirty years, numerous designers have become increasingly aware of the ability of the hand to function as a sensory device as well as a prehensile tool. In fact, Kant (2) referred to the hand as "man's outer brain." Thus, efforts to restore lost sensation in an amputated limb have begun to become more popular and have progressed along various fronts since 1951. In that year, Siehlow (3) reported that advances had been made in applying the phantom limb sensation (which persists following an upper extremity amputation) to the control of and sensory feedback from his "Electro-prosthesis." Fifteen years later, Beeker (4) described a prosthetic arm which used pressurized crystal deformation to generate an electric charge. When amplified, it delivered electric shock stimuli to the wearer's skin. In the late 1960's Rhode and Fabric (5) developed a prosthesis which delivered auditory feedback to a hearing aid earpiece. Since the middle of the 1960's investigators [Alles (6,7), and Mann (8)] at the Massachusetts Institute of Technology have been concerned with feedback systems that incorporate and transmit electrical stimuli to remote body areas or, through the use of audio signals to a hearing aid earpiece. Another system designed by Kawamura and Sueda (9) of Japan, utilized a mechanical vibrator which is controlled by a strain gauge transducer located on a voluntary opening Dorrance hook.

A more comprehensive discussion of the theory and principle of sensory feedback to limb prosthesis wearers can be found elsewhere (10), but the variety of different approaches that have been attempted is evident from the brief summary presented here. Note, however, that little, if any, work has been done to exploit the versatility of using fluids as transmitting media. Thus, the following device has been designed to allow for this possibility.

THE DESIGN

In designing a prosthetic system to incorporate sensory feedback, an attempt should be made to include at least information concerning the space orientation of the hand and the normal and shear forces associated with any given grasping operation. Moreover, regardless of the method of transduction used to restore lost sensory information, the artificial limb should also approach the anthropometric, kinematic and kinetic characteristics of the hand it is replacing. Extensive literature (11, 12, 13, 14) has evolved to describe what is lost with amputation and therefore, what needs to be replaced. Minimum design requirements (extracted largely from these works) are as follows:

1) a "finger" motion which allows for anatomic prehensile patterns while maximizing both mechanical efficiency (i.e., cable force compared with prehensile force) and mechanical advantage (i.e., cable excursion compared with finger movement);

2) a "wrist" which allows the anatomic ranges of extension, flexion, adduction and abduction, together with the corresponding limits of forearm pronation and supination;

3) operational simplicity within amputee limitations for control and power, (for example, one and one half inches of cable excursion and twenty-five pounds of cable force);

4) maximum comfort, including minimum weight; and,

5) maximum functionality to increase the prospect of amputee acceptance.

In attempting to satisfy these design criteria, two configurations were adopted: First, a hook-type terminal device was selected. The hook offers less bulk, such that finer work can be accomplished in tighter corners. It also permits greater visibility of terminal device activity, thereby increasing visual feedback. Second, the voluntary-closing device is preferred because (a) it gives the operator the capability of varying the prehensile force commensurate with the delicacy and weight of the object being grasped; (b) voluntary-closing is a "natural" action for the amputee, in that he/she exerts effort to provide a prehensile force and relaxes to release the force; and (c) with the voluntary-closing device, the prehensile force is limited only by the amputee's strength and not by the number of elastic bands on the terminal device (as is the case with the voluntary-opening hook). For a further definition of "voluntary-opening" and "voluntary-closing" devices, refer to the Appendix.

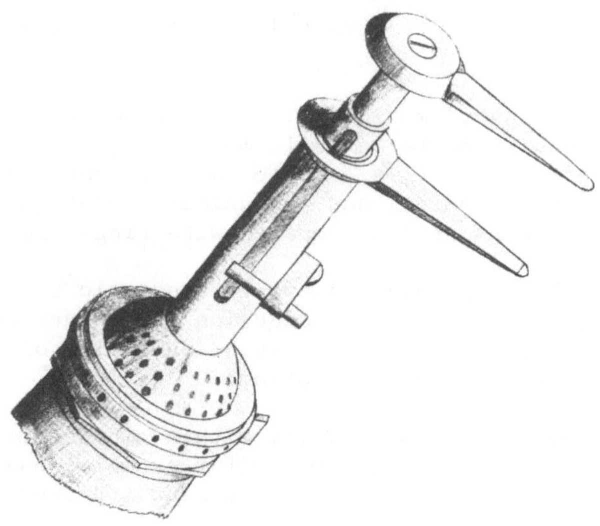

Figure 1 Center Push/Pull Terminal Device, (CPPTD) (Anterior Isometric View)

With the aforementioned considerations in mind, the Center Push/Pull Terminal Device (CPPTD) shown in Figures 1 through 6 was designed. Figures 1-3 are isometric views of the device, which actually consists of two distinct units: the hand (parts 1-12) and the wrist (parts 15-21).

The grasping components, or fingers 1 and 2, (see Figure 4) move co-axially in the positive or negative z-direction (up to 2.9 inches apart) to create the desired grasping modes. Finger 1 is attached with screw 5 to center support rod 3 and constitutes the "active" grasping member. That is, it telescopes in and out of outer support rod 4 which, in turn, slides in and out of the teflon sleeve 6 and outer sleeve 7. Finger 2 is pinned to outer sleeve 7 and represents the "stationary" finger, i.e., finger 1 moves toward and away from finger 2. Both fingers are tapered distal to the z-axis in order to increase their usefulness and both contain fluid-filled compartments.

The motion of center support rod 3, outer support rod 4, and finger 1 is controlled by cable 8 attached to center support rod 3 via ball and socket coupling 9. The fingers are kept normally open by compression springs 10 and 11 which tend to force the outer support rod 4 and center support rod 3, respectively, out of outer sleeve 7. This provides the voluntary-closing mode or operation. The telescoping effect of center support rod 3 and outer support rod 4 is maintained by a difference in spring constants for springs 10 and 11. That is, the spring constant for spring 11 is less than that for spring 10, and it is so chosen that the force which exists in spring 11 when the wearer is at rest is precisely balanced by the tension which normally remains in cable 8 under these same conditions (i.e., wearer at rest). Moreover, the displacement in spring 11 corresponsing to this resting tension is such that inner support rod 3 remains retracted into outer support rod 4. This gives the wearer a functional grasp width (the distance between fingers) which is normally one to one and a half inches (sufficient for most daily tasks). It is only when the operator allows the tension in the cable to approach zero that the inner support rod 3 is extended, thereby generating a grasp width of nearly three inches for more specific applications.

Pin 12, screwed into center support rod 3, travels in guide slot 23 milled into outer support rod 4, teflon

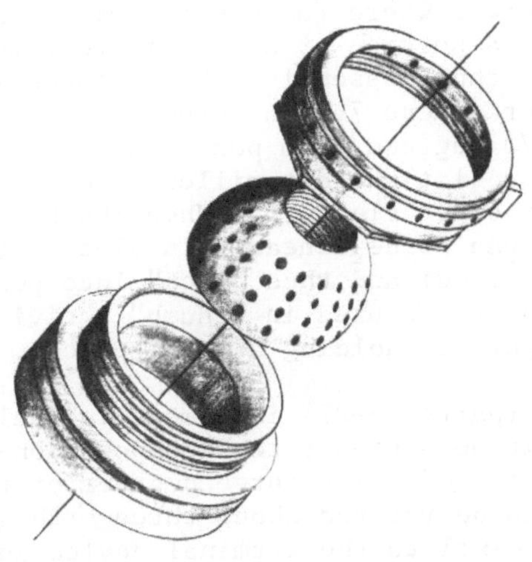

Figure 2 CPPTD Wrist Exploded

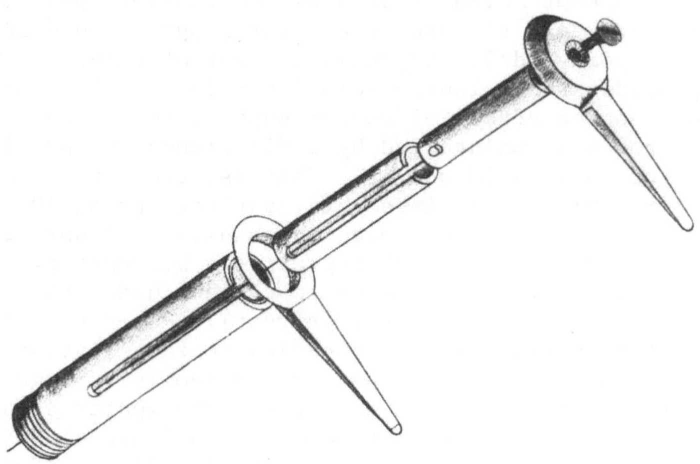

Figure 3 CPPTD Fingers Exploded

sleeve 6, and outer sleeve 7. During the linear motion of the rod this pin prohibits rotation of the rod-finger 1 configuration around the z-axis (centerline of the rod). Another pin (not shown), screwed into outer support sleeve 4, travels in guide slot 23, likewise prohibiting the rotation of the rod-finger configuration. Thumb 13 rotates about lateral y-axis. This movement is illustrated more clearly in Figure 5 where the profile view (looking along the y-axis), of a portion of outer sleeve 7 is seen together with the thumb assembly. Here, thumb 13 is shown affixed to outer sleeve 7 with screw 14. Thumb 13 rotates approximately 70 degrees from position "B" to "A", and vice versa, on a flat surfact milled into the cylindrical continuation of outer sleeve 6. When thumb 13 is in position "A", it can slide linearly in slot 25 (through which screw 14 is inserted) and thus "lock" into position "A". Here it is ready to be used in "thumb" activities, such as aiding the fingers in holding a fork, pencil, etc.

The wrist (parts 15-21) is a modified ball-and socket joint in that it possesses ratcheted-position-locking capabilities while offering three degrees of rotational freedom. It can be rotated about three orthogonal axes, one (z) longitudinal to the terminal device and two (x,y) lateral, allowing a wide variation in positioning. The ranges of rotation about the axes include: a full 360

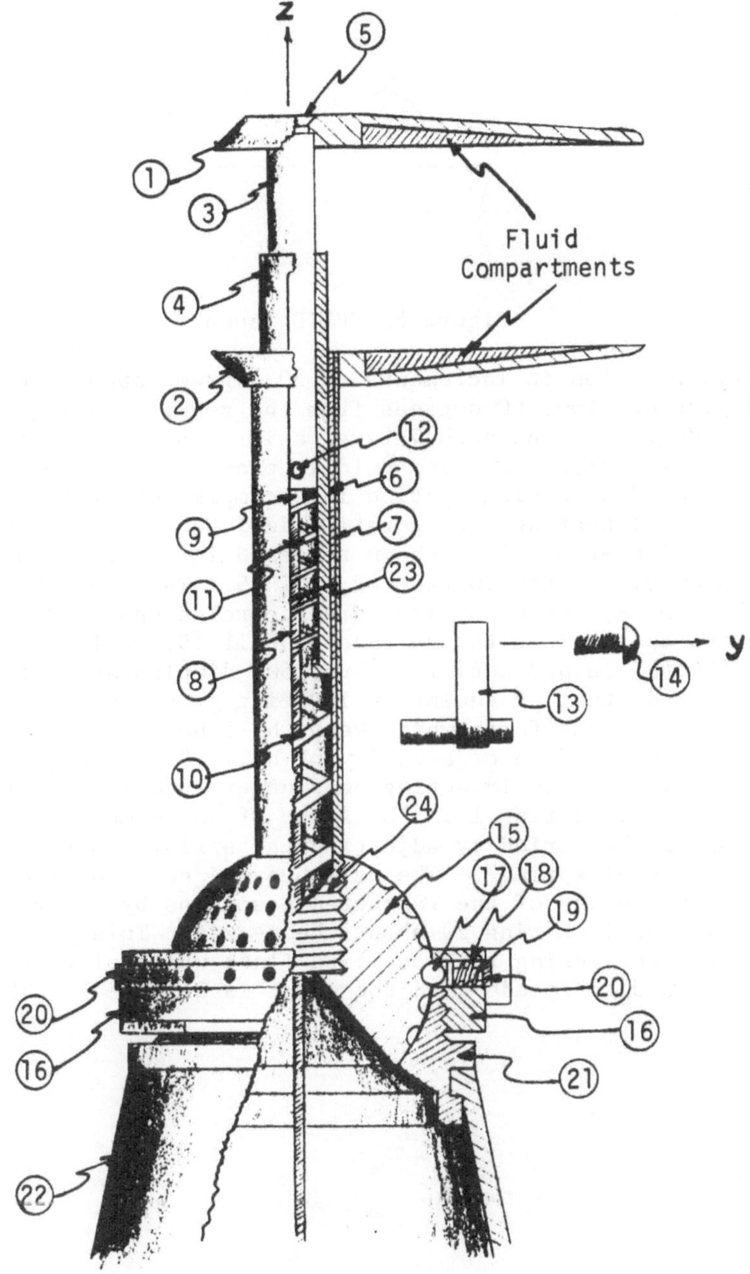

Figure 4 CPPTD Assembly Cross-Section

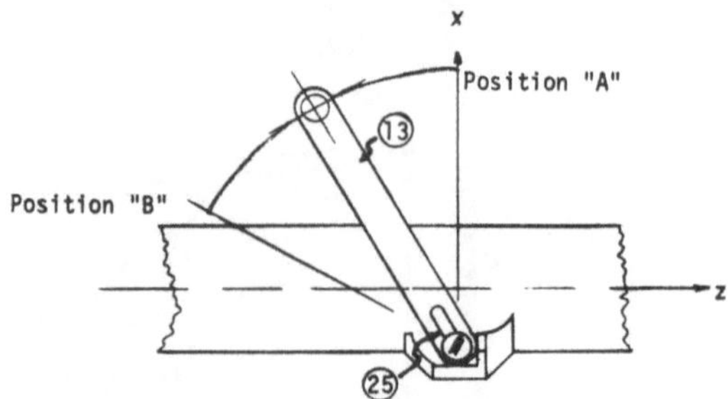

Figure 5 CPPTD Thumb

degree rotation in increments of 30 degrees about the z-axis and plus or minus 60 degrees from the z-axis, in increments of 15 degrees about each lateral axis. Ball 15, rests in ball socket 21 while being held in place by ring 16. The indexing or "ratcheting" action is accomplished by the spring-loaded ball bearing configuration depicted in Figure 6. Here ball bearing 17 is shown recessed in a mating socket, or "dimple," on the surface of ball 15. There are sixty such dimples, ten each along great circles spaced 30 degrees apart along the surface of ball 15. Ball bearing 17 "floats" in and out of the various dimples as ball 15 is rotated, thus producing an indexing or ratchet-positioning effect. The force which keeps ball bearing 17 resting against the surface or a cavity of ball 15 is supplied by compression spring 18 acting on plunger 19 which, in turn, acts on ball bearing 17. The force of the compression spring can be varied by adjusting the spring constant (i.e., the choice of spring). The ball bearing configurations may be locked in any of the ratcheted positions by a five degree rotation of lock-ring 20 about the z-axis. This rotation offsets the locking ring cavities (into which plungers 19 are allowed to recede when the wrist is unlocked). Thus,

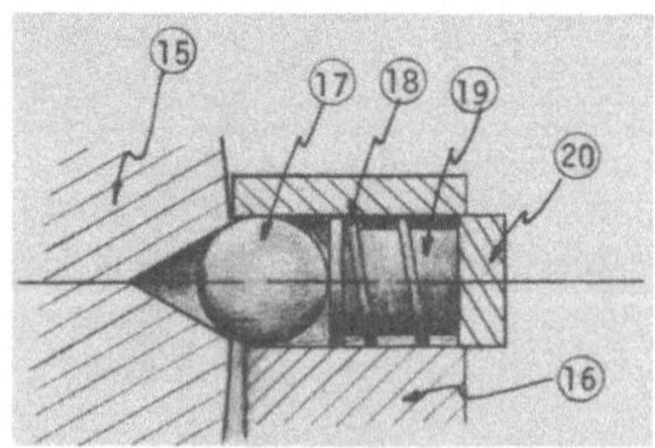

Figure 6 Ball Bearing Configuration

the plungers 19 now bottom up against the lock-ring and prevent the ball bearings 17 from retreating out of the depressions they are occupying, thereby locking the wrist in place. Finally, it can be seen from Figure 4 that ball socket 21 is fastened to the conventional, upper extremity, prosthetic forearem 22, using a plastic laminant. Consequently, the attached wrist components are simultaneously mounted to the forearm as well.

The interface between wrist unit and hand unit is the standard 1/2 inch threaded stud (24 in Figure 4) affording the wearer the option of utilizing the CPPTD hand component or any other conventional terminal device. When the CPPTD finger unit is disengaged from the wrist unit, the control cable 8 may be uncoupled from the center support rod 3 and withdrawn from the wrist and forearm, ready to be fastened to any eccentric-pull device. It should be noted, however, that when using any eccentric, or side pull terminal device, the CPPTD wrist component must be locked in place to avoid any involuntary and undesired rotation of the wrist.

DISCUSSION

Examination of the CPPTD as a whole reveals that the design fulfills a substantial number of the criteria suggested earlier, as well as providing the desired fluid-filled compartments for sensory feedback purposes. For example, the ratio of cable excursion to corresponding finger movement is unity. This relationship produces the desired 1-1/2 inches of grasp width while providing the capability for opening the fingers to a maximum separation of almost three inches. Similarly, it can be shown that the ratio of prehensile force to cable tension is close to one. That is, summing the forces in the z-direction on the cable: tension, T, approximately equals the restoring force, S, generated by the springs plus the prehensile force, P, (neglecting sliding friction which has been minimized by using the teflon inner sleeve 6). Thus, one can write:

$$T \approx S + P \qquad [1]$$

or

$$1 \approx \frac{S}{T} + \frac{P}{T} \qquad [2]$$

Now, it has been found elsewhere (15) that the maximum force, S, required for operation is between 2-1/2 and 3 pounds. Therefore, for T of order 25 pounds and up, the ratio of prehensile force to cable tension, P/T, is approximately equal to one, since, for higher cable tensions, the quantity S/T becomes progressively less significant. Finally, the range of motion of the CPPTD wrist unit actually exceeds anatomical values for wrist flexion/extension and adduction/abduction, and forearm pronation/supination. In addition, the tapered shape of the fingers and their fluid-filled compartments allow the desired variation in grasping mode.

The prototype terminal device described here can be instrumented for sensory feedback by proper choice of the transmitting fluid and associated transducers. This phase of the investigation is still in its earliest stages of development and will be reported upon in a later communication.

REFERENCES

1. Friedman, L.W., "Amputations and Prostheses in Primitive Cultures," Bulletin of Prosthetics Research, Vol. 10-17, 1972, pp. 105-138.

2. Alpenfels, E.J., "The Anthropology and Social Significance of the Human Hand," Artificial Limbs, Vol. 2, 1955, pp.4-21.

3. Siehlow, K., "Phantom Controlled Electre Prothesis for BE and AE Amputees," Translated from Orthopedic Technic, Number 3, 1951.

4. Beeker, T.W., During, J., and DenHertob, A., "Artificial Touch in a Hand Prosthesis," Medical and Biological Engineering, Vol. 5, 1967, pp. 47-49.

5. Pfeiffer, E.A., Rhode, C.M., and Fabric, S.J., "An Experimental Device to Provide Substitute Tactile Sensation From the Anaesthetic Hand," Medical and Biological Engineering, Vol. 7, 1969, pp. 191-199.

6. Alles, D.S., "Kinesthetic Feedback System For Amputees Via the Tactile Sense," presented at the Second Canadian Medical and Biological Engineering Conferences, Toronto, 1968.

7. Alles, D.S., "Information Transmission by Phantom Sensations," IEEE Transactions on Man-Machine Systems, Vol. MMS-11, 1970.

8. Mann, R.W. and Reimer, S.D., "Kinesthetic Sensation for the EMG Controlled 'Boston Arm,'" IEEE Transactions on Man-Machine Systems, Vol. MMS-11, 1970, pp.110-115.

9. Kawamura, Z. and Sueda, O., "Sensory Feedback Device for the Artificial Arm," monograph presented at the Fourth Pan-Pacific Rehabilitation Conference, 1969.

10. Davis, R.B., and Schneck, D.J., "Design of a Center Push/Pull Terminal Device and Quantitative Evaluation Technique For An Upper Extremity Prosthetic Limb," Virginia Polytechnic Institute and State University Technical Report, #VPI-E-79-8, January, 1979, Available NTIS, Accession # PB 290 783/AS.

11. Mason, C.P., "Design of a Powered Arm System for the Above-Elbow Amputee," Bulletin of Prosthetics Research, Vol. 10-18, 1972, pp. 10-24

12. Napier, J.R., "The Prehensile Movements of the Human Hand," Journal of Bone and Joint Surgery, Vol. 38B, 1956, pp. 902-913.

13. Taylor, C.L. and Schwartz, R.J., "The Anatomy and Mechanics of the Human Hand," Artificial Limbs, Vol. 2, 1955, pp. 22-35.

14. Taylor, C.L., "The Biomechanics of the Normal and of the Amputated Upper Extremity," Human Limbs and Their Substitutes, eds. Klopsteg, P.E. and Wilson, P.D., Hafner Publishing Company, New York, 1968, pp. 169-221.

15. Davis, R.B., notes from research at Duke University, 1976, unpublished.

16. Sanschi, W.R., ed., Manual of Upper-Extremity Prosthetics, Department of Engineering, University of California at Los Angeles, 1958, pp. 42-44.

17. Fletcher, M.J., "New Developments in Hands and Hooks," Human Limbs and Their Substitutes, eds. Klopsteg, P. E. and Wilson, P.D., Hafner Publishing Company, New York, 1968, pp. 222-238.

18. Childress, D.S., "Neural Organization and Myoelectric Control," Neural Organization and Its Relevance to Prosthetics, ed. Fields, W.S., Symposia Specialists, Miami, 1973, pp. 117-126.

19. Childress, D.S., Holmes, D.W., and Billock, J.N., "Ideas on Myoelectric Prosthetic Systems for Upper Exremity Amputees," The Control of Upper Extremity Prostheses and Orthoses, eds. Herberts, P., Kadefors, R., Magnusson, R., and Peterson, I., Thomas Books, Springfield, Illinois, 1974, pp. 86-106.

Appendix

The influx of artificial hand designs in the 1950's featured both body-powered and externally powered devices. Generally, the former are controlled and powered by distal muscle movement, e.g., shoulder and/or back motion. Externally powered devices utilize electric servomotors and/or hydraulic/pneumatic systems as power sources.

The majority of body-powered, upper extremity terminal devices resemble not the human hand, but the more task-oriented split hook consisting of two opposing, hook-shaped levers. A prehensile force is generated by rotating one lever relative to the other against a restoring torque which may act to keep the split-hook normally closed (the voluntary opening, VO hand) or normally open (the voluntary closing, VC hand). Lever rotation is accomplished by the wearer, who applies a force to cause excursion of a cable attached to a fixed "thumb" that provides the moment arm. The operator positions the terminal device passively by rotating it about the longitudinal axis of his/her forearm.

The VO hand is kept normally closed. When the wearer applies a force to the cable it acts through the thumb to generate a moment against the elastic restoring torque. The moment opens the VO hand until the patient discontinues applying tension to the cable, whereupon the hooks close upon the object to be grasped. A popular example of this type of terminal device is the Dorrance hook (16). This prosthesis utilizes heavy rubberbands to supply the prehensile force in the form of a torque. The wearer of the VC hand exerts muscular effort to close the normally open hook around what is to be grasped. The most popular of this type of VC device is the Army Prosthetics Research Laboratory (APRL)-Sierra Hook (16, 17), which offers a controllable prehensile force.

Externally-powered devices may be categorized into those which are electromechanical, hydraulic/pneumatic, or a combination of the two. The externally-powered hands are controlled either with manual switches (the switches are operated by distal body motions) or with myoelectric control. For more information regarding the details of myoelectric control, the reader is referred to the works of Childress (18, 19).

HYDRODYNAMICS OF PREY CAPTURE BY TELEOST FISHES

George V. Lauder

The Museum of Comparative Zoology

Harvard University, Cambridge, MA 02138

SUMMARY

The dominant mode of prey capture in teleost fishes is inertial suction: rapid expansion of the mouth cavity creates a negative (suction) pressure relative to the surrounding water. This pressure differential results in a flow of water into the mouth cavity carrying in the prey. Previous models of the suction feeding process have predicted the pattern and magnitude of pressure change in the mouth cavity based on kinematic profiles of jaw bone movement and the application of the Bernoulli equation and the Hagen-Poiseuille relation. These models predict similar pressure magnitudes and waveforms in both the buccal and opercular cavities, and rely on the assumption of a unidirectional steady flow. *In vivo* simultaneous measurement of buccal and opercular cavity pressures during feeding in sunfishes shows that (1) opercular cavity pressures average one-fifth buccal pressures (which may reach -650 cm H_2O),(2) the opercular and buccal cavities are functionally separate with distinct pressure waveforms, (3) a flow reversal (opercular to buccal flow) probably occurs during mouth opening, and (4) the kinetic energy of the water and inertial effects must be considered in hydrodynamic models of suction feeding.

INTRODUCTION

Despite the dramatic advances in our understanding of the hydrodynamics of fish locomotion in the last decade (Lighthill, 1969; Webb, 1975; Weihs, 1972, 1973), very little work has been done on the hydrodynamics of fish prey capture. This may in part be due to experimental difficulties involved in studying feeding behavior. Water-tunnel respirometers allow the study of locomotion under controlled circumstances in a fixed location. The process of locomotion is cyclical and allows repeated measurements over an experimental trial. Investigators of fish locomotion have also greatly benefited from the input of hydrodynamic engineers and theoretical physicists who have applied a large body of relevant experimental and theoretical work to problems of fish locomotion. In contrast, prey capture by teleost fishes occurs extremely rapidly (often within 50 ms), is not cyclical, and the fish cannot be excessively restrained or subjected to experimental trauma without eliminating the feeding response.

The difficulties of studying the hydrodynamics of feeding in fishes have been ably summarized by Holeton and Jones (1975: 547) (in the context of respiration). " The analysis of the breathing mechanics of fish is difficult because it involves the measurement of an unsteady flow of a dense fluid through a non-uniform system which is ill-defined. The compliance of the respiratory tract is variable, both spatially and temporally, and certain resistive elements (such as the gill filaments) are mobile, both actively and passively, throughout a breathing cycle." These difficulties are all compounded during feeding by the extremely short duration of the prey capture event.

In spite of these formidable problems, a number of investigators have modeled the process of prey capture using simple hydrodynamic equations and the kinematics of jaw bone movement to predict the pattern of pressure change in the mouth cavity. In this paper I will review these models and examine the few experimental studies with actual pressure measurements from the mouth cavity during feeding. I will then present new experimental data on the suction feeding mechanism in sunfishes and propose a new model of fluid flow and pressure change in the teleost mouth cavity.

II. ANATOMICAL BASIS OF THE SUCTION FEEDING MECHANISM

Prey capture in most teleost fishes occurs by <u>inertial suction feeding</u>. Mouth cavity volume is rapidly expanded by the contraction of certain jaw muscles (see Lauder and Liem,

1980; Liem, 1978), and this expansion results in the creation of a negative pressure (relative to the surrounding water) in the mouth cavity. This pressure differential creates a flow of water into the mouth from the region directly in front of the head and draws the prey in. The jaws are then closed trapping the prey in the mouth cavity while the water flows out over the gills.

The mouth cavity may be divided into an anterior buccal cavity and two posterolateral opercular cavities (Fig. 1B), separated from the buccal cavity by the gill curtain. The gills are supported on four gill arches and form a resistance to fluid flow within the mouth cavity. Changes in volume of the buccal and opercular cavities for the most part do not occur independently: anatomically they are coupled. Expansion of the buccal cavity may occur by elevation of the neurocranium, opening of the front jaws, depression of the hyoid apparatus, and lateral expansion of the suspensory apparatus (Fig. 1; also see Lauder and Liem, 1980; Liem, 1970, for a more detailed account of anatomical couplings). These movements may also effect opercular cavity expansion. However, some bone movements (such as opercular adduction) (Fig. 1) do predominantly affect only one cavity. In general, the dorsal, ventral, and lateral walls of the mouth cavity all rapidly expand to create a low pressure center during the attack at a prey item.

The role of the gills as a resistant element separating the buccal and opercular cavities was first recognized by Woskoboinikoff and Balabai(1937) and van Dam (1938), and the concept of gill resistance to water flow has received considerable attention in recent studies of fish respiration (Ballintijn, 1972; Hughes and Morgan, 1973; Hughes and Shelton, 1958; Jones and Schwarzfeld, 1974; Pasztor and Kleerekoper, 1962; Shelton, 1970). The resistance of the gills to flow is not equal in both directions: flow directed anteroposteriorly (i.e., from the buccal to opercular cavity) encounters less resistance than reverse flow from the opercular cavity into the buccal cavity due to the orientation of the gill filaments (Fig. 1). While several attempts have been made to measure gill resistance to anteroposterior flow (e.g., Brown and Muir, 1970; Davis and Randall, 1973; Hughes and Umezawa, 1968; Jones and Schwarzfeld, 1974), no data exist on the values of gill resistance to reverse flow. It is well established, however, that gill configuration (and thus resistance) may be actively modified by intrinsic gill

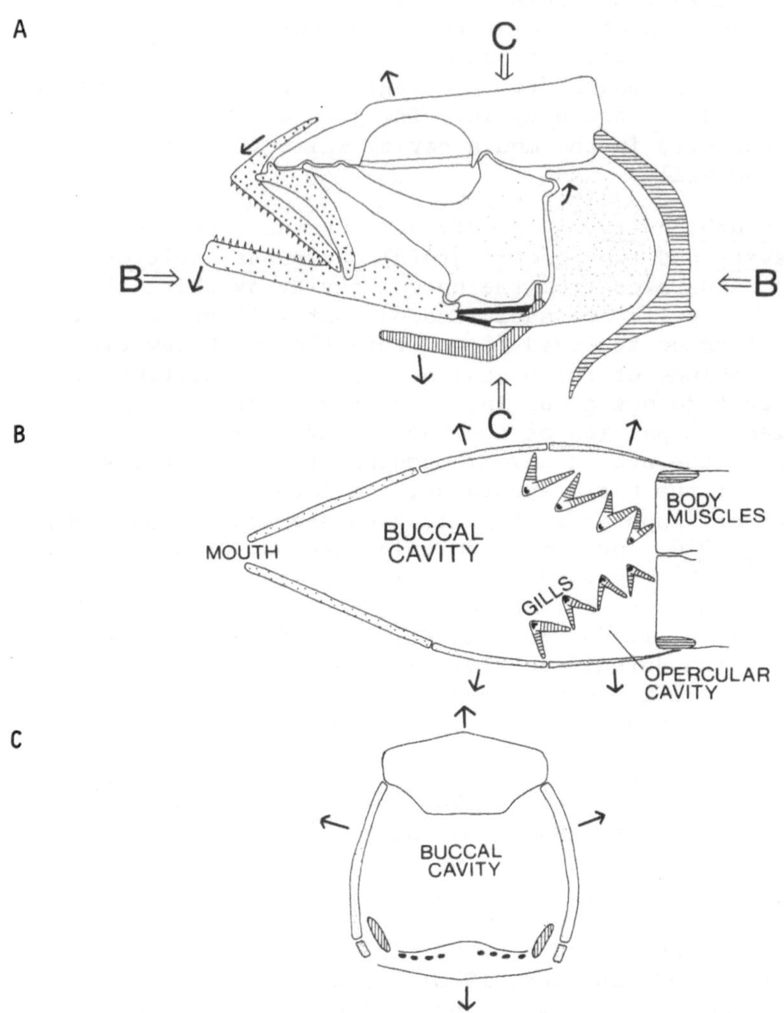

Fig. 1. Diagrammatic view of the head of an advanced teleost fish with protrusible jaws. B and C represent sections of the head at the level indicated in A. Arrows indicate major bony movements during prey capture. Key: white = neurocranium; vertical lines = hyoid apparatus; horizontal lines = pectoral girdle; dense stipple = opercular apparatus; fine stipple = suspensorium; large stipple = jaw apparatus.

arch musculature (Pasztor and Kleerekoper, 1962). Other resistance to flow occurs at the mouth opening and at the opercular and branchiostegal valves where water exits through a narrow slit of high resistance. Osse (1969: 371) and Alexander (1967) have suggested that gill resistance is very low during feeding.

III. RESPIRATORY HYDRODYNAMICS

Research on respiratory hydrodynamics has provided the conceptual basis for current models of fluid flow during feeding. The early work of Hughes (1960), Hughes and Shelton (1958), and Saunders (1961) established that water flow through the teleost mouth cavity is unidirectional and is regulated by two "pumps." An opercular suction pump draws water through the gill resistance by lateral expansion of the operculum which creates a pressure differential from the buccal to the opercular cavities. Shortly after opercular expansion has reached its peak, the buccal pressure pump is initiated by jaw closure and suspensorial adduction (Ballintijn and Hughes, 1965). This creates a positive buccal pressure (of 1-2 cm H_2O) which drives water through the gills and into the opercular cavity where it exits to the outside. Throughout this process buccal pressure is nearly always positive with respect to opercular pressure.

The key points established by studies of respiratory hydrodynamics are (1) that the gill cover functions as a fundamental element of the "opercular suction pump," drawing water over the gills, (2) that pressures in the opercular cavity are negative with respect to buccal cavity pressures, (3) that this pressure differential must exist if water is to flow unidirectionally through the mouth cavity (Saunders, 1961).

Holeton and Jones (1975) provided the first velocity measurements of flow during respiration and noted that water velocity varied within the buccal cavity. Velocities of up to 38 cm/sec were recorded during normoxic respiration.

IV. PREVIOUS MODELS OF SUCTION FEEDING IN FISHES

A. Pressure Waveforms and Magnitudes: Predictions

Osse (1969) first attempted to predict the magnitude of mouth cavity pressures in fishes using simple hydrodynamic relationships between velocity and pressure. The equation

$$\frac{P_1}{\rho g} + \frac{\frac{1}{2}V^2}{g} = \frac{P_o}{\rho g}$$

(where P_1 is the pressure near the mouth within the mouth cavity, P_o the pressure of the surrounding water, V the velocity of water entering the mouth, ρ the density of the liquid, and g the acceleration due to gravity) was applied to the fish head with V=200 cm/sec, and a buccal pressure of -20 cm H_2O was calculated. Velocity of water flow was calculated from the estimated change in buccal volume, the estimated rate of volume change, and the mean cross-sectional area of the mouth during mouth opening. This approach was indicated as a first approximation to problems of fluid flow in the mouth cavity and involved a number of assumptions. The most important of these is the assumption of steady flow in the Bernoulli equation, a condition that is certainly not met during feeding. Lauder (1979) also assumed steady flow conditions during his consideration of the effect of mouth geometry on flow rate. Osse (1969: 371) concluded that expansion of both the buccal and opercular cavities contributes to suction feeding: "The suction force due to enlargement of the opercular cavity is directly applied to the water entering the buccal cavity, thus increasing the quantity of water and the velocity of the current."

More recently, Pietsch (1978) has applied the Bernoulli equation and the Hagen-Poiseuille relation to the tubular mouth of Stylephorus to calculate the buccal cavity pressure and flow velocity during feeding. Assumptions of the Hagen-Poiseuille relation, none of which apply to fishes, include (1) a small pipe diameter, (2) steady flow, (3) absence of particles (i.e., prey) in the flow, and (4) that the relationship is not valid near the pipe entrance (see Prandtl, 1949; Streeter and Wylie, 1979). The predicted buccal pressure was -53 cm H_2O with a flow velocity of 325 cm/sec.

Muller and Osse (1978) and Osse and Muller (in press) have developed an elegant hydrodynamic model to predict the pattern of pressure and velocity change with time during feeding. The fish head is modeled as a radially symmetrical cone that expands to reduce the pressure inside. The timing of expansion of both the anterior and posterior bases of the cone can be varied to simulate the timing of mouth opening and opercular expansion respectively. Flow velocity is obtained from the equation of continuity

$$\frac{\partial u}{\partial x} + \frac{1}{r} \cdot \frac{\partial (vr)}{\partial r} = 0$$

where u is the component of velocity along the body axis, x the distance along the body axis, v is the velocity component perpendicular to the body axis (along the radius of the cone), and r is the radius of the cone at the point of interest. By solving this equation for velocity and substituting into the equation of motion (Navier-Stokes, for frictionless flow),

$$\frac{\partial u}{\partial t} + u \frac{\partial u}{\partial x} = \frac{-1}{\rho} \cdot \frac{\partial p}{\partial x}$$

where p is pressure and ρ is density, the pressures generated by the expanding cone can be calculated. This procedure does _not_ assume steady fluid flow through the mouth cavity.

Three major hydrodynamic assumptions have been made (Muller and Osse, 1978): (1) friction is neglected, (2) the fish head is assumed to be radially symmetrical, and (3) the prey is assumed to behave as an element of the water.

Elshoud-Oldenhave and Osse (1976: 411-412) have made the most specific predictions of pressure waveform in the teleost mouth cavity and correlated the hypothesized pressure changes with kinematic events to produce a theoretical model of suction feeding. Figure two summarizes the present hypothesis of pressure change in the buccal and opercular cavities and is drawn from discussions in Alexander (1969, 1970), Elshoud-Oldenhave and Osse (1976), Lauder (1979), Nyberg (1971), and Liem (1978).

A preparatory phase occurs first as the fish approaches the prey (Fig. 2:P). The volumes of both the buccal and opercular cavities are reduced and the pressure goes positive relative to the surrounding water. The mouth cavity then begins to expand (Fig. 2:mce) while the front jaws remain closed, and this results in a pressure decrease in both cavities. The mouth then opens (Fig. 2: mo), pressures reach their peak negative value, and compression of the

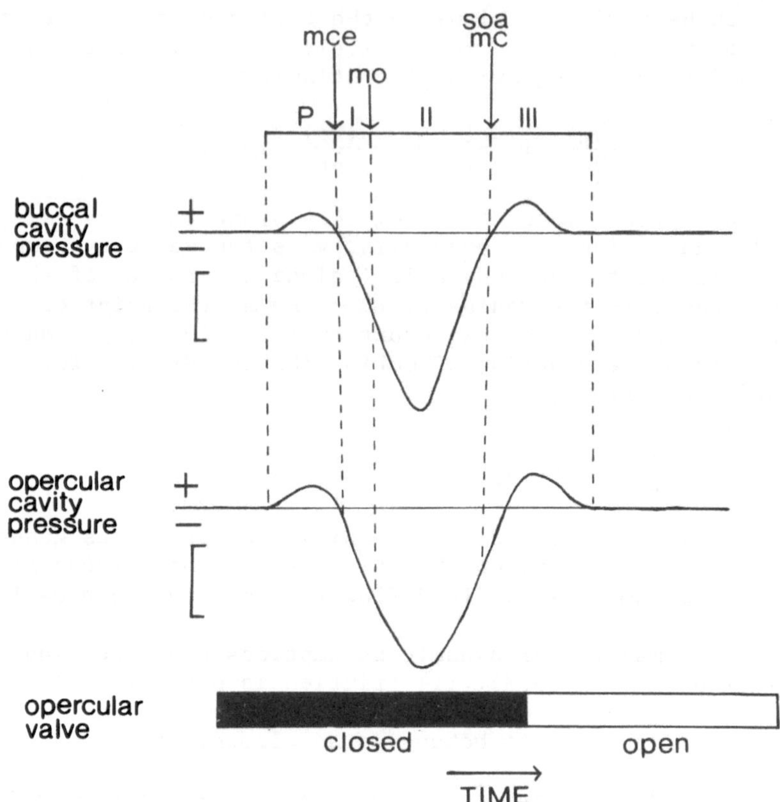

Fig. 2. Current model of buccal and opercular cavity pressure change with time during suction feeding. Phases P, I, II, and III are defined after Elshoud-Oldenhave and Osse (1976), as are the kinematic correlates of pressure change: mce, mouth cavity expansion; mo, mouth opening; soa, suspensorial and opercular adduction; mc, mouth closing. Note the close similarity in both waveform and magnitude (see arbitrary scale bar on left) between buccal and opercular cavity pressures.

mouth cavity occurs. Finally, as the buccal pressure reaches zero, suspensorial and opercular adduction commences and the mouth closes (Fig. 2: soa, mc), resulting in a positive pressure as water is forced out the opercular slit.

The key elements of this model are (1) the close similarity between buccal and opercular pressure waveforms and magnitudes, (2) the role of opercular abduction in the generation of a negative opercular cavity pressure, (3) pressure decrease before the mouth begins to open, and (4) unidirectional flow through the mouth cavity. O'Brien (1979:579) has also emphasized the importance of opercular expansion in contributing to the unidirectional flow of fluid through the mouth.

B. Experimental Data

Alexander (1969, 1970) provided the first direct measurements of pressures in the teleost mouth cavity. He used a pressure transducer attached to a nylon tube which was fixed in the aquarium. A small piece of food was attached to the tube and the fishes were trained to suck off the food by placing their mouths around the tube. Pressures were measured during the feeding act.

A survey of nine different species showed that the maximum negative pressure varied from -80 cm H_2O to -400 cm H_2O in the buccal cavity. Pressure waveforms typically showed a sharp negative pressure drop shortly after the mouth opened and a slight positive pressure pulse of +1 to 9 cm H_2O as " water which has been sucked in with the food is ... driven out through the opercular openings" (Alexander, 1969). These pressure traces agree well with the pattern of buccal pressure change hypothesized from kinematic analyses (Fig. 2), although data on the occurrence of a preparatory phase were not available since the fish had to open its mouth before pressures could be recorded. Casinos (1977) using similar equipment recorded pressures of -150 cm H_2O in cod (*Gadus*).

Osse (1976) presented preliminary pressure measurements from the buccal and opercular cavities of *Amia calva* and reported pressures as low as -170 cm H_2O and -95 cm H_2O respectively. Most recently, Liem (1978) measured buccal pressure profiles in two cichlid fishes and found a preparatory pressure pulse corresponding to phase P in Fig. 2.

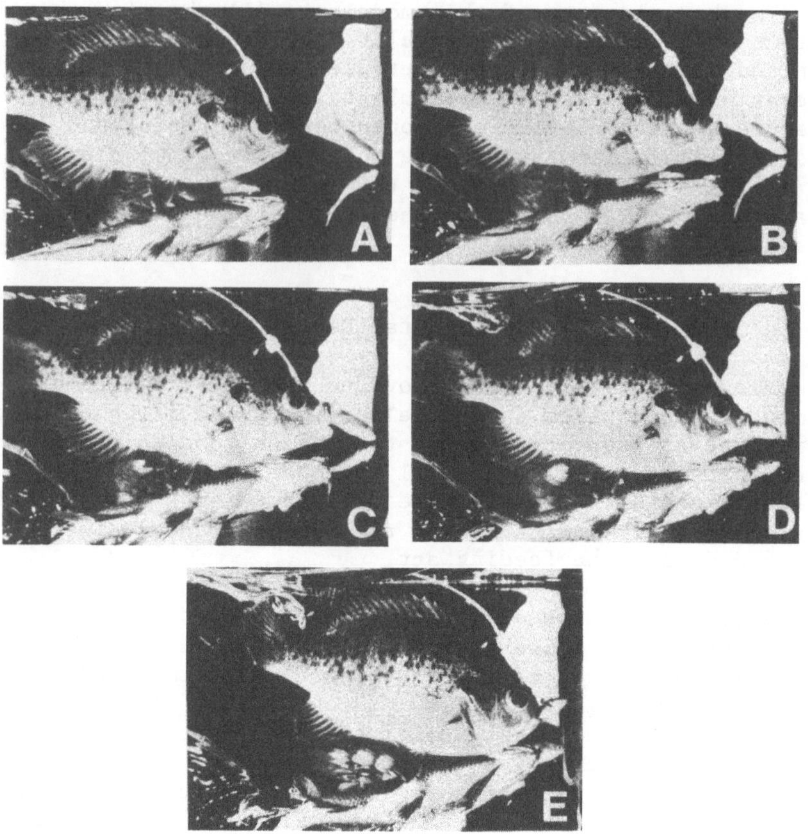

Fig. 3. Representative frames from a high-speed film (200 frames per second) of the bluegill (Lepomis macrochirus) capturing a goldfish. Note the plastic cannula leading into the buccal cavity and the attachment of the cannula to the clamp. Also note abduction of the gill filaments as seen in the ventral view of frame E. Frames A, B, C, D, and E correspond to frames 1, 4, 6, 8, and 15 from the film.

V. EXPERIMENTAL ANALYSIS OF FEEDING IN SUNFISHES

A. Materials and Methods

The suction feeding mechanism in the bluegill sunfish Lepomis macrochirus (Family Centrarchidae) was studied by the simultaneous recording of buccal and opercular cavity pressures together with a high-speed film (200 frames per second) of jaw movements. A detailed description of the recording apparatus and calibration technique may be found in Lauder (1980). Briefly, plastic cannulae (o.d. 1.52 mm, i.d. 0.86 mm) were chronically implanted in the buccal and opercular cavities (see Fig. 3) and attached to Statham P23 Gb pressure transducers filled with a mixture of 53% boiled (degassed) glycerine and 47% boiled distilled water. This mixture resulted in a transducer damping factor of 0.65 and a frequency response of 75 Hz. Films were then taken of the fish feeding over a mirror to allow accurate measurement of kinematic events. The fishes were fed a variety of prey types, from live goldfish (Carassius auratus) to earthworms and mealworms.

B. Results

The patterns of buccal and opercular cavity pressure recorded during feeding are shown in Fig. 4 and typical jaw movements occurring during capture of a goldfish in Fig. 3. There is tremendous variability in the pressure waveform between different feeding events and these variations correlate with specific kinematic patterns (Lauder, 1980).

Buccal pressures very rarely exhibit a preparatory phase. A pressure drop is recorded immediately after the mouth begins to open and peak gape occurs before the maximum negative pressure. The maximum recorded buccal cavity pressure was -650 cm H_2O. Pressure magnitudes correlate with prey type (goldfish elicit the greatest negative pressures, mealworms the least), and pressure varies inversely with the degree of satiation (Lauder, 1980). The most common buccal pressure waveform contains an initial large negative peak followed by a smaller positive pressure pulse and then by a final negative phase (see Fig. 4A: 1, 5, 7, 9). Occasionally the positive pulse or the second negative is absent (Fig. 4: 2, 10).

Opercular pressure waveforms exhibit an initial sharp positive phase which is followed by a negative pressure peak that may reach a maximum of about -130 cm H_2O (Fig. 4B). A positive pulse may follow the negative (Fig. 4B: 1, 2, 4, 5, 7) or it may be absent (Fig. 4B: 3, 8, 9). Feeding on stationary prey produced opercular pressures in the -10 to -40

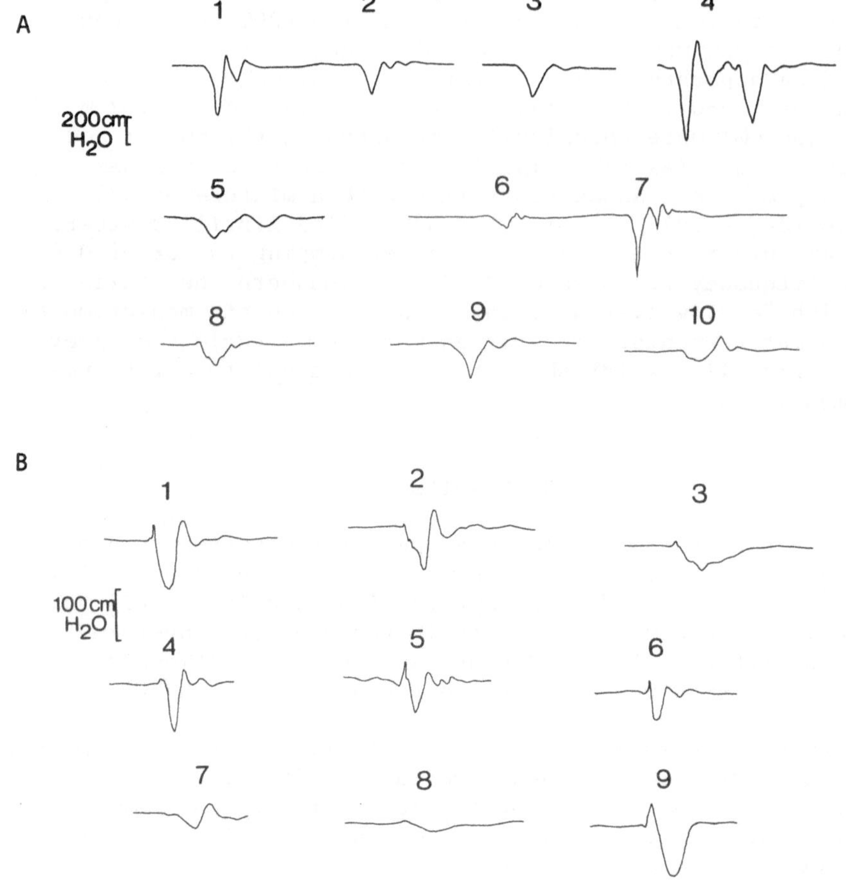

Fig. 4. Representative traces of pressure change in the buccal cavity (A) and the opercular cavity (B) in bluegill (Lepomis macrochirus) during feeding. Note the differing scales and the variation in pressure waveform and magnitude. See text for discussion.

cm H_2O range and tended to flatten out the pressure profile (Fig. 4B: 3, 8).

The temporal relationship between the buccal and opercular pressures is shown in Fig. 5A. Buccal cavity pressure begins to decrease immediately after the mouth starts to open (Fig. 5A: t_0 to t_1). During this same time interval, the operculum is adducted and opercular cavity pressure actually rises. Buccal pressure reaches its peak (usually 5 times the peak opercular pressure) 5 to 10 ms prior to the opercular pressure peak although the two peaks are occasionally temporally coincident. Buccal pressure then starts to rise and passes through zero while the opercular cavity pressure is still negative. The positive phase of the buccal waveform (Fig. 5: phase IV) occurs while the opercular cavity pressure is negative. In phase V, opercular pressure goes positive as the second negative buccal pressure pulse occurs. Opercular abduction is initiated at the peak in opercular cavity pressure (Fig. 5: oa); throughout the first third of the feeding sequence the operculum exhibits no lateral movement (see ventral view in Fig. 3B). Considerable opercular abduction occurs before the opercular and branchiostegal valves open (Figs. 3B; 5:om). Mouth closure, usually against partially protruded premaxillae, occurs before opercular pressure passes zero and at or near the peak of the positive buccal pressure pulse (Fig. 5:mc). The operculum often remains abducted after the mouth has closed and the pressures have returned to their ambient values (Fig. 5: t_7, t_8). At this point the gill filaments from adjacent arches are clearly seen to be abducted (Fig. 3E: ventral view) and gill resistance is presumably low.

C. New Model of Fluid Flow in the Mouth Cavity

A comparison of simultaneously recorded buccal and opercular cavity pressure waveforms and magnitudes (Fig. 5A) strongly suggests the hypothesis that flow is not unidirectional in the mouth cavity. Figure 5B illustrates the hypothesized flow pattern at representative stages of the feeding cycle.

During phase I, the period when opercular cavity pressure is positive (Fig. 5A), buccal cavity pressures may reach -150 cm H_2O. Between t_1 and t_2 (Fig. 5A) the ratio of buccal to opercular cavity pressure is about 8. This large pressure differential and the lack of opercular abduction indicate a reverse flow from the opercular to buccal cavity between t_0 and t_2 (Fig. 5). After the end of phase II, opercular abduction occurs and the direction of flow is hypothesized to be from the buccal into the opercular cavity. This change is due both to opercular cavity volume increase and the momentum of water entering the mouth. The branchiostegal membrane opens at t_4 and this allows flow between the opercular cavity and the exterior. If opercular abduction were delayed beyond t_3, then the anteroposterior flow pattern would likely not be established by t_4, and opening of the branchiostegal valve (by the hyohyoideus inferioris muscle) should actually result in water flow into the opercular cavity from the outside. This anterior flow would be temporary because by t_5 (Fig. 5B) the anteroposterior flow is well established as buccal pressure becomes positive with respect to that of the opercular cavity. At this point,

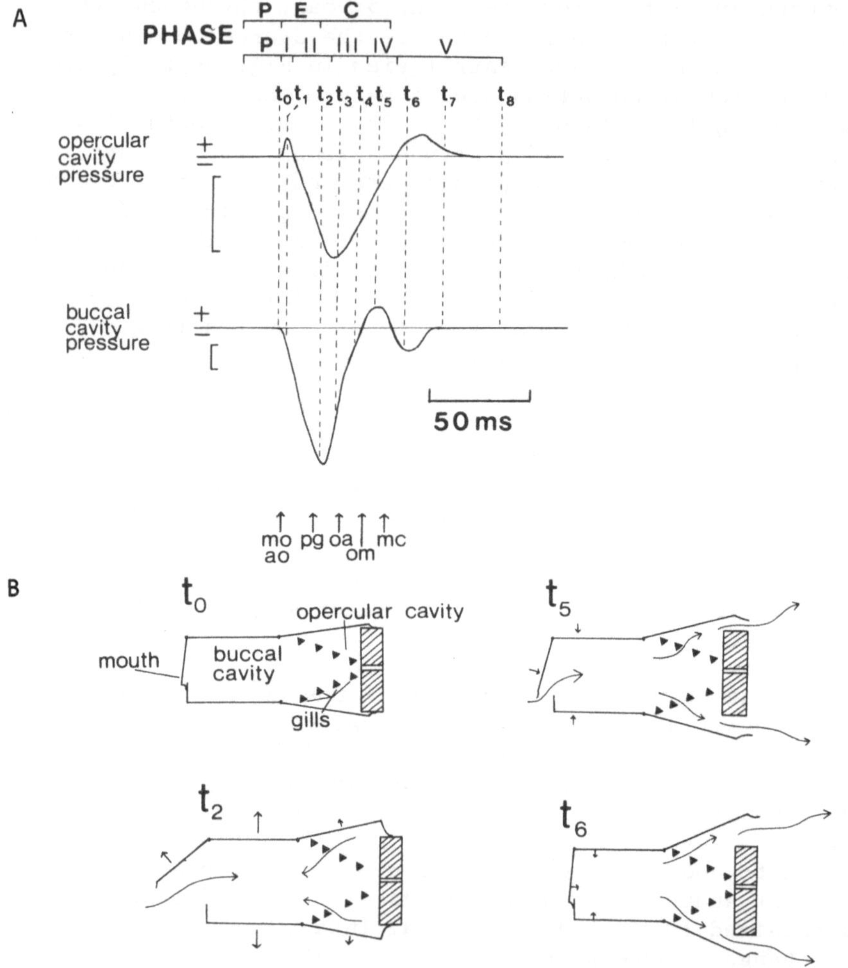

resistance of the opercular slit is high because of its small cross-sectional area.

The mouth closes during the buccal positive pulse (phase IV) and this event is followed by a rapid pressure decrease in the early part of phase V. This second buccal negative pressure is hypothesized to be due to the water hammer effect. Rapid closing of the mouth acts like the closing of a valve in a pipeline during flow. On the upstream side of the valve the pressure rapidly increases and a high pressure wave is propagated upstream. On the downstream side, the pressure is rapidly reduced (a cavity forms and the fluid returns with the same velocity) and a low pressure wave travels downstream. This tends to reduce the velocity of fluid flow and to contract the pipe downstream of the valve. The analagous situation during feeding is depicted in Fig. 5B: t_6). The mouth rapidly closes, water tends to continue flowing posteriorly causing a pressure reduction just inside the mouth (early phase V). Positive pressures are often recorded as water flows anteriorly after the pressure reduction (Fig. 4A: 1, 7, 9, 10). This phenomenon is analagous to events causing the dichrotic notch in the mammalian cardiac pressure waveform. Finally, by t_8 both the buccal and opercular cavities have returned to ambient pressure.

Fig. 5. A, simultaneous recordings of buccal and opercular cavity pressures during a typical strike at a goldfish. Scale bar equals 100 cm H_2O. P, E, and C refer to the preparatory, expansive, and compressive phases of the strike as conventionally defined (see Liem, 1978). Phases below are those proposed in this paper. Note the dissimilarity of pressure waveforms and magnitudes in the two cavities: e.g., the lack of a preparatory phase and the two negative phases in the buccal waveform. B, proposed pattern of fluid flow through the mouth cavity during feeding. t_0, t_2, t_5, and t_6 correspond to the times in A. Small arrows indicate movements of the mouth cavity. Note the hypothesized reverse flow between t_1 and t_2. Kinematic events are: mo, mouth opening; ao, opercular adduction; pg, peak gape; oa, opercular abduction; om, branchiostegal valve opens; mc, mouth closure.

VI. DISCUSSION

The assumptions and predictions of previous models of pressure change and fluid flow in the teleost mouth cavity during feeding are not supported by the experimental analysis of suction feeding in sunfishes presented here; no previous simultaneous buccal and opercular cavity pressure measurements exist. Current conceptions of the hydrodynamics of teleost feeding have been framed by the large body of data on respiratory mechanics and hydrodynamics. Thus, flow is assumed to be unidirectional, inertial effects have been generally neglected (but see Holeton and Jones, 1975; Muller and Osse, 1978), and the process of creating suction during feeding is viewed as a modification of the respiratory two-pump system. In particular, the operculum is suggested to be of key importance in creating negative mouth cavity pressures (Alexander, 1967; Muller and Osse, 1978; Nyberg, 1971; O'Brien, 1979; Osse, 1969), in a manner analagous to the opercular suction pump during respiration. Additional elements of current concepts of feeding hydrodynamics are the close similarity between buccal and opercular cavity pressure waveforms and magnitudes, the correlated view that the buccal and opercular cavities are a functional unit, and the assumption that gill resistance is low during feeding.

None of these assumptions appear to be true. Buccal cavity pressures in sunfishes consistently average five times the opercular pressures (Fig. 5). In addition, pressure waveforms from the two cavities differ significantly and do not agree with expected patterns (Fig. 2). Flow reversal also appears to occur while the mouth is opening.

Inertial effects play a fundamental role in the hydrodynamics of feeding. The process of creating suction is best viewed as being composed of a powerful _buccal suction pump_ that draws water into the buccal cavity from both the area in front of the mouth and from the opercular cavity. The operculum functions only as a passive element at this stage, preventing water influx from the outside. Flow from in front of the mouth is much greater than from the opercular cavity because the mouth opening is much less resistant to flow than the gill curtain. The inertia of the water drawn in through the mouth is primarily responsible for the transition to the anteroposterior flow pattern and the exit of water out over the gills to the exterior. Opercular abduction appears to contribute relatively little to the direction of fluid flow, the magnitude of opercular cavity negative pressure, or flow velocity.

The asymmetry of gill resistance plays a key role in this model. In the early stages of feeding, the drop in opercular cavity pressure is due both to the buccal cavity pressure reduction and perhaps also to expansion of opercular cavity volume as a result of anatomical couplings between the two cavities, not to opercular abduction. Opercular cavity pressures do not equal those in the buccal cavity because of gill resistance, and the filaments of adjacent arches may be adducted. As the inertia of water sucked in through the mouth results in flow into the opercular cavity, the gill filaments are abducted and resistance becomes low.

Based on the synchronously recorded buccal and opercular cavity pressures and the hydrodynamic considerations outlined above, a number of kinematic correlates of pressure waveform attributes may be predicted (Table I). The correspondence between the occurrence of different kinematic patterns and variations in pressure waveform will be considered in detail elsewhere (Lauder, 1980), but variations during phases IV and V (Figs. 4, 5A) may be correlated with the timing of opercular abduction and mouth closing.

The large negative pressures recorded in the mouth cavity (up to -650 cm H_2O) invite considerations of the structural demands imposed on the teleost head. Lauder and Lanyon (1979) have considered the morphology of the sunfish operculum to be primarily a response to deformation induced by negative opercular cavity pressures. Two prominent orthogonal bony struts on the operculum were hypothesized to resist bending and twisting moments imposed by the pressure reduction. This view of the role of the operculum is consistent with the model of suction feeding presented here: the gill cover acts primarily as a passive element preventing fluid influx from the exterior.

A number of clearly defined areas may now be outlined for future work. Of particular interest is a characterization of the velocity field, both anterior to the mouth in the vicinity of the prey, and within the buccal and opercular cavities. Opercular cavity flow velocity determinations would provide a test of the reverse flow hypothesis. The pressure -- velocity relationship during feeding is also of importance. Because of the prominence of inertial effects

TABLE I

Predicted Kinematic Correlates of Pressure Waveform Attributes

	Pressure Waveform Characteristic (See Fig. 5A)	Predicted Kinematic Correlate
BUCCAL PRESSURE	Phase P: Positive	Hyoid protraction; suspensorial adduction.
	Phase I, II: Negative	Mouth opening; suspensorial abduction; hyoid depression.
	Phase III: Negative → 0	Mouth closing; start of hyoid and suspensorial adduction.
	Phase IV: Positive	Delay in opercular abduction relative to mouth closing; hyoid, suspensorial adduction.
	Phase V: Negative	Rapid closing of jaws relative to mouth cavity expansion.
OPERCULAR PRESSURE	Phase P: Positive	Opercular adduction; suspensorial adduction; hyoid protraction.
	Phase I: Positive	Opercular adduction.
	Phase II: Negative	Mouth opening; suspensorial abduction; hyoid depression.
	Phase III: Negative	Opercular abduction; mouth closing.
	Phase IV: Negative → 0	Opercular abduction; mouth closing; hyoid and suspensorial adduction.
	Phase V: Positive	Hyoid and suspensorial adduction.

and changing gill resistance during the strike, calculation of flow velocity from measured pressures is unlikely to yield satisfactory results. Finally, correlation of attributes of the suction feeding mechanism (such as volume flow rate, pressure, velocity) with morphological features, feeding efficiency, and prey type in a number of closely related taxa, may provide insights into the evolutionary mechanisms governing changes of shape and function in teleost fishes.

ACKNOWLEDGMENTS

I was supported during the preparation of this paper by an NIH Predoctoral Fellowship in Musculoskeletal Biology (5T32GM077117) and by a Junior Fellowship in the Society of Fellows, Harvard University. A grant from the Bache Fund of the National Academy of Sciences provided the research funds. Preliminary work was partially supported by a Raney Fund Award from the American Society of Ichthyologists and Herpetologists. Professor F.A. Jenkins very kindly supplied funds for a pressure transducer. I thank K. Hartel, N. Bigelow, S. Norton, T. Chisholm, and especially Marty Fitzpatrick and Charlie Gougeon for their help with experiments. K.F. Liem, T. McMahon, M. Patterson, and P. Elias kindly reviewed the manuscript.

LITERATURE CITED

Alexander, R. McN. 1967. Functional design in fishes. Hutchinson and Co.: London.

Alexander, R. McN. 1969. Mechanics of the feeding action of a cyprinid fish. J. Zool., Lond. 159:1-15.

Alexander, R. McN. 1970. Mechanics of the feeding action of various teleost fishes. J. Zool., Lond. 162:145-156.

Ballintijn, C.M. 1972. Efficiency, mechanics, and motor control of fish respiration. Resp. Physiol. 14:125-141.

Ballintijn, C.M. and Hughes, G.M. 1965. The muscular basis of the respiratory pumps in the trout. J. exp. Biol. 43:349-362.

Brown, C.E. and Muir, B.S. 1970. Analysis of ram ventilation of fish gills with application to skipjack tuna (Katsuwonus pelamis).J. Fish. Res. Bd. Can. 27:1637-1652.

Casinos, A. 1977. El mechanisme de deglucio de l'aliment a Gadus callarius, Linnaeus 1758 (Dades preliminars).

Butelleti Soc. Cat. de Biol. 1:43-52.

Davis, J.C. and Randall, D.J. 1973. Gill irrigation and pressure relationships in rainbow trout (*Salmo gairdneri*). J. Fish. Res. Bd. Can. 30:99-104.

Elshoud-Oldenhave, M.J.W. and Osse, J.W.M. 1976. Functional morphology of the feeding system in the ruff--*Gymnocephalus cernua* (L. 1758) - (Teleostei, Percidae). J. Morph. 150:399-422.

Holeton, G.F. and Jones, D.R. 1975. Water flow dynamics in the respiratory tract of the carp (*Cyprinus carpio* L.) J. exp. Biol. 63:537-549.

Hughes, G.M. 1960. A comparative study of gill ventilation in marine teleosts. J. exp. Biol. 37:28-45.

Hughes, G.M. and Morgan, M. 1973. The structure of fish gills in relation to their respiratory function. Biol. Rev. 48:419-475.

Hughes, G.M. and Shelton, G. 1958. The mechanism of gill ventilation in three freshwater teleosts. J. exp. Biol. 35:807-823.

Hughes, G.M. and Umezawa, S-I. 1968. On respiration in the dragonet *Callionymus lyra* L. J. exp. Biol. 49:565-582.

Jones, D.R. and Schwarzfeld, T. 1974. The oxygen cost to the metabolism and efficiency of breathing in trout (*Salmo gairdneri*). Resp. Physiol. 21:241-254.

Lauder, G.V. 1979. Feeding mechanics in primitive teleosts and in the halecomorph fish *Amia calva*. J. Zool., Lond. 187:543-578.

Lauder, G.V. 1980. The suction feeding mechanism in sunfishes: an experimental analysis. (in press)

Lauder, G.V. and Lanyon, L.E. 1979. Functional anatomy of feeding in the bluegill sunfish, *Lepomis macrochirus*: in vivo measurement of bone strain. J. exp. Biol.

Lauder, G.V. and Liem, K.F. 1980. The feeding mechanism and cephalic myology of *Salvelinus fontinalis*: form, function, and evolutionary significance. Chapter 10, In: Chars: salmonid fishes of the genus *Salvelinus*. E.K. Balon Ed., Junk Publishers, The Netherlands.

Liem, K.F. 1970. Comparative functional anatomy of the Nandidae (Pisces: Teleostei). Fieldiana, Zool. 56:1-166.

Liem, K.F. 1978. Modulatory multiplicity in the functional repertoire of the feeding mechanism in cichlid fishes. I. Piscivores. J. Morph. 158:323-360.

Lighthill, M.J. 1969. Hydromechanics of aquatic animal propulsion. Ann. Rev. Fluid Mech. 1:423-446.

Muller, M. and Osse, J.W.M. 1978. Structural adaptations to suction feeding in fish. Proc. Zodiac. Symp. on

Adaptation, Wageningen, The Netherlands.
Nyberg, D.W. 1971. Prey capture in the largemouth bass. Am. Midl. Nat. 86:128-144.
O'Brien. W.J. 1979. The predator-prey interaction of planktivorous fish and zooplankton. Amer. Sci. 67:572-581.
Osse, J.W.M. 1969. Functional morphology of the head of the perch (<u>Perca fluviatilis</u> L): an electromyographic study. Neth. J. Zool. 19:289-392.
Osse, J.W.M. 1976. Mechanismes de la respiration et de la prise des proies chez <u>Amia calva</u> Linnaeus. Rev. Trav. Inst. Peches Marit. 40:701-702.
Osse, J.W.M. and M. Muller. (in press). Feeding by suction in fish and some implications for ventilation. In: Environmental Physiology of fishes, 1979 NATO/ASI conference. M.A. Ali, Ed. Plenum Press.
Pasztor, V.M. and Kleerekoper, H. 1962. The role of the gill filament musculature in teleosts. Can. J. Zool. 40:785-802.
Pietsch, T.W. 1978. The feeding mechanism of <u>Stylephorus chordatus</u> (Teleostei: Lampridiformes): functional and ecological implications. Copeia 2:255-262.
Prandtl, L. 1949. Essentials of Fluid Dynamics. Hafner Publ. Co., New York. Trans. from German by W.W. Deans.
Saunders, R.L. 1961. The irrigation of the gills in fishes. I. Studies of the mechanism of branchial irrigation. Can J. Zool. 39:637-653.
Shelton, G. 1970. The regulation of breathing. In, Fish Physiology, Vol. 4. W.S. Hoar and D.J Randall Eds. Academic Press, New York.
Streeter, V.L. and Wylie, E.B. 1979. Fluid Mechanics, 7th edition. McGraw Hill: New York.
Van Dam, L. 1938. On the utilization of oxygen and regulation of breathing in some aquatic animals. Dissertation, Groningen. (Cited in Saunders, 1961).
Webb, P.W. 1975. Hydrodynamics and energetics of fish propulsion. Bull. Fish. Res. Bd. Can. 190:1-158.
Weihs, D. 1972. A hydrodynamical analysis of fish turning manoeuvers. Proc. R. Soc. Lond. B. 182:59-72.
Weihs, D. 1973. The mechanisms of rapid starting of slender fish. Biorheology. 10:343-350.
Woskoboinikoff, M. and Balabai, D. 1937. Comparative experimental investigations of the respiratory apparatus of bony fishes. II. Acad. Sci. Ukrain. S.S.R. trav. Inst. Zool. Biol. 16:77-127. (Cited in Saunders, 1961).

HYDROMECHANICAL ADAPTATIONS IN *ALCYONIUM SIDEREUM* (OCTOCORALLIA)

Mark R. Patterson

Museum of Comparative Zoology Laboratories 105

Harvard University, Cambridge, MA 02138

ABSTRACT

The relation between colony size and shape *vs*. flow regime was investigated in a boreal species of octocoral, *Alcyonium sidereum* with particular regard to two hydrodynamic problems faced by these organisms: (1) maximizing the area of feeding surfaces presented to flow while (2) limiting potentially damaging drag forces induced by water movement.

The results are summarized as follows: (1) *A. sidereum* exhibits three general morphologies depending upon flow regime: A) a stubby lobed morphology (C_D = 0.37) is found in areas characterized by turbulent flow of high velocity ($U > 25$ cm/s), B) an ellipsoidal plate morphology (C_D = 0.76) is found in areas of predictable bi-directional flow ($10 < U < 25$ cm/s), and C) an elongate pencil-like morphology (C_D = 0.60), occasionally arborescent, is characteristic of calm areas ($U < 10$ cm/s). (2) Contraction of *A. sidereum* colonies during hydromechanical stress reduces their height above the substrate as well as their greatest horizontal dimension by 40-80%, significantly reducing drag. The normalized reduction is significantly greater in the height above the substrate indicating the importance of seeking the wall boundary layer during periods of strong flow. (3) Peak drag forces experienced by finger-like, lobed, and ellipsoidal colonies (isovolumetric) during storms were estimated to be 17 N, 9 N, and 23 N respectively. (4) Colony size (height above substrate) exhibits a positive correlation with water movement.

INTRODUCTION

Living in a fluid medium has important mechanical consequences for sessile marine invertebrates. These organisms typically rely on moving water to bring them food and to carry away waste products and sediments.

Many sessile planktivores show marked orientation to flow, e.g., sponges (Bidder 1923, Vogel 1977), brachiopods (LaBarbera 1977), hydroids (Riedl 1971, Svoboda 1976), octocorals (Théodor 1963, Théodor and Denizot 1965, Barham and Davies 1968, Wainwright and Dillon 1969, Grigg 1972, Leversee 1972, Rees 1972, Velimirov 1976, Muzik 1978), scleractinians (Abbott 1974, Chamberlain and Graus 1975a,b), crinoids (Meyer 1973, Macurda and Meyer 1974), and ophiuroids (Warner 1971, Warner and Woodley 1975). In most cases, orientation occurs in a manner which maximizes the feeding surface area presented to flow; this strategy effectively maximizes the rate at which plankton are removed from the water column. Changes in the orientation, size, and types of species existing in a community can occur between differing flow environments. Octocorals (Robins 1968, Kinzie 1973), zoanthids (Koehl 1977b), scleractinians (Jokiel 1978), sea anemones (Koehl 1977a) and crinoids (Meyer 1973) exhibit species replacement between areas characterized by different flow regimes. Water movement can thus play a significant role in the distribution of sessile marine invertebrates (Riedl 1971, Vosburgh 1978).

Many cnidarians are sessile planktivores; in the western north Atlantic as well as the Caribbean, members of the class Anthozoa are often the dominant space holders on subtidal substrates. Hence they are important consumers of plankton in shallow water habitats. I investigated biomechanical adaptations to flow in an octocoral, Alcyonium sidereum in order to better understand the general ecology of the species.

MATERIALS AND METHODS

Study Sites

A. sidereum is a boreal species of octocoral with a distribution from Cape Cod to the Bay of Fundy (Gosner 1971). Colonies of this species exhibit remarkable plasticity with respect to shape. This species is one of the dominant subtidal space holders in vertical rocky substrate communities (K. Sebens, personal communication) often blanketing large areas of rock wall (see Figure 1). A congeneric species, A. digitatum, is similar in morphology and habitat (Roushdy 1962, Robins 1968) and feeds on zooplankton (Pratt 1905) and phytoplankton (Roushdy and Hansen 1961). While the ecology of A. sidereum is virtually unknown, it is assumed to be similar in many respects to A. digitatum.

Studies of A. sidereum were conducted at several shallow water sites (< 15 m depth) near East Point, Nahant, Massachusetts (42°25'N, 70°54'W). Scuba dives were made year round in order to obtain data under a variety of current conditions. The surface water temperature at Nahant ranges from +15° to -1°C annually. Two study sites were established: one was located approximately 60 m offshore

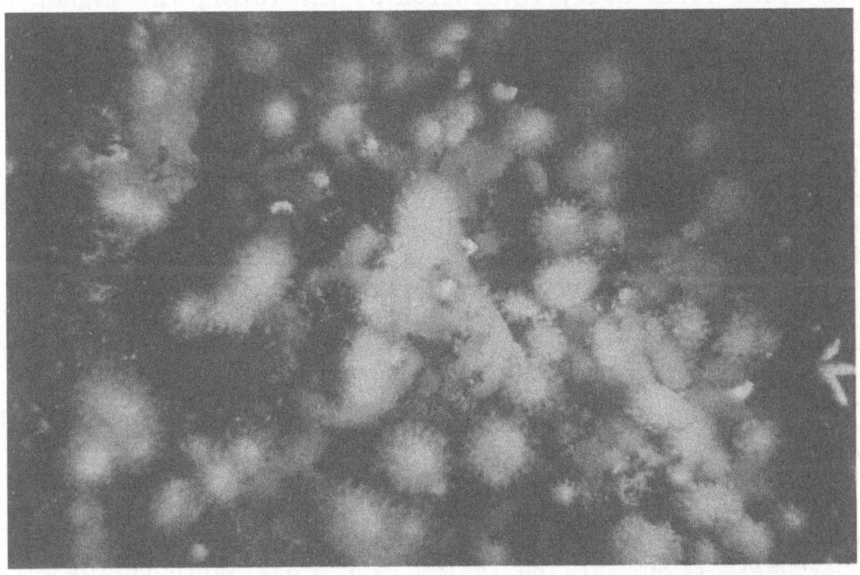

Figure 1. Colonies of A. sidereum in an expanded feeding condition on a vertical rock wall at 5 m depth, Nahant, Massachusetts. Colonies are 3-7 cm in their greatest dimension.

(high tide) on a vertical rock wall in 8 m of water; the other was located approximately 0.7 km offshore on a vertical rock wall in 12 m of water, and was accessible only by boat. Both sites were located adjacent to property owned by Northeastern University's Marine Science Institute. The property is an ecological preserve and these sites were largely free from disturbance by other divers.

Flowmeter

Flow regimes were characterized in various microhabitats with a flowmeter built in collaboration with K. Sebens, Harvard University. The flow transducer consisted of a two-tiered plexiglas (10 cm radius) Savonius rotor mounted on pin bearings in an aluminum bracket. The rotor and bracket were attached to an underwater housing (Ikelite No. 5210) which housed a digital electronic counter with LED display. A small permanent magnet was glued to the bottom of the rotor and counterweighted on the diametrically opposite side. With every revolution of the rotor, the magnet closed a magnetic reed switch imbedded in epoxy underneath the rotor. Leads from the magnetic reed switch were routed through holes in the housing and sealed with epoxy. In addition to an externally mounted "count" reed switch, an internally mounted reed switch was wired to clear the display. This reed switch was placed near the side wall of the housing and could be externally activated by the diver with a small magnet. The device was powered by a 9V transistor radio battery; with the display continuously running, battery life was about eight hours. The transducer was also easily interfaced to any of a variety of micro-cassette tape recorders currently available. A 1.5V AA penlite battery in series with the microphone input and the reed switch recorded a sharp click on the tape with every rotor revolution. Such an arrangement allowed continuous recording of flow data for periods as long as one-half hour.

The threshold of the prototype was 10 cm/s and its response to flow above 11 cm/s was linear ($r = 0.98$). This inexpensive flowmeter was tested over a six month period and found to be reliable and rugged. While it is possible to use the device to calculate flow values for short time intervals of less than one minute, it has great utility when used as a flow averager. Emplacing the device in areas for periods of 15 minutes or longer allows one to detect differences in integrated water flow which might not be seen on a shorter time scale.

Fluorescein Dye Release

Gross patterns of flow at the study sites were determined by releasing packets of uranine ($C_{20}H_{12}O_5$) dissolved in seawater, with a plastic syringe. Uranine exhibits intense fluorescence in an alkaline solution such as seawater, appearing a brilliant green. I also injected dye near colonies to permit an *in situ* visualization of flow. Dye release was useful for both detecting the presence of turbulent eddies in the lee of the colony, and determining the depth of the boundary layer over substrates.

Photography

I documented colony morphologies in various flow regimes, a volume contraction response to fluid mechanical stress in *A. sidereum*, and fluorescein dye movements around colonies using a Nikonos III underwater camera.

In situ Morphometry

Measurements of the sizes of *Alcyonium* colonies in expanded and contracted states were made using a metric ruler (1 mm divisions). Data were recorded on a submersible magnetic tape recorder (Wet-Tape) for later transcription ashore.

Wind Tunnel Studies

The relationship between colony morphology and induced drag force was investigated using Harvard University's one-meter wind tunnel and associated force balance. Colonies fabricated out of plasticine were mounted flush with the roof of the wind tunnel to simulate a substrate, and could be rotated to change the orientation of the colony with respect to flow. Koehl (1977a) found that the presence or absence of wart-like bumps (verrucae) on models of sea anemones made very little difference in the induced drag measured. Hence no attempt was made to simulate surface texture in the model colonies.

RESULTS AND DISCUSSION

The Effect of Flow on Morphology

A. sidereum exhibits a wide variety of morphologies. I observed subtidal morphology patterns and classified the shapes found into three distinct groups: 1) tall, thin fingers, occasionally arborescent, 2) ellipsoidal plates, and 3) short, stubby arborescent colonies. (See Figure 2.)

Thin, finger-like colonies are always found in areas of low ambient flow (see Figure 2A). During dives at Nahant, such colonies were observed to occur in crevices,

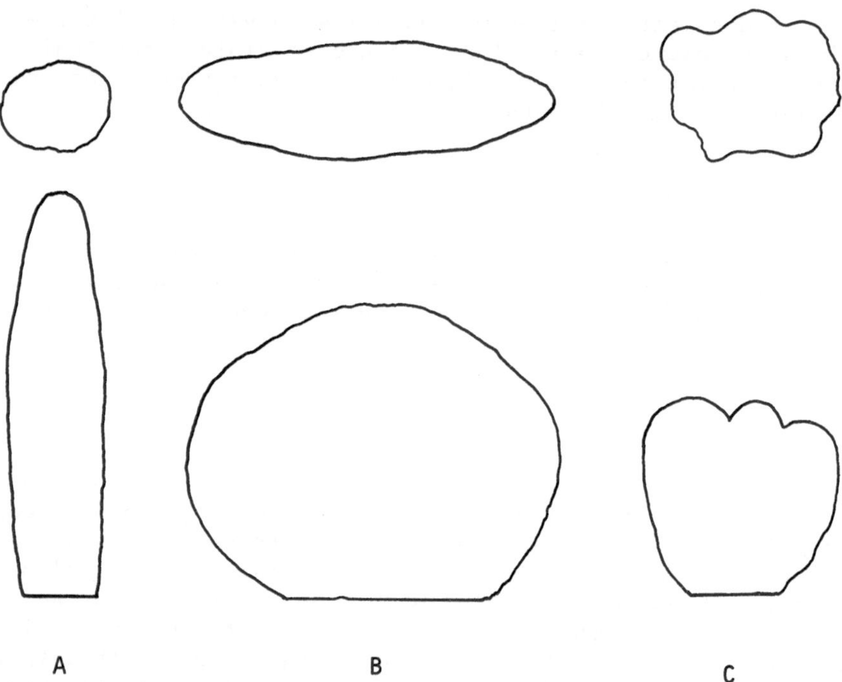

Figure 2. Projected surface areas of colony morphologies of *A. sidereum*. Upper tracing: top view; lower tracing: side view. A. Pencil-like. B. Ellipsoidal. C. Lobed.

deep under rock underhangs, and on vertical rock walls in protected areas. In many areas where such colonies occurred, large amounts of drift kelp and fine sediment had gathered on the bottom, good evidence that the area was chronically stagnant. Flow was measured 15 times (½ hour/area, 7.5 hours of flow data) in areas where finger morphologies were present and it was always below the threshold of the flowmeter, (10 cm/s). Fluorescein dye release confirmed that stagnation was indeed occurring.

In areas of predictable bi-directional flow, ellipsoidal morphologies were found (see Figure 2B). At a site off Dive Beach, Nahant, a small cove recedes into the vertical rock walls characteristic of the area. A. sidereum colonies and the colonial tunicate Aplidium constellatum are abundant here. Much of the cove is shaded during

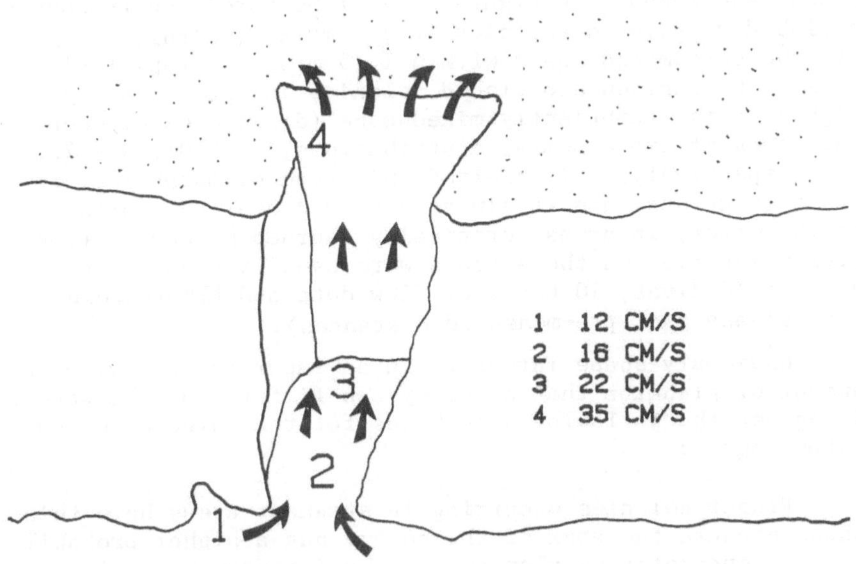

Figure 3. Flow direction and velocity at Dive Beach site. Velocities are average flows for the points indicated: 10 dives; 2 hours of data each for points 1, 2, 3, and 1.5 hours for point 4. Height of vertical walls approximately 5 m. Pencil-like colonies of A. sidereum occur in a crevice at 1, ellipsoidal colonies occur along the vertical wall to the left of 2 and 3, and lobed colonies occur at 4.

the day; the lack of light effectively excludes the kelp *Laminaria saccharina* and other brown and red algae everywhere, except for a small zone near the top of the rock wall. As waves pass over the site, the water within the cove is entrained and drawn through the top of the cove. This effect was observed by releasing packets of fluorescein dye, and consistently occurred during the 25 dives made at the site over a one year period. Dye paths allowed me to map the predominate direction of flow, while flowmeter readings permitted measurements of flow speed (see Figure 3).

At the rear of the cove at Dive Beach, I observed that a gradient of morphology types occurred between (1) colonies near the bottom of the rock wall that were subjected to bi-directional flow, and (2) those colonies located near the top of the wall where the water was turbulently mixed. Those colonies nearer the surface are stubby and lobed (see Figure 2C). I measured the frequency of lobed *vs* unlobed colonies in a 1 m^2 area within 1.5 m off the bottom and again within 1.25 m of the top of the wall. The incidence of lobed colonies is significantly higher in the turbulently mixed zone (difference between the means of two binomial distributions, $p < 0.01$, N = 22, 38, respectively). I observed this type of morphology in other locations, and it always occurred near the surface of the water, in areas chronically churned by wave action. Flow velocities in these areas were usually greater than 25 cm/s (6 dives, 10 hours of flow data and fluorescein dye release over pre-measured distances).

Obviously shape interacts in an important way with the amount of plankton that a colony can filter from the water. I suggest the following hypotheses for the three observed morphologies:

Finger colonies occurring in stagnant areas have this shape because the apex of the colony has a higher probability of encountering plankton in the faster-moving flow away from the substrate. Most *small* colonies (<3 cm in height) of *A. sidereum* found in all three flow regimes possess a spire morphology. There are two possible reasons for this growth pattern: A) growth away from the substrate decreases the probability that a young colony will be overgrown by the tunicate *Aplidium constellatum*; overgrowth by this colonial tunicate is a significant cause of mortality in young A. sidereum (K. Sebens, personal communication), and B) the growing apex of the colony enjoys an increasing probability of successful plankton capture as it moves away from the boundary layer of the wall. Growth out of the boundary layer allows the tip of the colony to sample a greater volume of water per unit time; this water has a smaller probability of being plankton-depleted due to filtering by other organisms growing on the wall. Indeed, A. sidereum is often found attached to mussel shells (Modiolus modiolus), effectively elevating the colony several cm away from the substrate. Because of the stagnant flow environment, the high surface-to-volume ratio compensates for the fact that a smaller volume of water passes over the colony per unit time. This growth mechanism has been proposed as an explanation for the distribution of a species of Alcyonium possessing a similar morphology (Robins 1968).

Ellipsoidal morphologies occur in areas of predictable bi-directional flow; their geometry indicates that broadside orientation is a natural response in a predictable flow regime. That is, the projected surface area (= feeding area) of the colony presented to flow is maximized by the differential growth of the colony.

Stubby lobed forms show no preferred orientation; their roughly spherical symmetry may indicate that all parts of the colony capture prey with equal probability, given the unpredictable direction and speed of the ambient flow near the surface where they occur.

The Effect of Flow on Colony Size

The height of a colony above the substrate is a measure of colony size. Colony size might be expected to reflect the colony's success in capturing plankton; larger colony size implies a higher filtration rate/unit volume, all other variables such as plankton availability, net circulation, etc. being equal for a given microhabitat.

To test this hypothesis, I measured heights of colonies in 1 m^2 areas in three microhabitats. In each instance, the microhabitat experienced a gradient of flow velocities as indicated by fluorescein dye movements and flowmeter readings.

A 1 m^2 plot was chosen in a stagnant area of the microhabitat and another 1 m^2 plot picked in an active flow area. Measurements of (1) colonies in crevices *vs*. nearby colonies not in crevices, (2) colonies near the edge of an underhang *vs*. those recessed, and (3) colonies near the top of a wall *vs*. colonies lower on the same wall, show the same trend of increased height in areas of higher flow velocity. All colonies surveyed were in an expanded feeding condition and care was taken to select areas where the velocity gradient was low enough to prevent significant mixture of morphologies due to drastic changes in flow regime. The mean height of colonies located in the faster flow regimes (out of crevices, near the lips of under hangs, and near the tops of protected walls) is significantly larger than the mean height of colonies located in adjacent stagnant areas (in crevices, recessed in underhangs, and near the bottom of protected walls). [Student's t-test, $p < 0.01$, d.f. = 41, 57, 45, respectively.]

Drag Reduction During Hydromechanical Stress

A. *sidereum* colonies living in flow are subjected to induced drag forces which must be transmitted to the substrate. The Reynolds' number (Re) for a "typical" colony *in situ* is calculated as follows:

$$Re = \frac{\rho U L}{\eta} = 2.21 \times 10^4$$

where: U = velocity of the seawater = 0.5 m/s (strong flow)
L = length of the colony = 0.06 m
ρ = density of seawater at 10°C = 1026 kg/m^3 (Zerbe and Taylor 1953)
η = dynamic viscosity of seawater at 10°C = 1.39 x 10^{-3} kg/m/s (Sverdrup, Johnson, and Fleming 1942)

Drag under conditions of high Re and steady flow is given by:

$$F_D = \tfrac{1}{2} C_D \rho U^2 A$$

where: F_D = induced drag force (N)
C_D = drag coefficient (dimensionless)
ρ = density of seawater (kg/m^3)
U = velocity of the fluid (m/s)
A = projected surface area of the colony (m^2)

If a colony is to persist, it must be able to withstand periodically severe hydromechanical stresses such as would be induced by storms. Typical flow environments of these organisms are on the order of 5-50 cm/s. Flow velocities during storms are reasonably estimated as being several m/s. Thus, flow velocities during storms are approximately one order of magnitude greater than non-storm flow velocities. Since F_D increases as the square of velocity, F_D experienced by the colony during a storm will be two orders of magnitude higher than the non-storm mean. Thus, colonies might be expected to possess some mechanism of drag reduction during periods of high F_D.

Observations of colonies in the field revealed that they do possess a mechanism of drag reduction: size change. During hydromechanical stress, water is pumped out of the gastrovascular cavities of the colony, effecting a considerable reduction in volume. Hydromechanical stress was artifically induced in the field by gently fanning a colony with a diving fin. Contraction of the colony typically occurred over a period of several minutes. The normalized percent changes in both "height above the substrate" and "greatest dimension parallel to the substrate" are -66.3% ($s = 10.3$, $N = 20$) and -42.3% ($s = 11.3$, $N = 20$), respectively. The mean value of percent contraction in the direction towards the substrate is significantly greater than the mean value of percent lateral contraction [student's t-test, $p < 0.01$, d.f. = 19]. This pattern of asymmetrical contraction may indicate the importance of seeking the boundary layer during periods of high F_D, as well as reducing the projected surface area of the colony presented to flow.

Drag Coefficients of Model Colonies

Since direct measurements of the drag forces experienced by colonies in flow were not feasible, I conducted wind tunnel studies on plasticine models of _A. sidereum_ in the Harvard one-meter wind tunnel. F_D was measured on a sensitive force balance at a variety of air speeds appropriate for the model size in order to preserve dynamic similarity. Projected surface areas of the models were determined by tracing an outline of the colony on graph paper. A regression was computed for area of graph _vs_. weight of paper and the projected surface area calculated by weighing the cut-out tracings.

Calculated data for the various morphologies are presented in Figures 4 and 5. Drag on colonies in situ was calculated using the C_D values obtained from the wind tunnel trials. The three models were isovolumetric; the plasticine was remodeled after each experiment. The data show that for a given amount of biomass, the ellipsoidal configuration experiences the highest drag. Notice the effect of orientation evident in Figure 5; an ellipsoidal colony oriented parallel to flow decreases its drag by 50% relative to a colony oriented broadside.

For "typical" colonies of all three morphology types, the drag experienced in situ during non-storm conditions ($U < 1$ m/s) is less than 1 N. Riedl and Machan (1972) claim that 5 m/s is the maximum velocity likely to be

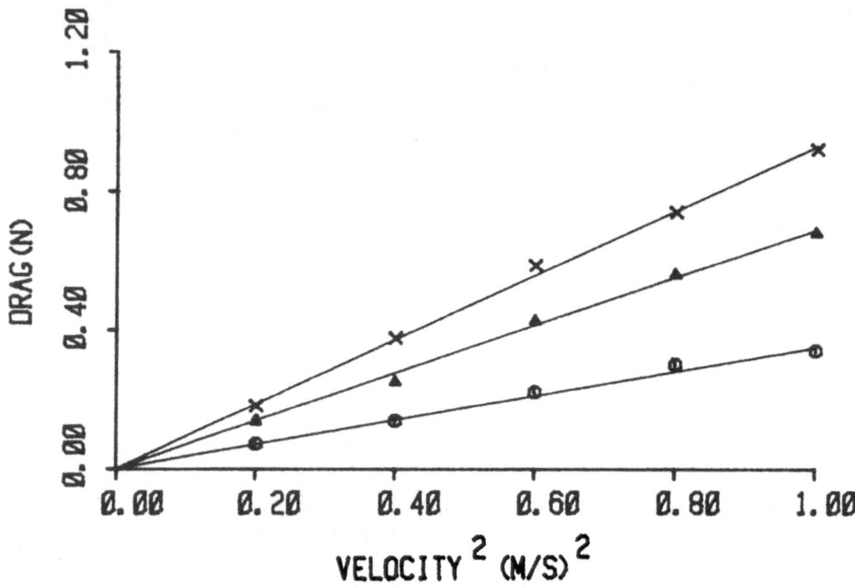

Figure 4. Graph of drag against velocity^2 for three colony morphologies. Values for F_D were calculated using C_D values obtained from wind tunnel testing of models. Lobed (⊕)(C_D = 0.37), pencil-like (▲)(C_D = 0.60), and ellipsoidal (X) (C_D = 0.76) colonies are isovolumetric.

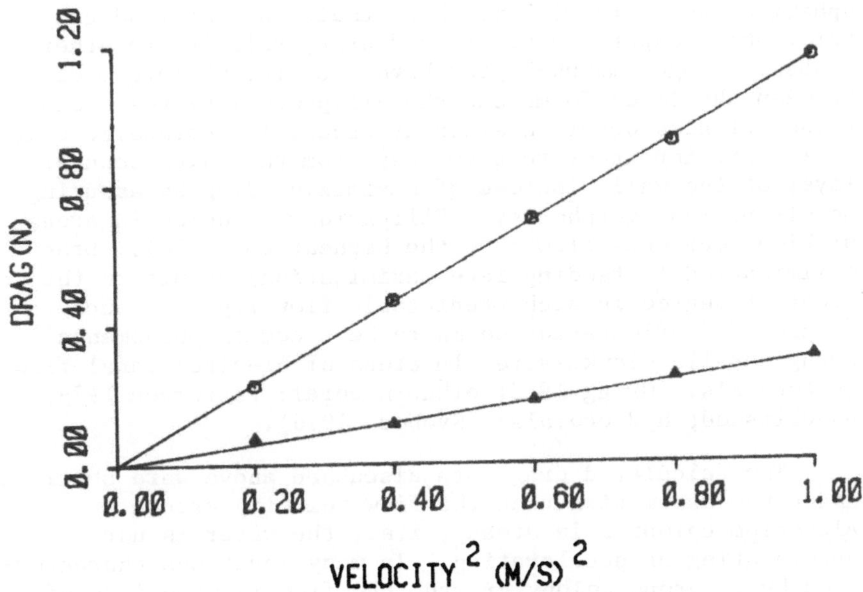

Figure 5. Graph of drag against velocity^2 for an ellipsoidal colony showing the effect of broadside orientation (⊕) (C_D = 0.55) vs. parallel (▲)(C_D = 0.30) orientation to flow.

reached in coastal channels during storms. Assuming the colonies don't contract, and are exposed to mainstream velocities during storms, a worst-case analysis reveals drag forces of 17 N, 9 N, and 23 N for the isovolumetric finger-like, lobed, and ellipsoidal colonies respectively.

Examination of the drag coefficients for the three types of morphologies shows an expected correlation between flow regime and drag-inducing character due to shape. Lobed, globular colonies characteristic of high velocity turbulently mixed areas have the lowest C_D (0.37). With respect to the drag coefficient, such geometries are very similar to spheres; both possess the same C_D at an Re of 2.21×10^4 (F. Eisner cited in Li and Lam 1964). Flow direction continually changes in turbulently mixed areas; it might be reasonably expected that colonies possessing a spherical geometry will maximize their rate of food capture, while experiencing reduced drag, relative to other shapes. Finger morphologies have a C_D (0.60) intermediate between the lobed forms and the ellipsoidal plates. Since these colonies occur in stagnant areas, I hypothesize that it is more important to grow away from the thick boundary layer of the wall, instead of maximizing drag by assuming an ellipsoidal morphology. Ellipsoidal colonies in areas of bi-directional flow have the highest C_D (0.76). Drag maximization (= feeding rate maximization) occurs to the greatest degree in such predictable flow regimes; such "broadside" orientation seems to be a common phenomenon among sessile planktivores in areas of bi-directional flow (octocorals: Grigg 1972; elkhorn coral: Patterson 1979, unpublished; hydrocorals: Svoboda 1976).

The calculated drag data discussed above were obtained under the assumption that the flow velocity around *Alcyonium* colonies is steady, i.e., the water is not accelerating or decelerating. In many locations characterized by vigorous colony growth, the flow is wave-induced oscillatory surge. As Koehl (1977a) points out, oscillatory flow exerts a higher drag force on marine organisms than steady flow does due to the acceleration of the fluid. Thus, my estimates of F_D are apt to be lower than the actual values experienced by colonies in flow.

CONCLUSION

Previous investigators have found orientation to flow in many sessile planktivores. In many cases, orientation occurs in such a way as to maximize the projected feeding surface area of the colony or reduce mechanical stresses. Hence, I believe a useful generalization follows: passive sessile planktivores are drag maximizers. However, the advantage of increased feeding efficiency exacts a heavy toll during periods of hydromechanical stress; drag forces experienced during periods of high flow must be transmitted to the substrate. Thus, all sessile planktivores should possess some mechanism of reducing drag during periods of high flow.

A. sidereum exhibits trends found in other sessile planktivores: it is a drag maximizer in flow regimes where the direction of flow is predictable, yet it possesses a mechanism of drag reduction which occurs during hydromechanical stress. The colony shrinks; by becoming small during periods of strong flow, it can afford to maintain a much larger size during feeding without a concomitant increase in the cost of materials normally required for a rigid organism of the same size to successfully resist high drag. Size as well as shape is strongly determined by the ambient flow of the microhabitats in which these colonies occur.

ACKNOWLEDGEMENTS

This research was supported in part by National Science Foundation Grant #OCE 78-08482 to K. Sebens, Harvard University, and undergraduate research funds from the Department of Biology, Harvard University. The author was supported during the preparation of this manuscript by a National Science Foundation Graduate Fellowship in Ecology. K. Sebens gave valuable advice and encouragement and read preliminary drafts of this work, as did T. McMahon, G. Lauder, R. Olson, S. Sharar, and S. Johnson. F. Abernathy kindly allowed me use of the Harvard one-meter wind tunnel, while R. Rex provided valuable technical assistance. N. W. Riser allowed me unlimited access to the facilities of Northeastern University's Edwards Laboratory at Nahant, Massachusetts, where this study was conducted. A special debt of thanks is due S. Johnson for encouragement and support, and to my father, R. C. Patterson, who once taught a little boy of six to snorkel.

LITERATURE CITED

Abbott, B. M. 1974. Flume studies on the stability of model corals as an aid to quantitative paleoecology. Paleogeogr. Paleoclimatol. Paleoecolog., 15: 1-27.

Barham, E. G., and I. E. Davies. 1968. Gorgonians and water motion studies in the Gulf of California. Underwat. Nat., 5: 24-28.

Bidder, G. P. 1923. Relation of the form of a sponge to its currents. Quart. Journ. Micros. Sci., 67(2): 293-325.

Chamberlain, J. A. and R. R. Graus. 1975a. Water flow and hydromechanical adaptations of branched reef corals. Bull. mar. Sci., 25: 112-125.

__________. 1975b. Adaptation in corals: How do corals withstand waves and currents? Abstracts with Programs, U.S. Geol. Soc. Ann. Meetings, Salt Lake City, Utah, p. 1024.

Gosner, K. L. 1971. Guide to Identification of Marine and Estuarine Invertebrates. Wiley-Interscience, New York.

Grigg, R. W. 1972. Orientation and growth form of sea fans. Limnol. Oceanogr., 17: 185-192.

Jokiel, P. L. 1978. Effects of water motion on reef corals. J. exp. mar. Biol. Ecol., 35: 87-97.

Kinzie, R. A., III. 1973. The zonation of West Indian gorgonians. Coral reef project-papers in memory of Dr. Thomas F. Goreau. Bull mar. Sci., 23: 93-155.

Koehl, M.A.R. 1977a. Effects of sea anemones on the flow forces they encounter. J. exp. Biol., 69: 87-105.

_____. 1977b. Water flow and the morphology of zoanthid colonies. Proc. Third Int. Coral Reef Symp., 1: 437-444.

LaBarbera, M. C. 1977. Brachiopod orientation to water movement 1: Theory, laboratory behavior and field observations. Paleobiol., 3(3): 270-287.

Leversee, G. J. 1972. Field and laboratory studies of the effect of water currents on morphology and feeding in the seawhip, *Leptogorgia*. Am. Zool., 12: 719.

Li, W.-H. and S-H. Lam. 1964. *Principles of Fluid Mechanics*. Addison Wesley, Reading, MA.

Macurda, D. B. and D. L. Meyer. 1974. Feeding posture of modern stalked crinoids. Nature (Lond.), 247: 394-396.

Meyer, D. L. 1973. Feeding behaviour and ecology of shallow-water unstalked crinoids (Echinodermata) in the Carribean Sea. Mar. Biol., 22: 105-129.

Muzik, K. 1978. A bioluminescent gorgonian, *Lepidisis olapa*, new species (Coelenterata, Octocorallia), from Hawaii. Bull. mar. Sci., 28(4): 735-741.

Patterson, M. R. 1979. Hydromechanical investigations in the Cnidaria: investigations involving *Alcyonium sidereum* (Octocorallia) and *Acropora palmata* (Scleractinia). Unpublished MS.

Pratt, E. M. 1905. The digestive organs of the Alcyonaria and their relation to the mesogleal cell plexus. Quart. Journ. Micros. Sci., 49: 327-362.

Rees, J. T. 1972. The effect of current on the growth form in an octocoral. J. exp. mar. Biol. Ecol., 10: 115-124.

Riedl, R. J. 1971. Water movement. In: *Marine Ecology, Vol. I, Part 2*. O. Kinne, editor. Wiley Interscience, London, pp. 1085-1088, 1124-1156.

Riedl, R. J. and R. Machan. 1972. Hydrodynamic patterns in lotic intertidal sands and their biological implications. Mar. Biol., 13: 179-209.

Robins, M. W. 1968. The ecology of *Alcyonium* species in the Scilly Isles. Underwat. Ass. Rept., pp. 67-71.

Roushdy, H. M. 1962. Expansion of Alcyonium digitatum (Octocorallia) and its significance for the uptake of food. Vidensk. Medd. bansk natur. Foren. Kjobenhaun, 124: 409-420.

Roushdy, H. M. and V. K. Hansen. 1961. Filtration of phytoplankton by the octocoral Alcyonium digitatum L. Nature (London), 190: 649-650.

Sverdrup, H. U., M. W. Johnson, and R. H. Fleming. 1942. The Oceans: Their Physics, Chemistry, and Biology. Prentice-Hall, New York.

Svoboda, A. 1976. The orientation of Aglaophenia fans to current in laboratory conditions (Hydrozoa, Coelenterata). In: Coelenterate Ecology and Behavior. G. O. Mackie, editor. Plenum, New York, pp. 41-48.

Théodor, J. 1963. Contribution à l'étude des Gorgones III. Trois formes adaptives d'Eunicella stricta en fonction de la turbulence et du courant. Vie et Milieu, 14: 815-818.

Théodor, J. and M. Denizot. 1965. Contribution à l'étude des Gorgones. I. A porpos de l'orientation d'organismes marins fixés végétaux et animaux en fonction du courant. Vie et Milieu, 16: 237-241.

Velimirov, B. 1976. Variation in forms of Eunicella cavolini Koch (Octocorallia) related to intensity of water movement. J. exp. mar. Biol. Ecol., 21: 109-117.

Vogel, S. 1977. Current-induced flow through living sponges in nature. Proc. Nat. Acad. Sci., 74(5): 2069-2071.

Vosburgh, F. 1977. The response to drag of the reef coral Acropora reticulata. Proc. Third Int. Coral Reef Symp., 1: 477-482.

Wainwright, S. A. and J. Dillon. 1969. On the orientation of sea fans (genus Gorgonia). Biol. Bull., 136: 130-139.

Warner, G. F. 1971. On the ecology of a dense bed of the brittle-star Ophiothrix fragilis. J. mar. biol. Ass. U.K., 51: 267-282.

Warner, G. F. and J. D. Woodley. 1975. Suspension-feeding in the brittle-star Ophiothrix fragilis. J. mar. biol. Ass. U.K., 55: 199-210.

Zerbe, W. B. and C. B. Taylor, 1953. Sea water density reduction tables. In: Coast and Geodetic Survey Special Publication, no. 298, Wash., D.C., U.S. Dept. of Commerce, pp. 18-19.

Control of the Output of the Heart:

A Historical Perspective

Matthew N. Levy

Mt. Sinai Hospital

Cleveland, Ohio 44106

The circulation of the blood was first described by William Harvey about 4 centuries ago. Subsequently, there was a considerable lag before some of the major factors that governed the circulatory system were elucidated. Most of that work has been accomplished in the present century. The various components of the circulatory system have usually been studied in isolation, under "open-loop" conditions. Studies of the function of the heart have also usually been done in the "open-loop" mode, and only relatively recently have investigators made a serious attempt to analyze the system in its natural "closed-loop" condition.

In this paper, I will direct attention to some of the highlights in the evolution of our understanding of the function of the heart. I will discuss the information derived from studies of isolated hearts. I will then trace the development from experiments on intact anesthetized animals subjected to extensive surgical procedures, to more modern technics involving chronically instrumented, awake and active animals. I will also describe some of the recent attempts to model the closed-loop operation of the circulatory system, wherein the interactions between the cardiac and vascular components of the system are given their proper consideration.

At the time of the pioneering studies of the German physiologist, Otto Frank, near the end of the last century the relationship between the resting tension of skeletal muscle and the extent of its subsequent contraction had

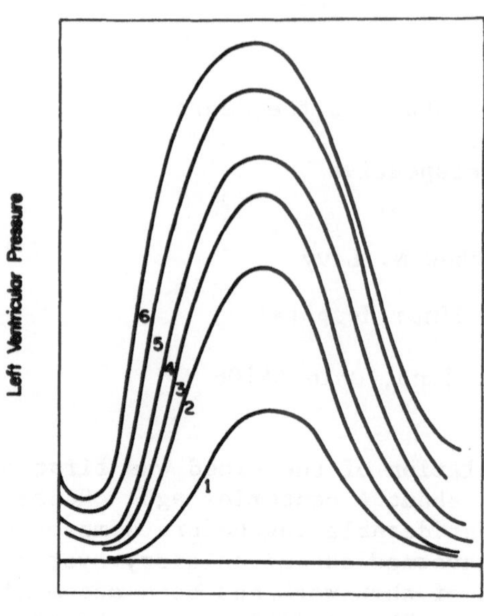

Fig. 1. The effect of increased filling on the response of the frog ventricle. The intraventricular pressure at the beginning of each contraction increases as the filling volume is raised (denoted by the numbers along each curve), and this is accompanied by a progressively greater peak tension during systole. (Redrawn from Frank, O.: Z. Biol., 32:370, 1895, with permission.)

already received considerable attention. Frank's studies[1] on the isolated frog heart constituted the first important transition from the observations on skeletal muscle to those on cardiac tissue. His principle conclusion was that the initial tension on the cardiac muscle, just prior to its contraction, was a critical determinant of the extent of that contraction. As shown in Figure 1, as he increased the volume of fluid (curves 1 to 6) in a heart made to contract isovolumetrically, the end-diastolic pressure increased accordingly. During contraction, there was a disproportionately great augmentation of the peak systolic pressure as the initial tension was progressively raised. He also found that if the initial tension was increased excessively, there was a decline in the peak pressure attained during systole. This constituted a form of acute heart failure.

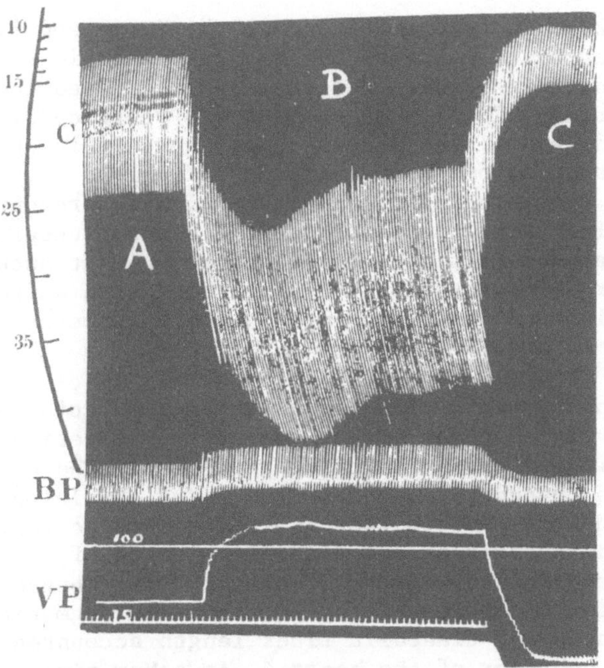

Fig. 2. The effect of a rise in filling pressure on the ventricular volume and stroke volume in a canine heart-lung preparation. The blood reservoir connected to the right atrium was suddenly elevated at the time denoted by the sudden rise in venous pressure (VP). Arterial blood pressure (BP) was limited experimentally to only a slight increase. The changes in ventricular volume are denoted by the cardiometer (C) tracing; increasing volume is registered as a downward deflection. Note that as venous pressure is elevated, the ventricular end-diastolic volume (lower border of the cardiometer tracing) and the stroke volume (amplitude of the cardiometer tracing) both increase. (From Patterson et al., J. Physiol. (London), 48:465, 1914, with permission.)

The English physiologist, Ernest Starling, extended the observations of Frank. Starling and his collaborators used the isolated, canine heart-lung preparation, in which it was possible to vary the effects of preload and afterload independently, and to examine their effects on the cardiac output.[2] They showed that, within limits, the stroke volume varied directly with the cardiac preload

(Fig. 2). The accommodation to the increased filling was achieved by an increase in the end-diastolic volume of the ventricles, which reflected an increased length of the myocardial fibers that make up the ventricular walls. Starling and his colleagues also showed that the ventricular adaptation to an increased afterload similarly involved a stretching of the myocardial fibers during diastole. Subsequent work on the ultrastructure of both skeletal and cardiac muscle have shown that such intrinsic cardiac adaptations involve changes in alignment of the thick and thin filaments in the sarcomeres of the muscle fibers.

The Frank-Starling mechanism had a profound influence for several decades on the concepts of experimental physiology and of clinical cardiology.[3] There was a pronounced tendency to explain the major physiological and pathological factors that influence cardiac function by invoking changes in end-diastolic length of the ventricular myofibrils. Starling himself contributed to the confusion, because at times he asserted dogmatically that changes in end-diastolic fiber length accounted for the entire behavior of the heart.[2] At other times, however, he acknowledged that other factors, such as myocardial oxygen supply and tissue pH, also played a critical role.

Sarnoff and his collaborators[4,5] contributed to a more rational appreciation of the role of the Frank-Starling mechanism by investigating the interaction of this intrinsic mechanism with a variety of extrinsic and certain other intrinsic factors. They emphasized that the Frank-Starling relationship could not properly be represented by a single characteristic curve, but that a family of such curves was the appropriate representation (Fig. 3).

The dramatic advances in instrumentation technology have facilitated the collection of data from healthy, unanesthetized animals under a variety of experimental conditions. Rushmer was one of the pioneers in the development of such studies. He showed that during physical exercise, for example, the neural influences on the heart far outweighed the effects ascribable to the Frank-Starling mechanism.[6] Refinements in such technics have led Braunwald and Ross[7] and their collaborators, and

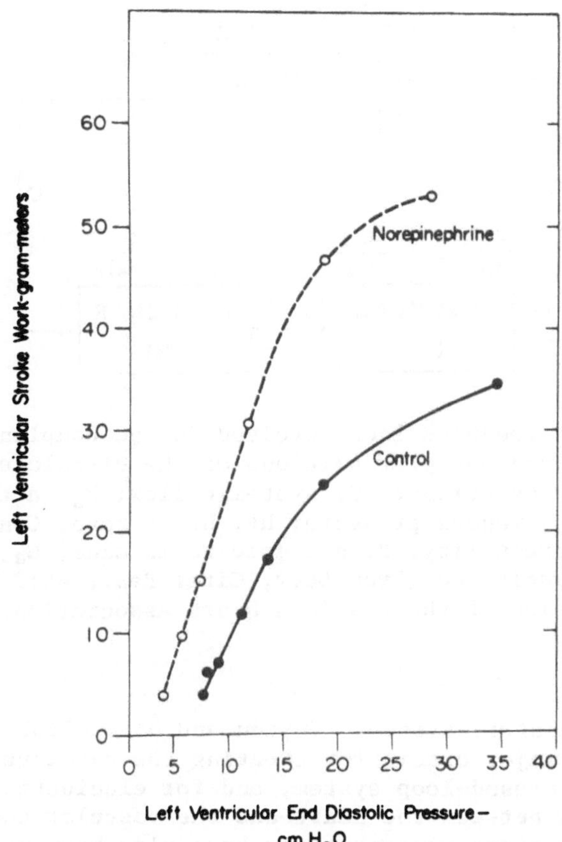

Fig. 3. The effect of a norepinephrine infusion on the relation between left ventricular stroke work and left ventricular end-diastolic pressure (Redrawn from Sarnoff et al., Circ. Res., 8:1108, 1960, with permission from the American Heart Association.)

numerous other groups as well, to carry out sophisticated experiments on healthy, chronically instrumented animals, and on animals with a variety of abnormal cardiovascular conditions as well.

Cardiac physiologists have characteristically focused their attention on the heart itself, and have neglected the coupling between the heart and the vascular component

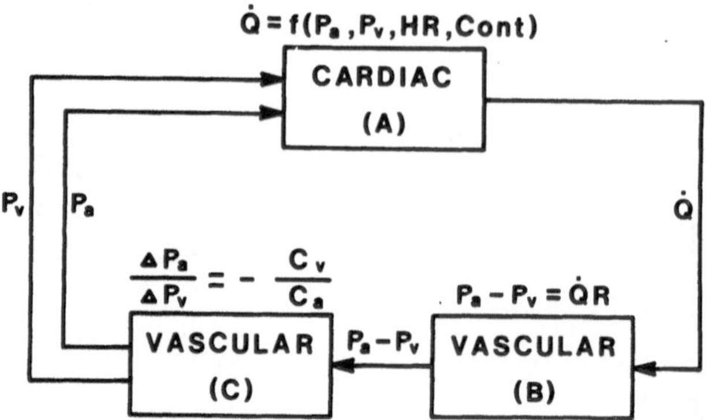

Fig. 4. The feedback loop involved in the coupling of the cardiac and vascular portions of the circulatory system. Abbreviations: $\dot{Q}$, systemic flow; P_a, arterial pressure; P_v, venous pressure; HR, heart rate, Cont, cardiac contractility; R, systemic resistance; C_a, arterial capacitance (From Levy, Circ. Res., 44:739, 1979, with permission of the American Heart Association.)

of the circulatory system. Guyton and his collaborators[8] deserve the major credit for treating the cardiovascular system as a closed-loop system, and for elucidating the interactions between the heart and the vascular components. However, important contributions have also been made in this area by Grodins et al.[9] and by Sagawa.[10] The major factors affecting cardiac performance are usually considered to be the preload, afterload, heart rate, and contractility. This is depicted in block (A) of Figure 2. Depending on the levels of these critical factors, the heart delivers a certain flow, $\dot{Q}$, to the vascular component of the system. The resistive and capacitive characteristics of the vasculature, in turn, determine the preload and afterload of the heart.[8-11] In addition, reflexes invoked by changes in pressure on the arterial and venous sides of the circuit also result in changes in heart rate and myocardial contractility. There are also important interactions between the cardiovascular system and other organ systems, notably the renal, endocrine, and pulmonary systems. These interactions have been treated in great detail by Guyton et al.[8] Considerably more work must still be done to unravel the extent of such interactions during various types of activities in normal subjects and in a variety of pathological states.

1. Frank, O.: Zur Dynamik des Herzmuskels. Z. Biol. 32:370-447, 1895

2. Patterson, S. W., Piper, H., and Starling, E.H.: The regulation of the heart beat. J. Physiol. (London) 48:465-513, 1914

3. Chapman, C. B.: Impact of the Starling concepts on clinical cardiology. JAMA 183:352-357, 1963

4. Sarnoff, S.J., and Berglund, E.: Ventricular function I. Starling's law of the heart studied by means of simultaneous right and left ventricular function curves in the dog. Circulation 9:706-718, 1954

5. Sarnoff, S.J., and Mitchell, J.H.: The control of the function of the heart. In: Handbook of Physiology, 1962, eds: Hamilton, W.F., and Dow, P., Waverly Press, Baltimore, pp. 489-532

6. Rushmer, R.F.: Effects of nerve stimulation and hormones on the heart; the role of the heart in general circulatory regulation. In: Handbook of Physiology, 1962, eds: Hamilton, W.F., and Dow, P., Waverly Press, Baltimore, pp. 533-550

7. Braunwald, E., and Ross, J., Jr.: Control of cardiac performance. In: Handbook of Physiology, 1979, eds: Berne, R.M., Sperelakis, N., and Geiger, S.R., Waverly Press, Baltimore, pp. 533-580

8. Guyton, A.C., Jones, C.E., and Coleman, T.G.: Circulatory Physiology: Cardiac Output and its Regulation. 1973, W. B. Saunders Company, Philadelphia, pp. 3-549

9. Grodins, F.S., Stuart, W.H., and Veenstral, R.L.: Performance characteristics of the right heart bypass preparation. Am. J. Physiol. 198:552-560, 1960

10. Sagawa, K.: The use of control theory and systems analysis in cardiovascular dynamics. In Cardiovascular Fluid Dynamics, Vol. 1, 1972, ed. Bergel, D.G., Academic Press, London, pp. 115-171

11. Levy, M.N.: The cardiac and vascular factors that determine systemic blood flow. Circ. Res. 44:739-745, 1979

BLOOD FLOW DISTURBANCES IN THE CARDIOVASCULAR SYSTEM: SIGNIFICANCE IN HEALTH AND DISEASE

Paul D. Stein, Hani N. Sabbah and Frederick J. Walburn

Departments of Medicine and Surgery, Henry Ford Hospital, Detroit, Michigan, 48202

The characteristics of blood flow in the cardiovascular system are unique in that they involve an array of variables not commonly encountered in engineering applications. Among these factors are: pulsatile flow, non-Newtonian fluid, compliant arterial walls, complex structure of entrance regions, vessel tapering, curvature and bifurcations. All of these factors interact in the determination of flow characteristics of the arterial system. Many may undergo changes in various disease states and thereby alter the characteristics of flow. The following discussion will relate to: 1) The nature of flow disturbances in normal states. 2) Its alteration under abnormal conditions. 3) Factors that tend to augment or minimize disturbances, and 4) The implications of the presence of flow disturbances in physical diagnosis, and 5) its potential in the perpetuation of cardiovascular abnormalities.

The medical community has recognized for many years that turbulence may occur in the cardiovascular system because of its likely association with cardiac murmurs, particularly that of aortic stenosis. It was many years, however, before in vivo measurements of turbulence or disturbed flow were conducted in laboratory animals (1-3), in human beings during open heart surgery (4) and during cardiac catheterization (5,6).

MEASUREMENTS OF BLOOD FLOW DISTURBANCES IN THE REGION OF SEMILUNAR VALVES

Direct measurement of point velocity near the aortic

valve of normal subjects with a hot-film probe showed flow disturbances during ejection (6). Fluctuations in velocity persisted during the deceleration phase (Fig. 1). In the mid-ascending aorta and the proximal aortic arch, the velocity signal was smooth with no apparent disturbances present. The maximal aortic velocity in these subjects with normal aortic valves ranged between 55 and 120 cm/sec (6).

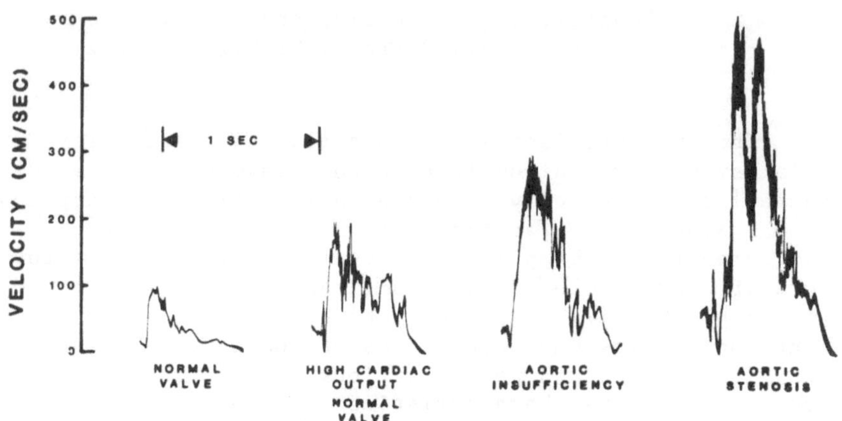

Fig. 1 Blood velocity above the aortic valve in a patient with a normal aortic valve, a patient with high cardiac output of 12.9 L/min and a heart rate of 144 beats/min, a patient with aortic insufficiency and a patient with severe aortic stenosis. These tracings were redrawn from original data in order to show velocities at the same amplification.

Higher magnitude flow disturbances in the region of the aortic valve were observed in a patient with an elevated heart rate and stroke volume (6). In this patient, disturbances of flow started earlier during ejection and continued throughout the ejection phase (Fig. 1). As the probe was moved farther away from the aortic valve into the mid-ascending aorta, velocity fluctuations persisted during peak velocity and into the deceleration phase (6). These observations are consistent with those made in laboratory animals (7-9). It is clear from these observations that an increased stroke volume contributes to an increased blood velocity. An increased blood velocity would augment the Reynolds number, and an increased heart rate would augment the unsteadiness parameter, α, (9, 10). The presence of increased flow disturbances is consistent with an augmentation of these parameters.

Disturbed or turbulent blood flow in the cardiovascular system has been related to the transitional and critical Reynolds number (4, 9), and the unsteadines parameter (α number) (9). In addition, physical factors not included in these parameters also modify turbulent or disturbed flow. Roughness or irregularity of the vessel walls, stenoses, bifurcations, and projection into the stream of flow are factors that contribute to disturbances of flow. Compliant walls of blood vessels (11), a favorable branch-to-trunk area ratio (12), vessel tapering (13), normal flexibility of red cells (14), and a normal number of red cells (15) are factors that tend to minimize disturbances.

In vitro studies of the effect of semilunar leaflets upon the production of disturbed or turbulent blood flow in the region of a stent mounted normal porcine aortic valve indicated that the localization of maximal turbulence above the aortic valve was related to the location of the free margin of the leaflets during ejection (16). The presence of maximal turbulence occurred in close proximity to the location of the free margins of the semilunar cusps during systole. This can be explained by the tripping of flow at the free margin of the cusps. Fluttering of the free margin of the leaflets may contribute to this affect, and we have observed such fluttering on high speed cine in vitro. These observations are consistent with the speculations made by McDonald (17). The observed disturbances can be considered to result from the cusps acting as natural projections into the stream of flow. In pulsatile flow in tubes, large instantaneous Reynolds numbers reported during systole do not result in transition to turbulence everywhere, but only at sources of disturbances (18). Two conditions were given for turbulence to occur in systole. These were the presence of a large shear rate and a source of instability (18). This led to the conclusion that it is insufficient to rely on the instantaneous peak Reynolds number when attempting to determine when the transition from laminar to turbulent flow occurs in a pulsatile system (18). It is likely that the valve cusps act as a source of instability producing blood flow disturbances in the region of normal semilunar valves. An increased Reynolds number would augment disturbances that occur at the site of such projections.

In the presence of aortic valvular disease, disturbed or turbulent blood flow is markedly increased in comparison to that observed in normal subjects (6) (Fig. 1). In patients with pure aortic insufficiency, the increased flow disturbances may be attributed to an increased forward flow across the valve productive of high ejection velocities. In such patients, peak velocity in the region of the aortic valve was twice that observed in normal subjects. Significant levels of flow disturbances were measured in the proximal aortic arch region (6). In patients with aortic stenosis the level of turbulence was approximately ten times that observed in normal subjects due to the turbulent jet. Peak velocities in such patients ranged from 300 to 450 cm/sec (6).

Turbulence in the region of aortic prosthetic valves has also been shown to be present both in vivo (6) and in vitro (19). The structural configuration of these valves contributes to flow separation which results in the breakdown of flow distal to the prosthesis; and the amount of turbulence depends upon the configuration of the valve (19). Some recently designed prosthetic valves have taken this problem under consideration, and designs of prosthetic valves have been introduced which minimize turbulence (20).

Blood flow disturbances have also been shown to occur in the region of normally functioning pulmonary valves in patients (21) (Fig. 2). However, the absolute intensity of turbulence in the main pulmonary artery, 1.6 ± 0.2 cm/sec (mean ± SEM), was significantly lower than in the ascending aorta, 4.2 ± 0.4 cm/sec, and the relative intensity of turbulence was also lower in the pulmonary artery (0.06 ± 0.01 vs 0.04 ± 0.004) (21).

One factor that may have contributed to the higher turbulence in the aorta may be the higher velocity in the aorta (21). Even though volumetric flow is identical in the pulmonary artery and aorta, maximal velocity is lower in the pulmonary artery due to several factors. Among these are the larger diameter of the main pulmonary artery (22, 23) and pulmonary valve (24), and a longer systolic ejection period within the pulmonary artery (25, 26). On this basis alone, mean velocity would be approximately 29 percent lower in the pulmonary artery than in the aorta, and a lower velocity has been measured (21, 25, 27). In addition, the greater distensibility of the pulmonary artery (22, 23) may have a damping effect upon turbulence that is independent of velocity, since compliant surfaces themselves appear to have a damping effect upon turbulence (11, 28).

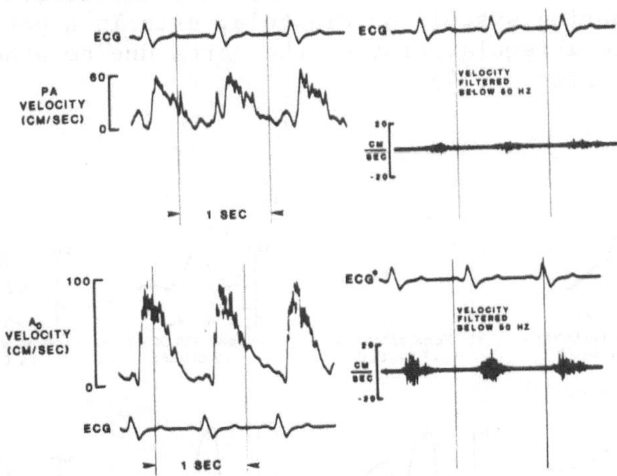

Fig. 2 Pulmonary arterial (PA) velocity (top) and aortic (Ao) velocity (bottom). To the right of each velocity tracing is a velocity signal that has been filtered below 50 Hz. Random fluctuating velocities occur during mid-systole, and are shown superimposed upon pulsatile velocity (left). To the right, the pulsatile velocity has been eliminated by filtering. The root-mean-square value of these fluctuating velocities defined the turbulence intensity. It is particularly apparent on the filtered tracings that velocity disturbances in the aorta were greater than in the pulmonary artery.

FLOW CHARACTERISTICS IN THE ABDOMINAL AORTA AND ILIAC BIFURCATION

Studies in glass tubes have indicated that turbulent or disturbed flow would be expected at bifurcations, and turbulence would occur at lower velocities than expected with comparable fluids in straight tubes of similar dimensions (29-31). Thus, the occurrence of flow disturbances on the basis of in vitro studies is predicted, and mechanical damage due to such disturbances conceivably could contribute to atherosclerosis. In view of this, we conducted measurements of instantaneous point velocity in the descending aorta and iliac artery in eight patients during diagnostic cardiac catheterization (32). Our measurements showed no disturbances of flow in the abdominal aorta or in the common iliac artery. No high frequency velocity fluctuations were observed on any of the recorded velocity signals during systole or diastole, even in a patient with noticeable irregularities of the aorta due to atherosclerosis (32) (Figs. 3, 4).

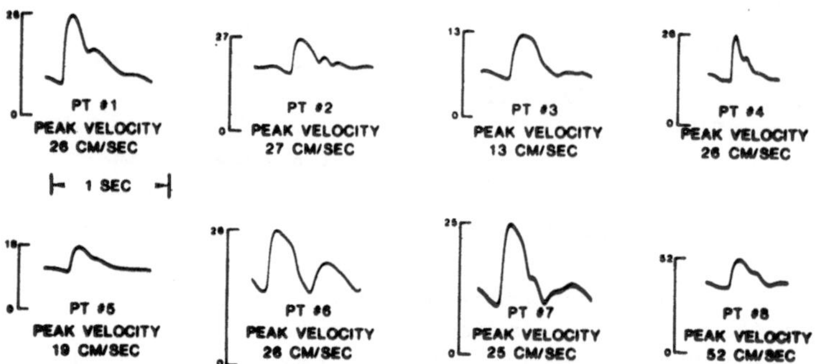

Fig. 3 Velocity recorded just proximal to the bifurcation of the aorta at the common iliac artery in each of the patients studied. None of the recorded velocities showed flow disturbances during systole or diastole. Reproduced from Biorheology (32) with permission of Pergamon Press.

Peak velocity recorded in the aorta in the region of the renal artery was 28 ± 3 cm/sec (mean ± SEM). Just proximal to the aortic bifurcation it was 27 ± 4 cm/sec. Velocity was only somewhat lower in the proximal portion of the common iliac artery (24 ± 4 cm/sec). The Reynolds number at the aortic bifurcation was 730 (range 400-1100). In the common iliac artery the average Reynolds number in four patients was 540 (range 390-620). The unsteadiness parameter (α number) measured just proximal to the iliac bifurcation was 9.3 (range 8.3-10.9).

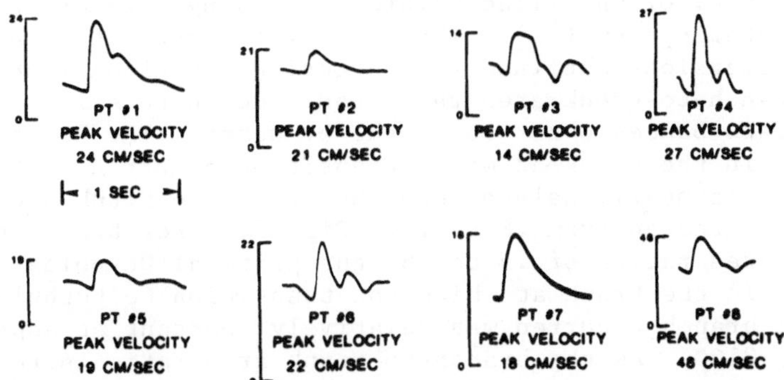

Fig. 4 Velocity recorded in the common iliac artery in each of the patients studied. In each instance, velocity showed no flow disturbances during systole or diastole. Reproduced from Biorheology (32) with permission of Pergamon Press.

The angle measured between the common iliac arteries, as they branched from the aorta, ranged between 52° and 80° in the four patients in whom it was measured. The branch-to-trunk area ratio (measured in a different group of patients) was .71 ± .03 (range .5 to .95) (12) and the angle of tapering was 0° to 3° (13).

The absence of blood flow disturbances in the abdominal aorta and iliac bifurcations of human subjects, in view of its conjectured presence suggested by others based upon studies in glass models led to investigations of the factors that may be responsible for the attenuation of disturbances in these regions. These included studies of 1) the effects of the branch-to-trunk area ratio; 2) the effects of vessel tapering; and 3) the effects of vessel distensibility.

Effect of Branch-To-Trunk Area Ratio on Flow Disturbances

The effect of the branch-to-trunk area ratio on the transition to turbulent flow was studied in an in vitro glass model of the iliac bifurcation using a laser Doppler anemometer system (12). The model was constructed based upon dimensions obtained from arteriograms of patients. The branch-to-trunk area ratio had a definite effect upon the disturbances observed in the branches. The Reynolds number in the trunk at which transition occurred in the branch was nearly halved when the branch-to-trunk area ratio increased from .8 to 1.4 (Fig. 5). For branch-to-trunk area ratios of .4 to .8, the critical Reynolds number in the trunk at which the transition to turbulence in the branch occurred was relatively constant at approximately 2100. As the branch-to-trunk area ratio increased beyond .8, the critical Reynolds number decreased until, at a ratio of 1.4, the critical Reynolds number was 1200 (12).

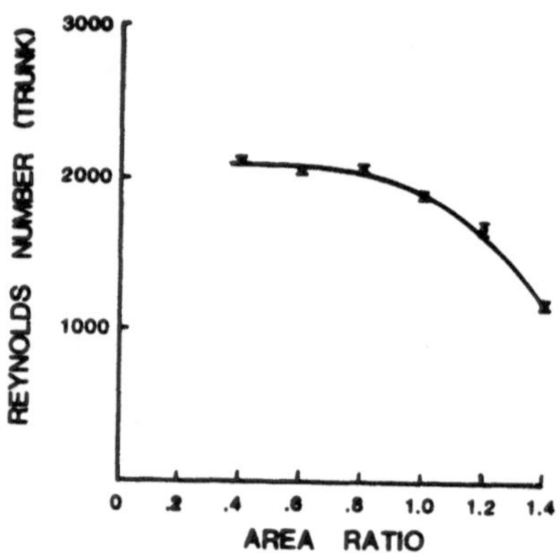

Fig. 5 The Reynolds number in the trunk at which the transition to turbulence occurred in the branch is shown as a function of the branch-to-trunk area ratio. Reproduced from Biorheology (12) with permission of Pergamon Press.

A number of investigators have shown that the critical Reynolds number is diminished due to the presence of a branch (29, 30, 33). Based upon these studies, some investigators have been led to speculate that turbulence or at least non-laminar flow probably exists in all arteries greater than 1 or 2 mm diameter (33). However, all of these studies were performed with branch-to-trunk area ratios of one or more (29, 30, 33). Our study showed that the critical Reynolds number is indeed lower than in unbranched straight tubes for area ratios greater than one. Conversely, in the range of physiological branch-to-trunk area ratios observed in the abdominal aorta, there was no reduction of the critical Reynolds number.

Effect of Vessel Tapering on Flow Disturbances

The effect of vessel tapering on the transition to turbulent flow was studied in a glass model of the abdominal aorta. Physiological angles of tapering had a definite stabilizing effect upon disturbances (13). The transition Reynolds number at which the flow was fully turbulent was nearly doubled when the angle of tapering was increased from 0.5° to 2.5° (Fig. 6).

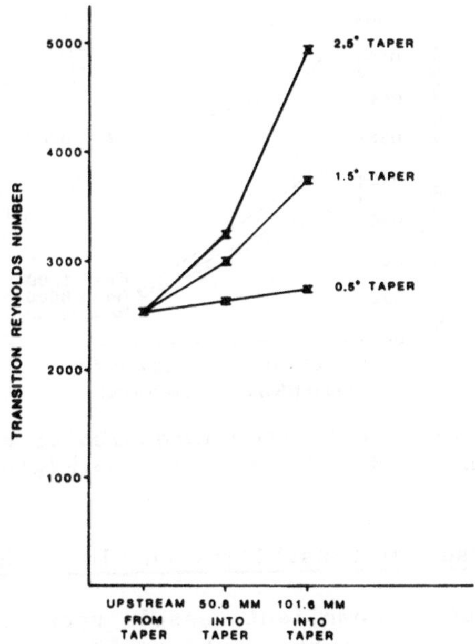

Fig. 6 Transition Reynolds number at various sites with different angles of tapering.

Derivations based upon Bernoulli's equation for steady flow of an incompressible fluid suggest that the dimensionless pressure drop would increase with an increase in the angle of tapering. The similarity of increases in the transition Reynolds number and dimensionless pressure drop implies that the dimensionless pressure drop and the transition Reynolds number are related and a close correlation was shown (r = .99). The pressure gradient exerts an overwhelming influence on the stability of a laminar boundary layer; a decrease in pressure in the downstream direction has a stabilizing effect (34). Therefore, one may conclude that the dimensionless pressure drop is primarily responsible for stabilizing the flow. On the basis of these results, we concluded that tapering tends to promote laminar flow in the descending aorta of human beings.

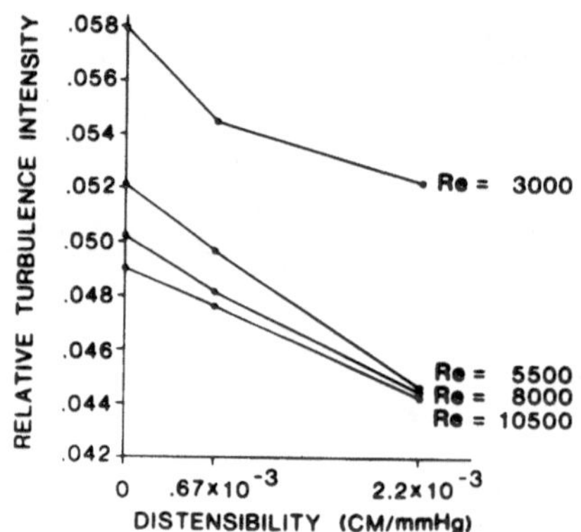

Fig. 7 Relative turbulence intensity at various Reynolds numbers (Re) shown as a function of the distensibility of the tubes.

Effect of Vessel Distensibility on Flow Disturbances

The effects of compliant vessels upon turbulent flow were studied in latex rubber tubes. In a tube with distensibility characteristics comparable to the human aorta, the absolute intensity of turbulence was 10% to 14% lower than in a rigid tube, depending upon the Reynolds number (11). The relative intensity of turbulence was also 10% to 14% lower than in a rigid tube (11). The more compliant rubber tube showed a lower relative intensity of turbulence than the less compliant tube, and the latter showed less turbulence than the rigid tube (Fig. 7). These observations suggested that the compliance of arteries may be a physiological factor that participates in reducing turbulence in the cardiovascular system.

Effect of Red Cell Concentration on Turbulent Flow

An assessment of factors that may affect blood flow stability in the cardiovascular system requires an evaluation of the effects of blood upon disturbances of flow, since blood is a non-Newtonian fluid. We studied the effects of the erythrocytes themselves upon turbulence, and attempted to separate the effects from those produced by differences of viscosity and density. Blood of various hematocrits and plasma of comparable viscosity and density were caused to flow in a turbulent fashion through an in vitro flow system (15). The viscosity of the plasma was made equal to that of blood by the addition of dextrose, and the density of the mixture was within 0.5% of that of blood. The intensity of turbulence of the blood was compared to that of the equally viscous plasma at equal Reynolds numbers. At hematocrits between 20 and 30 ml/100 ml (levels indicative of anemia) the intensity of turbulence of the blood was greater than that of comparably viscous plasma (15) (Fig. 8). At a normal hematocrit (40 ml/100 ml) blood was not more turbulent than comparably viscous plasma (Fig. 8). This indicates that at normal hematocrits, blood does not tend to cause greater disturbances than expected with a Newtonian fluid; but at hematocrits observed in anemic patients, disturbances are augmented by the red cells (15).

The observed augmentation of turbulence may be partially explained by the slip postulate (35). Newtonian non-slipping plasma is constrained to the wall, which promotes the stability of flow. In the case of blood, on the other hand, the red cells may skid along the wall (35), which would suggest a less stable condition according to this theory. Blood would therefore be more prone to turbulence than plasma under equivalent conditions of flow until a higher hematocrit tends to stabilize the flow, perhaps by virtue of the inertia of the red cell mass. Other factors, also may contribute to this effect. For example the size of the particles in suspension plays a role (36). Smaller particles tend to destabilize the flow of some fluids (gases); whereas larger particles tend to stabilize it (36).

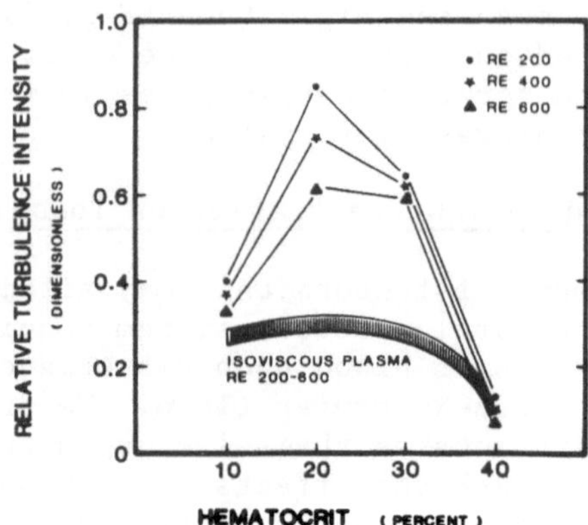

Fig. 8 Relative turbulence intensity snown as a function of the hematocrit of blood at Reynolds numbers 200, 400, and 600. For purposes of comparison, the relative intensity of turbulence of equally dense and viscous plasma at these hematocrits is also shown. Reproduced from Biorheology (15) with permission of Pergamon Press.

Effect of Red Cell Deformability on Turbulent Flow

The effects of red cell deformability upon turbulent blood flow were measured in vitro using plasma containing fresh red cells and plasma containing red cells that were hardened by glutaraldehyde (14). The intensity of turbulence of blood composed of hardened cells was consistently higher than that of blood composed of an equal number of fresh normal cells (14) (Fig. 9). In hereditary spherocytosis, the red cells are less deformable than normal erythrocytes (37) and mechanical fragility is increased (38). Shear stress in the presence of turbulent flow is known to be higher than in laminar flow because of additional stresses (Reynolds stresses) due to fluctuating velocities (39). Lysis of normal erythrocytes occurs when the critical strain of the membrane yield strength is exceeded (40). If the cells are abnormally fragile, a lower lethal shear stress would be expected. The increased turbulence in the presence of non-deformable cells associated with increased shear stresses perhaps would increase the rate of hemolysis in patients suffering from hereditary spherocytosis. Sickled erythrocytes in patients with sickle cell disease are also less deformable than normal red cells (41). Turbulent flow has been found to augment the sickling process (44). Therefore, in sickle cell disease a contribution to the pathological process perhaps would be made by the effects of the non-deformable cells on the intensity of turbulence.

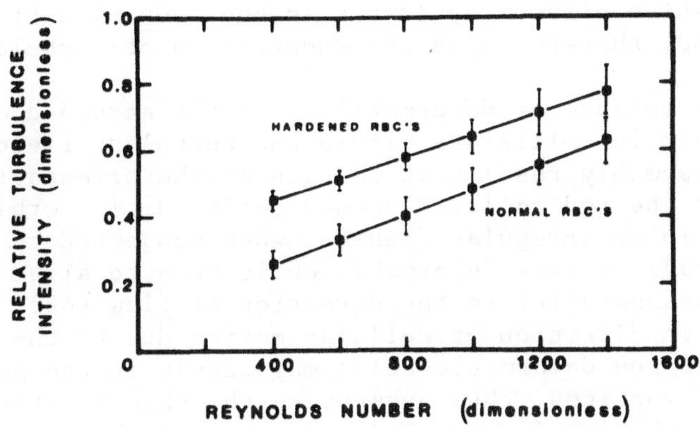

Fig. 9 Relative intensity of turbulence shown in relation to the Reynolds number, for normal red blood cells (rbc's) and hardened red blood cells. Reproduced from Biorheology (14) with permission of Pergamon Press.

Deformability of normal erythrocytes is known to be of functional importance. The viscosity of blood composed of normally deformable cells is lower than suspensions of non-deformable cells (43, 44). Susceptibility to mechanical trauma also appears to be diminished by erythrocytic deformability. The low resistance to bending of the normal erythrocyte allows the changes of shape or deformation necessary to prevent destruction in the crowded circulation (45).

The mechanism of a relative suppression of turbulence by normally deformable cells in comparison to hardened cells appears to be related to the viscoelastic properties of the membrane of the red cells. Turbulence suppression by viscoelastic fluids is a function of the ratio of elastic to viscous forces (46). This ratio is reduced in blood containing hardened cells, since the viscosity of the hardened cells is increased and the elasticity reduced. High rates of collision among red cells would invariably occur in turbulent flow. Deformation of human red cells upon collision has been observed (47). Deformation of the cells as a result of impact, would tend to cause the cells to act as an energy sink by absorbing some of the kinetic energy generated by the turbulent flow. This cushion effect may result in the reduction of vortex sizes and, therefore, in the suppression of turbulence.

The absence of deformability of the hardened cells may result in collisions within the turbulent field, and this presumably results in changes of the orientation and paths of the red cells. Hardened cells tumble, orbit, and collide in an irregular fashion, when subjected to high shear (43); whereas deformable cells tend to align their major axes parallel to the direction of flow (43). Changes in direction of cellular motion due to the collision of non-deformable cells may result in the generation of vortices, thus augmenting the chaotic state associated with turbulence. This effect would be even more pronounced with non-deformable cells because such cells may rotate and present their discoid surface at right angles to the velocity vector. Larger vortices, therefore, may be generated by non-deformable cells than would be generated by normal cells in the same turbulent field (14).

RELATION OF TURBULENT FLOW TO CARDIAC MURMURS

The observations of factors that produce disturbances of flow in the cardiovascular system have provided insight into the mechanism of the origin of cardiac murmurs (Fig. 10, 11, 12). We have shown that the energy of turbulence is directly related to the energy of intra-arterial sound pressure fluctuations (the intracardiac murmur) measured within the turbulent field (48). Also, a direct relation was observed between the turbulent power supply and the acoustic power output (48). The latter was related to the intensity of the murmur, as observed at the chest wall

(48). In view of this, one can now apply hydraulic theories of turbulent flow in pipes and in jets to better understand the clinical characteristics of murmurs, such as the time of their occurrence in the cardiac cycle, their duration, shape, intensity, location, and quality. During circumstances which increase flow (fever, exercise, hyperthyroidism, and anemia), turbulence would increase. Higher magnitudes of turbulent energy would be associated with higher levels of sound energy. With higher flows, turbulence would be present throughout a larger portion of the ejection period (6, 48).

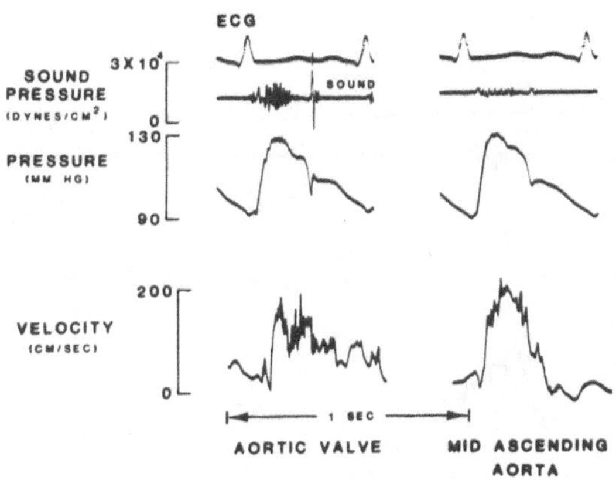

Fig. 10 Aortic velocity, pressure, and sound in a patient with normal aortic valve and high cardiac output (12.9 liters/ min). In the region of the aortic valve, an ejection murmur occurred coincidently with turbulence. In the mid-ascending aorta, the amplitude of sound diminished coincidently with reduction of flow disturbances. Reproduced from Circulation Research (48) with permission of the American Heart Association.

Anemia would increase the peak Reynolds number and turbulence, both by its effect upon the cardiac output (49) and by its effect upon viscosity (52). Anemia also augments turbulence by virtue of hemorheological effects related to the hematocrit (15).

Turbulent flow distal to a stenotic valve would occur at Reynolds numbers well below the critical Reynolds number which applies to steady flow in smooth pipes. Even so, the mean Reynolds number at the valve of patients with combined aortic stenosis and aortic regurgitation ranged between 10,000 and 20,000 (6). The intensity of turbulence of blood flowing through a stenotic orifice in a pulsatile fashion increases as the Reynolds number increases (51). Therefore, high magnitudes of turbulence would be expected distal to stenotic valves with the peak Reynolds numbers estimated to have occurred in these subjects.

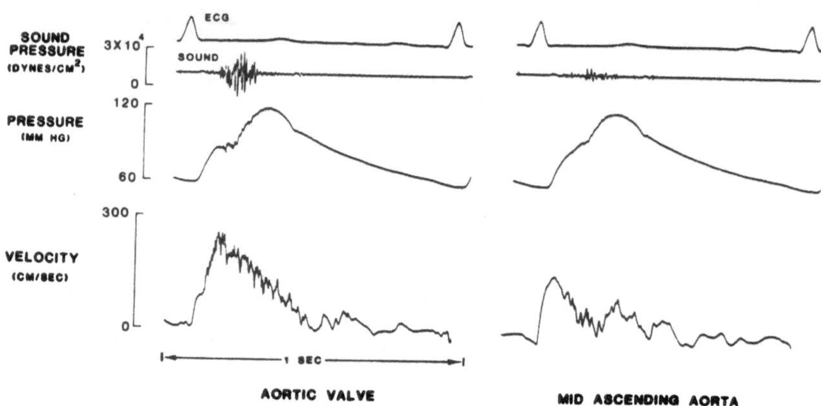

Fig. 11 Aortic velocity, pressure, and fluctuating pressures indicative of sound recorded in a patient with aortic insufficiency. In mid-ascending aorta, amplitude of intra-arterial sound diminished coincidently with reduction of turbulence density. Reproduced from Circulation Research (48) with permission of the American Heart Association.

The cause of the reduced intensity of organic ejection murmurs in subjects with congestive heart failure (52) can be explained in terms of fluid mechanics related to these observations. In congestive heart failure the maximal rate of flow diminishes (53) and therefore turbulence diminishes. A reduction of the issuing velocity caused by reduced ventricular performance (53) would cause a reduction of sound energy. In the region of unbounded jets, the acoustic efficiency itself diminishes with a reduction of velocity (54). Therefore, a reduction of velocity would certainly produce a reduction of sound.

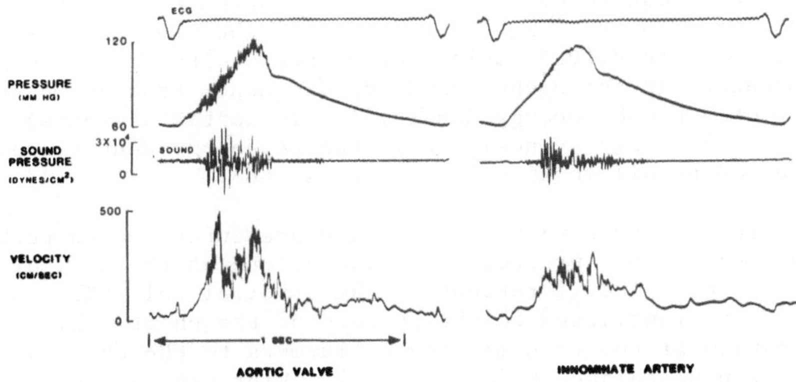

Fig. 12 Highly turbulent flow recorded distal to the aortic valve in a patient with predominant aortic stenosis. Fluctuations in pressure due to turbulence are apparent on the upstroke of the aortic pressure curve. Turbulence persisted in the innominate artery. A clearly defined ejection murmur was recorded at this site within the innominate artery. Reproduced from Circulation Research (48) with permission of the American Heart Association.

The cause of the diamond shape (55) of ejection murmurs can also be explained on the basis of the Reynolds number and magnitudes of turbulence. As velocity increases the turbulence energy density increases (6); and the turbulence energy density was shown to be linearly related to the sound energy density (48). Velocity increases during ejection, reaches a peak, then diminishes. Sound energy density, therefore, would be expected to follow the same curve, which results in a diamond-shaped murmur (48).

The quality of organic murmurs has been thought to reflect turbulence, as indicated by a wide spectrum of frequencies inherent in these murmurs (56). Murmurs caused by turbulent flow should contain a broad spectrum of frequencies, whether they are innocent murmurs or murmurs caused by stenotic valves. Some investigators postulated that the cause of innocent murmurs is different from that of organic murmurs. This was based on spectrograms of innocent murmurs which showed a compact type of output with the majority of frequencies below 250 Hz, whereas organic murmurs (mitral regurgitation) showed a higher range of frequencies (56). Our study showed that turbulence of a lower energy density (as in the presence of a normal aortic valve) would produce sound with lower energy components at the higher bands of frequency than turbulent flow of a higher energy density (as in aortic stenosis) (48). In aortic stenosis an increased energy density was observed at all measured frequencies (48).

The fluctuating velocities and pressures due to turbulence presumably produce local vibrations at the wall of the vessel. The vibrations of the arterial wall then would be transmitted to the surface of the chest. This mechanism of the transmission of murmurs to the chest wall has been postulated by others (57). Vibration of the walls of distensible tubes may occur with turbulent flow distal to an orifice (58). The failure of a phonocatheter to detect sound pressure vibrations upstream from a jet suggests that the conduction of sound through the blood by compression waves is negligible (59). Measurements of the rate of transmission of sound from the heart to the chest wall indicate that there is transmission by vibration of overlying tissues (60).

Among the postulated mechanisms for the genesis of cardiac murmurs is periodic vortex shedding (Aeolian tones) (61). If periodic vortex shedding were productive of murmurs in the patients that we studied, velocity measurements should have shown components of velocity which occurred periodically. However, spectral analyses of our data failed to show any periodicity (48).

Based upon the association of turbulence with ejection murmurs, one can speculate that the systolic murmur of mitral regurgitation, the diastolic rumble of mitral stenosis and the decrescendo murmur of aortic regurgitation are due to turbulence. Many of the characteristics of these murmurs can be explained on the basis of turbulent blood flow.

Measurement of blood flow velocity in the aorta and pulmonary artery of patients demonstrated greater flow disturbances in the ascending aorta than in the main pulmonary artery (21). These observations suggest that innocent ejection murmurs may originate in the aorta and we observed this in adult patients during diagnostic cardiac catheterization (62). Intracardiac sound measurements consistently showed higher amplitude systolic murmurs in the region of the normal aortic valve, in comparison to the pulmonary valve. The aortic murmur was intensified more than the pulmonary murmur with transient augmentations of flow in beats following premature ventricular contractions (62). Comparable observations were also made in dogs (62). Innocent murmurs are generally thought to originate in the pulmonary artery, and sometimes are termed pulmonary ejection systolic murmurs or innocent pulmonary systolic murmurs (63). The apparent pulmonary origin of innocent murmurs results from their precordial location and the demonstration, in intracardiac phonocardiographic studies, that systolic ejection murmurs originate within the pulmonary artery of most normal persons (64, 65). However, these intracardiac studies were confined to the right side of the heart and therefore did not exclude the possibility that murmurs heard at the wall of the chest may have originated from the aortic valve (66). A few intracardiac studies in the left side of the heart in normal subjects in addition to our study (62) demonstrated systolic ejection murmurs within the aorta (67, 68). Also, a specific type of innocent murmur, the "vibratory" variety, has been thought to originate near the aortic valve (69, 70).

If innocent murmurs originate in the region of the aortic valve, it is necessary to determine why such murmurs are not best heard at the same site as organic aortic murmurs. The explanation may be related to our observation that distal to a stenotic aortic valve, maximal turbulence occurred several diameters away from the orifice (48). This reflects the fact that in the first part of a jet, flow is laminar (39); a transition region follows and finally the flow becomes turbulent (71). The location on the chest wall that overlies the site of maximal turbulence distal to a stenotic aortic valve is in the region of the second right intercostal space. In contrast, turbulence distal to normal valves occurs with greatest intensity close to the valve (48). This is where disturbances initiated by the normal valve leaflets are maximal, and where instability due to inlet effects is most prominent. Because turbulence associated with normal valves occurs near the valve, intraarterial pressure fluctuations productive of murmurs would also be maximal near the valve (48). The location on the chest wall that corresponds to the anatomic location of the aortic valve is in the region of the third or fourth left intercostal space. This corresponds to the site at which innocent murmurs are usually heard (72). Because it approximates the site at which organic pulmonary murmurs are of greatest intensity such murmurs have been attributed to the pulmonary valve (66).

Although innocent ejection murmurs are a common auscultatory findings, particularly in children and adolescents (73, 74), the mechanism of such murmurs has not been determined. By definition, innocent ejection murmurs are murmurs heard at rest in subjects with no apparent cardiac abnormality (66). If one considers turbulent flow to be productive of ejection murmurs, then factors that augment turbulence such as viscosity, density, and blood velocity may provide a clue to the mechanism of innocent murmurs. We observed that the blood viscosity and plasma viscosity were lower in healthy, non-anemic young women with innocent murmurs in comparison to those with no audible murmurs. Small differences were observed between both groups in regard to the hematocrit, although in both groups, the hematocrit was normal (75). The diminished blood viscosity observed in subjects with an innocent ejection murmur increased the tendency for flow to become turbulent and may have contributed to the production of a murmur. High frequency pressure fluctuations are an intrinsic characteristic of turbulent flow (76) and such fluctuations define intracardiac sound-pressure (48). The decreased viscosity that we observed in these subjects would increase the sound-pressure sufficiently to cause approximately a 2 dB increase in sound intensity (75). This difference in sound intensity would be detectable with normal hearing (77). Thus, the increased turbulence calculated to have occurred due to the lower viscosity in the subjects we examined may have contributed to causing the intensity of inaudible ejection murmurs to exceed the threshold of audibility.

RELATION OF TURBULENCE TO THROMBUS FORMATION

Thrombus formation on prosthetic devices, particularly prosthetic cardiac valves, has been a major problem in prosthetic valve replacement. Turbulent flow in the region of prosthetic valves may contribute to thrombus formation (78). We showed in dogs with extracorporeal shunts that more thrombi, by weight, occurred within a turbulence-producing shunt than within a laminar flow shunt (51, 78). Thrombi invariably occurred in the region of the turbulence generator, and most of the thrombi were located just distal to the orifice (51). The mechanism of thrombus formation induced by turbulence is presumably related to the effects of turbulence on formed elements in the blood. Such effects could have been caused by 1) shear, 2) collision with the walls of the tubing, or 3) prolonged contact with the foreign surface. Both high shear and agitated random flow are intrinsic characteristics of turbulence (79). The possibility that high rates of shear contribute to the activation of platelets has been previously considered (80).

Turbulent blood flow, through its effects upon various blood elements may be implicated in the continuing disease process of aortic valvular stenosis. We observed microthrombi with evidence of organization on several calcified aortic valves obtained at surgery during aortic valve replacement (81). Subsequently, platelet depositions were also observed by scanning electron microscopy, but only on valves from occasional patients (82). We postulate that the repetitive deposition and organization of such thrombi may be an important cause of the continuing process of stenosis of previously deformed aortic valves, and turbulent flow may be a factor that contributes to the platelet deposition. Turbulent blood flow would be expected in the region of even mildly stenotic aortic valves. Hemolysis occurs in patients with aortic stenosis (83, 84) and is thought to be caused by mechanical effects related to turbulence. Sublethal hemolysis would cause the release of small amounts of adenosine diphosphate that produce platelet aggregation (85). Partial thromboplastin, also released by red cells, would form thrombin, thereby causing the aggregates to be irreversible (85). The hydrodynamic patterns of flow in the region of the aortic valve, including normal reversed flow, regurgitant flow (with aortic insufficiency) and captive annular eddies (with aortic stenosis), would cause various elements in the blood to interact with the valve surface and become deposited as a result of their increased adhesiveness. Abnormal valve surfaces would further enhance the deposition of platelets.

With the potential detrimental effects of turbulence and its associated high shear stress in mind, particularly in relation to prosthetic valve thrombosis, several investigators, ourselves included, have considered the potential use of turbulence reducing chemicals. Long chain polymers, such as polyethylene oxide have been shown to reduce turbulence in vitro (42, 86) as well as in laboratory animals (87). Polyethylene oxide, however, despite its effectiveness in quantities as low as one part per million, is a toxic substance. Desoxyribonucleic acid, is a long chain compound that may have comparable turbulence reducing effects, and conceivably could be used with less hazard (88). However, the use of chemicals for the reduction of turbulence is a remote therapeutic goal, since the detrimental effects of turbulence have not been fully evaluated.

In spite of the apparent absence of turbulence at the bifurcation of normal vessels, atherosclerosis is known to occur at such sites. Both high and low shear have been suggested to relate to the presence of atherosclerotic lesions in these areas and potentially detrimental levels of shear can occur in the absence of turbulence. Endothelial proliferation has been observed in experimentally induced areas of high shear stress (89). On the other hand, it has also been suggested that low shear rates may contribute to atherosclerosis through its effects upon the mass transport of cholesterol between the arterial wall and the intraluminal blood (90). Irrespective of the initial cause of atherosclerosis, once plaques or irregularities of the intima develop, turbulence may occur. One may conjecture that turbulent flow may accelerate the disease process and contribute to its proliferation.

In conclusion, physical mechanisms that relate to the characteristics of flow in the cardiovascular system appear to relate to a variety of physiological events and pathological disorders. As further information is derived related to the characteristics of flow in the human body, it is likely that some of this information will be applicable to the diagnosis and treatment of patients.

REFERENCES

1. McDonald, D.A.: The occurrence of turbulent flow in the rabbit aorta. J. Physiol. (London) 118: 340-347, 1952.

2. Seed, W.A., Wood, N.B.:Velocity patterns in the aorta. Cardiovasc. Res. 5: 319-330, 1971.

3. Nerem, R.M., Rumberger, J.A., Jr., Gross, D.R., Hamlin, R.L., Geiger, G.L.: Hot film anemometer velocity measurements of arterial blood flow in horses. Circ. Res. 34: 193-203, 1974.

4. Schultz, D.L., Tunstall-Pedoe, D.S., Lee, G. DeJ., Gunning, A.J., Bellhouse, B.J.: Velocity distribution and transition in the arterial system. In: Wolstenholme, G.E.W. and Knight, J., (eds.), Circulatory and Respiratory Mass Transport, (CIBA), Little, Brown & Company, Boston, 1969, pp. 172-202.

5. Seed, W.A., Thomas, I.R.: The application of hot-film anemometry to the measurement of blood flow velocity in man. In: Cockrell, D.J. (ed.), Fluid-Dynamic Measurements in the Industrial and Medical Environments, Leicester University Press, Leicester, England, 1972, pp. 298-304.

6. Stein, P.D., Sabbah, H.N.: Turbulent blood flow in the ascending aorta of humans with normal and diseased aortic valves. Circ. Res. 39: 58-65, 1976.

7. Seed, W.A., Wood, N.B.: Development and evaluation of a hot-film velocity probe for cardiovascular studies. Cardiovasc. Res. 4: 253-263, 1970.

8. Seed, W.A., Wood, N.B.: Velocity patterns in the aorta. Cardiovasc. Res. 5: 319-330, 1971.

9. Nerem, R.M., Seed, W.A.: An in vivo study of aortic flow disturbances. Cardiovasc. Res. 6: 1-14, 1972.

10. Wormersly, J.R.: An elastic tube theory of pulse transmission and oscillatory flow in mammalian arteries. Technical Report WADC TR 56-614, Dayton, Ohio, Wright Air Development Center, 1957, p. 1.

11. Stein, P.D., Walburn, F.J.: Damping effect of distensible tubes on turbulent flow: Implications in the cardiovascular system. Advances in Bioengineering, American Society of Mechanical Engineers 1979 Winter Annual Meeting, New York, pp. 117-119, 1979.

12. Walburn, F.J., Blick, E.F., Stein, P.D.: Effect of the branch-to-trunk area ratio on the transition to turbulent flow: Implications in the cardiovascular system. Biorheology (In Press).

13. Walburn, F.J. and Stein, P.D.: Effect of vessel tapering on the transition to turbulent flow: Implications in the cardiovascular system. Federation Proceedings (In Press).

14. Sabbah, H.N., Stein, P.D.: Effect of erythrocytic deformability upon turbulent blood flow. Biorheology 13: 309-314, 1976.

15. Stein, P.D., Sabbah, H.N., Blick, E.F.: Contribution of erythrocytes to turbulent blood flow. Biorheology 12: 293-299, 1975.

16. Sabbah, H.N., Stein, P.D.: Contribution of semilunar leaflets to turbulent blood flow. Biorheology 16: 101-108, 1979.

17. McDonald, D.A.: Blood Flow in Arteries. Williams and Wilkins Co., 2nd ed., Baltimore, 1974, pp. 95-97.

18. Yellin, E.L.: Laminar-turbulent transition process in pulsatile flow. Circ. Res. 19: 791-804, 1966.

19. Wieting, W.D., Hall, W.C., Liotta, D., DeBakey, M.E.: Dynamic flow behavior of artificial heart valves. In: Brewer, L.A., III (ed.), Prosthetic Heart Valves, Thomas Publisher, Springfield, Ill., 1969, pp. 34-49.

20. Reif, T.H.: A preliminary flow study of a two-dimensional model of a concave-convex pivoting disc prosthetic heart valve. Proc. 7th Annual New England Bioengineering Conference, 1979, pp. 209-211.

21. Stein, P.D., Sabbah, H.N., Anbe, D.T.: Comparison of disturbances of flow in the pulmonary artery and aorta of man. Biorheology 16: 357-362, 1979.

22. Greenfield, J.C., Jr., Griggs, D.M., Jr.: Relation between pressure and diameter in main pulmonary artery of man. J. Appl. Physiol. 18: 557-559, 1963.

23. Greenfield, J.C., Jr., Patel, D.J.: Relation between pressure and diameter in the ascending aorta of man. Circ. Res. 10: 778-781, 1962.

24. Davies, M.J., Pomerance, A., Lamb, D.: Techniques in examination and anatomy of the heart. In: Pomerance, A., and Davies, M.J. (eds.), The Pathology of the Heart, Blackwell Scientific Publications, Oxford, 1975, p. 21.

25. Murgo, J.P., Altobelli, S.A., Dorethy, J.F., Logsdon, J.R., McGranahan, G.M.: Normal ventricular ejection dynamics in man during rest and exercise. In: Physiologic Principles of Heart Sounds and Murmurs, AHA Monograph No. 46, 1975, pp. 92-101.

26. Shaver, J.A., Nadolny, R.A., O'Toole, J.D., Thompson, M.E., Reddy, P.S., Leon, D.F., Curtiss, E.I.: Sound pressure correlates of the second heart sound. Circulation 49: 316-325, 1974.

27. Franklin, D.L., Van Citters, R.L., Rushmer, R.F.: Balance between right and left ventricular output. Circ. Res. 10: 17-26, 1962.

28. Fisher, K.H., Blick, E.F.: Turbulent damping by flabby skins. J. Aircraft. 3: 163-164, 1966.

29. Stehbens, W.E.: Turbulence of blood flow. Quart. J. Exptl. Physiol. 44: 110-117, 1959.

30. Roach, M.R., Scott, S., Ferguson, G.G.: The hemodynamic importance of the geometry of bifurcations in the circle of Willis (glass model studies). Stroke 3: 255-267, 1972.

31. Stehbens, W.E.: Flow in glass models of arterial bifurcations and Berry aneurysms at low Reynolds numbers. Quart. J. Exper. Physiol. 60: 181-192, 1975.

32. Stein, P.D., Sabbah, H.N., Anbe, D.T., Walburn, F.J.: Blood velocity in the abdominal aorta and common iliac artery of man. Biorheology 16: 249-255, 1979.

33. Korvetz, L.J.: The effect of vessel branching on haemodynamic stability. Phys. Med. Biol 10: 417-428, 1965.

34. Schlichting, H.: Boundary-Layer Theory. McGraw-Hill Book Co., Ed. 6, New York, 1968, pp. 123, 443, 445, 468-469.

35. Nubar, Y.: Blood flow slip and viscometry Biophysical J. 11: 252-264, 1971.

36. Michael, D.H.: The stability of plane Poiseuille flow of a dusty gas. J. Fluid Mech. 18: 19, 1964.

37. Murphy, J.R.: Erythrocyte shape and blood viscosity. Hemorheology, Pergamon Press, Oxford, 1968, pp. 469-478.

38. Young, L.E., Izzo, M.J., Platzer, R.F.: Hereditary Spherocytosis I. Clinical, Hematologic and Genetic Features in 28 Cases, with Particular Reference to the, Osmotic and Mechanical Fragility of Incubated Erythrocytes. Blood 6: 1073-1098, 1951.

39. Pai, S.E.: Fluid Dynamics of Jets. D. Van Nostrand, New York, 1954, pp. 98, 121, 122, 132, 133.

40. Blackshear, P.L., Jr., Dorman, F.D., Steinbach, J.H., Maybach, E.J., Singh, A., Collingham, R.E.: Shear, wall interaction and hemolysis. Trans. Am. Soc. Artif. Int. Organs 12: 113, 1966.

41. Chien, S., Usami, S., Bertles, J.R.: Abnormal rheology of oxygenated blood in sickle cell anemia. J. Clin. Invest. 49: 623-634, 1970.

42. Stein, P.D., Sabbah, H.N., Mandal, A.K.: Augmentation of sickling process due to turbulent blood flow. J. Appl. Physiol. 40: 60-66, 1976.

43. Schimd-Schönbein, H., Wells, R., Goldstone, J.: Influence of deformability of human red cells upon blood viscosity. Circ. Res. 25: 131-143, 1969.

44. Chien, S., Usami, S., Dellenback, R.J., Gregersen, M.L.: Blood viscosity: influence of erythrocyte deformation. Science 157: 827-829, 1967.

45. Burton, C.A., Shrivastava, B.B., Hemorheology, Pergamon Press, Oxford, 1968, p. 479.

46. Metzner, A.B., Park, M.G.: Turbulent flow characteristics of viscoelastic fluids. J. Fluid Mech. 20, 291-303, 1964.

47. Fung, J.S.K., Cahman, P.B.: The mode and kinetics of the human red cell doublet formation. Biorheology 11: 241-251, 1974.

48. Sabbah, H.N., Stein, P.D.: Turbulent blood flow in humans: Its primary role in the production of ejection murmurs. Circ. Res. 38: 513-525, 1976.

49. Sproule, B.J., Mitchell, J.H., Miller, W.F.: Cardiopulmonary physiological responses to heavy exercise in patients with anemia. J. Clin. Invest. 39: 378-388, 1960.

50. Burch, G.E., DePasquale, N.P.: Hematocrit, viscosity and coronary blood flow. Dis. Chest 48: 225-232, 1965.

51. Stein, P.D., Sabbah, H.N.: Measured turbulence and its effect upon thrombus formation. Circ. Res. 35: 608-614, 1974.

52. Levine, S.A., Harvey, W.P.: Clinical Auscultation of the Heart. W. B. Saunders, ed. 2, Philadelphia, 1959, pp. 196-198, 325.

53. Stein, P.D., Sabbah, H.N.: Ventricular performance measured during ejection: Studies in patients of the rate of change of ventricular power. Am. Heart J. 91: 599-606, 1976.

54. Olson, H.F.: Acoustical Engineering. D. Van Nostrand, Princeton, N.J., 1957, p. 15.

55. Humphries, J.O., McKusick, V.A.: The differentiation of organic and "innocent" systolic murmurs. Prog. Cardiovasc. Dis. 5: 152-171, 1962.

56. Harris, T.N., Friedman, S., Tuncali, M.T., Hallidie-Smith, K.A.: Comparison of innocent cardiac murmurs of childhood with cardiac murmurs in high output states. Pediatrics 33: 341-355, 1964.

57. Rushmer, R.F., Morgan, C.: Meaning of murmurs. Am. J. Cardiol. 21: 722-730, 1968.

58. Meisner, J.E., Rushmer, R.F.: Production of sounds in distensible tubes. Circ. Res. 12: 651-658, 1963.

59. Yellin, E.L.: Hydraulic noise in submerged and bounded liquid jets. In: Biomedical Fluid Mechanics Symposium, New York, American Society of Mechanical Engineers, 1966, pp. 209-221.

60. Hotta, S.: The mechanism of transmission of the cardiovascular sound; an experimental study of the conduction velocity of sound on the chest wall. Jap. Heart J. 8: 354-368, 1967.

61. Bruns, D.L.: A general theory of the causes of murmurs in the cardiovascular sy tem. Am. J. Med. 27: 360-374, 1959.

62. Stein, P.D., Sabbah, H.N.: Aortic origin of innocent murmurs. Am. J. Cardiol. 39: 665-671, 1977.

63. Fowler, N.O.: Cardiac Diagnosis. Hoeber Medical Division Harper and Row, New York, 1968, p. 153.

64. Lewis, D.H., Ertugrul, A., Deitz, G.W., et al.: Intracardiac phonocardiography in the diagnosis of congenital heart disease. Pediatrics 23: 837-853, 1959.

65. Segal, B.L, Novak, P., Kasparian, H.: Intracardiac phonocardiography. Am. J. Cardiol. 13: 188-197, 1964.

66. Tavel, M.E." Innocent murmurs. In: Leon, D.F., Shaver, J.A. (eds.), American Heart Association Monograph 46, Physiologic Principles of Heart Murmurs, American Heart Association, New York, 1975, pp. 102-106.

67. Liebman, J., Sood, S.: Diastolic murmurs in apparently normal children. Circulation 38: 755-762, 1968.

68. Luisada, A.A., Liu, C.K., Szatkowski, J., et al.: Intracardiac phonocardiography in 172 cases studied by left or right heart catheterization or both. Acta Cardiol. 18: 533-570, 1963.

69. Monchy, C.D.: Studies on functional heart murmurs in children. I. The external carotid tracing of children with a precordial vibratory murmur. Ann. Paediatr. 206: 356-362, 1966.

70. Wennevold, A.: The origin of the innocent "vibratory" murmur studied with intracardiac phonocardiography. Acta Med. Scand. 181: 1-5, 1967.

71. Hinze, J.O., Turbulence, McGraw-Hill, New York, 1959, p. 3,421.

72. Weaver, W.F., Walker, C.H.M.: Innocent cardiovascular murmurs in the adult. A 16-year follow-up. Circulation 29: 702-707, 1964.

73. Fogel, D.H.: The innocent systolic murmur in children: a clinical study of its incidence and characteristics. Am. Heart J. 59: 844-854, 1960.

74. Groom, D., Chapman, W., Francis, W.W., Bass, A., Sihvonen, Y.T.: The normal systolic murmur. Ann. Intern. Med 52: 134-144, 1960.

75. Sabbah, H.N., Lee, T., Stein, P.D.: Role of blood viscosity in the production of innocent ejection murmurs. Am. J. Cardiol. 43: 753-756, 1979.

76. Sabbah, H.N., Blick, E.F., Stein, P.D.: High-frequency pressure fluctuations: their significance in the documentation of turbulent blood flow. Cath. Cardiovasc. Diag. 3: 375-384, 1977.

77. Yost, W.A., Nielsen, D.W.: Fundamentals of Hearing. Holt, Rinehart and Winston, New York, 1977, pp. 130, 136.

78. Smith, R.L., Blick, E.F., Coalson, J., Stein, P.D.: Thrombus production by turbulence. J. Appl. Physiol. 32: 261-264, 1972.

79. Parker, J.D., Boggs, J.H., Blick, E.F.: Introduction to Fluid Mechanics and Heat Transfer. Addison-Wesley Publishing Company, Reading, Massachusetts, 1969, pp. 224-225.

80. Goldsmith, H.L.: Flow of model particles and blood cells and its relation to thrombogenesis. In: Spaet, T.H. (ed.), Progress in Hemostasis and Thrombosis, vol. 1, Grune & Stratton, New York, 1972, pp. 97-172.

81. Stein, P.D., Sabbah, H.N., Pitha, J. V.: Continuing disease process of calcific aortic stenosis. Role of microthrombi and turbulent flow. Am. J. Cardiol. 39: 159-169, 1977.

82. Riddle, J.M., Magilligan, D.J., Stein, P.D.: Surface topography of stenotic aortic valves: A scanning electron microscopic study. Circulation (In Press).

83. Eyster, E., Mayer, K., McKenzie, S.: Traumatic hemolysis with iron deficiency anemia in patients with aortic valve lesions. Ann. Intern. Med. 68: 995-1104, 1968.

84. Forshaw, J., Harwood, L.: Red blood cell abnormalities in cardiac valvular disease. J. Clin. Pathol. 20: 848-853, 1967.

85. Johnson, S.A.: Platelets in hemostasis. In: Seegers, W.H. (ed.), Blood Clotting Enzymology, Academic Press, New York, 1967, pp. 379-420.

86. Stein, P.D., Parsons, E.D., Blick, E.F.: Modifications of dynamic flow properties of turbulently flowing human blood by long chain polymers. Med. Res. Eng. 11: 6-10, 1972.

87. Mostardi, R.A., Greene, H.L., Nokes, R.F., Thomas, L.C., Lue, T.: The effect of drag reducing agents on stenotic flow disturbances in dogs. Biorheology 13: 137-141, 1976.

88. White, W.D., Hoyt, J.W.: The effect of linear high molecular weight polymers on turbulent flow properties of human blood. In: Proceedings of the 8th International Conference on Medical and Biological Engineering, Chicago, 1969, Session 11-11.

89. Fry, D.L.: Acute vascular endothelial changes associated with increased blood velocity gradients. Circulation Research 22: 165-197, 1968.

90. Caro, C.G., Fitz-Gerald, J.M., Schroter, R.C.: Atheroma and arterial wall shear: Observation, correlation and proposal of a shear dependent mass transfer mechanism for atherogenesis. Proc. Roy. Soc. London 177: 109-159, 1971.

PULSATILE PROSTHETIC VALVE FLOWS: LASER-DOPPLER STUDIES

P.G. Alchas, A.J. Snyder and Winfred M. Phillips

Artificial Heart Engineering Laboratory
312 Mechanical Engineering Bldg.
The Pennsylvania State Univ., Univ. Park, PA
16802

INTRODUCTION

Many of the factors related to cardiovascular disease seem to be significantly dependent upon the corresponding fluid dynamics of blood flow through the cardiovascular system. A knowledge of such fluid mechanical parameters as shear rate, turbulence intensities and recirculation areas in the entry region of the aorta, for example, would provide a better understanding of associated diseases (e.g., thrombus formation and endothelial damage). In addition, this knowledge would serve as a foundation for the design and testing of prosthetic heart valves.

The need for a detailed fluid mechanical description of the prosthetic valve-induced flow characteristics has initiated extensive research. Analytical models have proved to be of limited utility to date as the problem is complicated by the flow field being three-dimensional, time-dependent, transitory, and in the case of blood, non-Newtonian (1, 9, 16, 21, 24). Such a situation indicates that a primary emphasis must be placed on experimental verification of models and proposed analyses (4, 6, 11, 13, 14, 15, 20, 22, 25). However, very little successful experimental information has been obtained from *in vivo* or *in vitro* studies of cardiovascular flows. *In vivo* studies present obvious diagnostic difficulties and *in vitro* studies require a reasonable mechanical mock circulation system and pump that will produce the proper pressure and flow wave forms.

At The Pennsylvania State University, a left ventricular assist prosthesis (essentially one side of the total artificial heart being used experimentally in animals) has been developed along with a suitable systemic and pulmonary analog circulation system (18, 19) With this system detailed experimental fluid mechanical studies have been made via laser-Doppler anemometry in areas distal to prosthetic aortic valves. This quantitative laser anemometry data is complemented by flow visualization photographs. Only in this way, with a system which adequately simulates normal cardiovascular flow, can fluid mechanical studies contribute significantly to the cardiovascular blood flow problem.

This experimental study involves the fluid mechanics of a Bjork-Shiley #25 (20 mm orifice diameter) convexo-concave tilting disc aortic prosthesis during various time increments in the cardiac cycle. In addition, a comparative look is taken at the flow patterns found from a similar previous investigation in our lab using the Bjork-Shiley straight tilting disc aortic prosthesis (7).

EXPERIMENTAL METHOD

It was necessary to represent the *in vivo* environment as closely as possible when incorporating the Penn State prosthetic ventricle. The "mock" circulation system used (described in Ref. 19) consists of "lumped element" approximations for vascular compliance, resistance, and inertance. The systemic side of the loop used for this study is shown in Fig. 1. All connective tubing is 2.54 cm ID Tygon and the blood analog consists of a 60-40% (by volume) solution of saline-glycerine with a viscosity of 3.5 cp at 30°C.

The Penn State prosthetic ventricle provides pulsatile flow in the same manner as the *in vivo* situation. The diaphragm within the pump is driven by alternate positive (systolic) and negative (diastolic) pressure pulses provided by an airline connected to a Vitamek*

*Vitamek Corporation, Houston, Texas

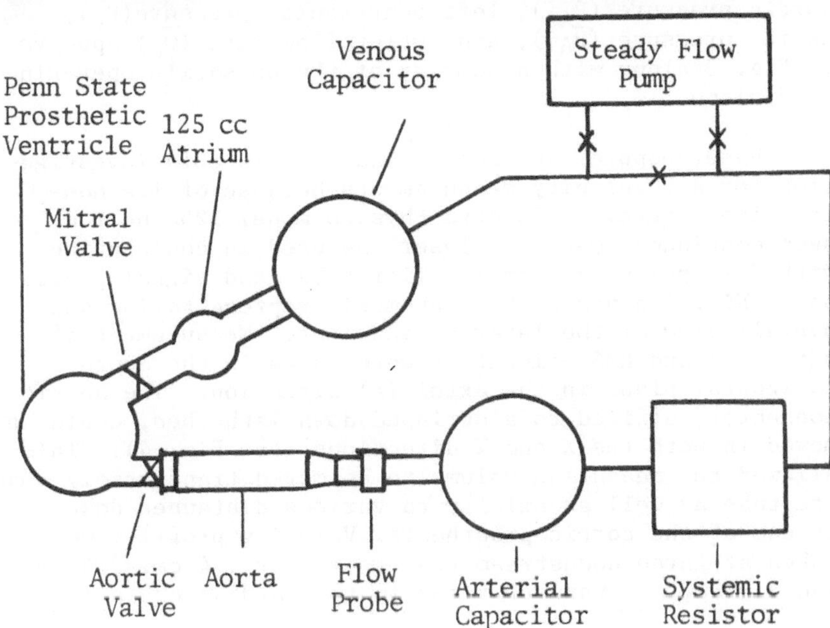

Fig. 1 Systemic side of the PSU mock circulatory loop.

pneumatic drive system. The unit has variable maximum and minimum drive pressures, systolic duration, and beat rate.

Interfacing the pump outlet with the mock circulation system is a rigid aortic connector within which velocity data is taken. The connector is machined from clear plexiglass and has an inner diameter of 2.54 cm (1.00"). Extreme optical transparency and planar outer surfaces minimize refractive effects when the laser beams converge upon the measuring volume during velocity measurements. The pump inlet is connected with the circulatory loop by a 125 cc artificial atrium constructed of segmented polyurethane Biomer*. The left prosthetic ventricle-valve-aortic connector components are shown in Fig. 2.

During pumping, the system is monitored with the necessary instrumentation required for recording pressure and flow waveforms. A sample oscillograph trace of aortic pressure ($\overline{P}_{AO}$), left ventricular pressure($\overline{P}_V$), atrial pressure ($\overline{P}_{AT}$), and outlet flow rate ($\overline{Q}_o$) appears in Fig. 3 along with a summary of the pulsatile operating conditions.

Laser-Doppler anemometry was used in this investigation for all velocity measurements because of its non-invasive nature. A Spectra Physics model 125A helium-neon continuous wave gas laser was used in conjunction with TSI optics, photomultiplier tube, and signal processor. Fig. 4 gives both a schematic representation and overall view of the laser arrangement. Measurement of both mean and RMS velocities were taken in the center horizontal plane in the axial (X) direction. The aortic connector, affixed to a stripped down lathe bed, could be moved in both the X and Y directions (see Fig. 4). This allowed the measuring volume to be moved transversely across the tube as well as axially to various distances downstream of the aortic prosthesis. Velocity profiles were taken at three downstream traverses, X = 1.4 cm, 2.0 cm, and 3.0 cm, for various times in the cardiac cycle (for a

*Ethicon Corp.

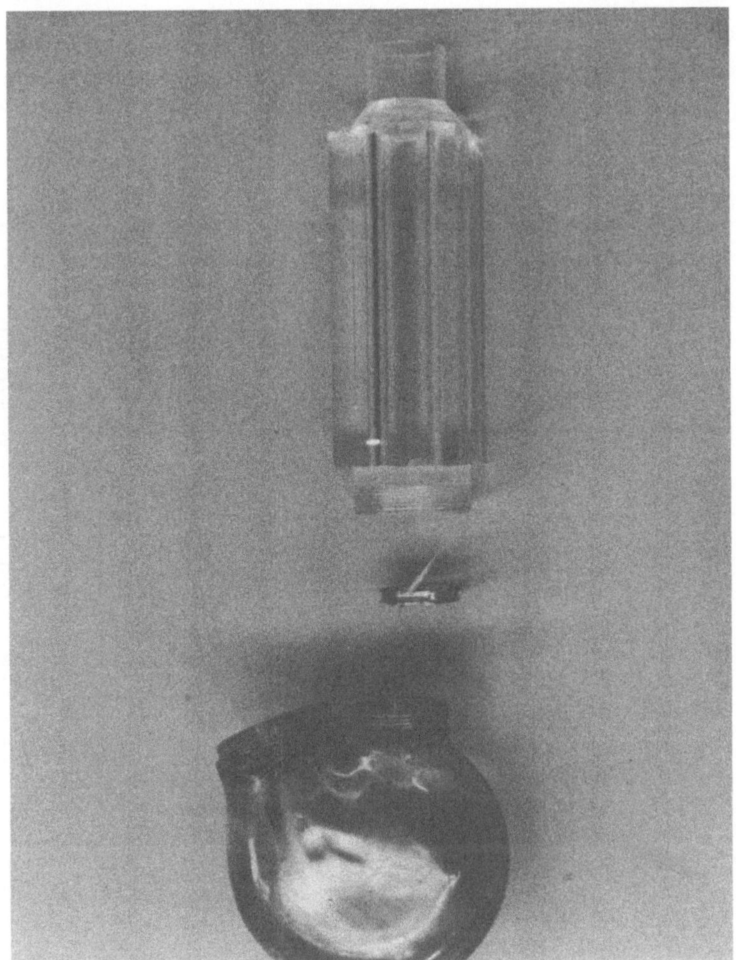

Fig. 2 Photograph of the left prosthetic ventricle-valve-aortic connector components.

beat rate of 100/min, the cycle duration is .60 sec).

Flow visualization pictures were taken within the aortic connector to provide qualitative flow pattern information. The fluid was seeded with 100 micron diameter aluminum dust. A high intensity mercury vapor lamp was used as a light source (12, 17).

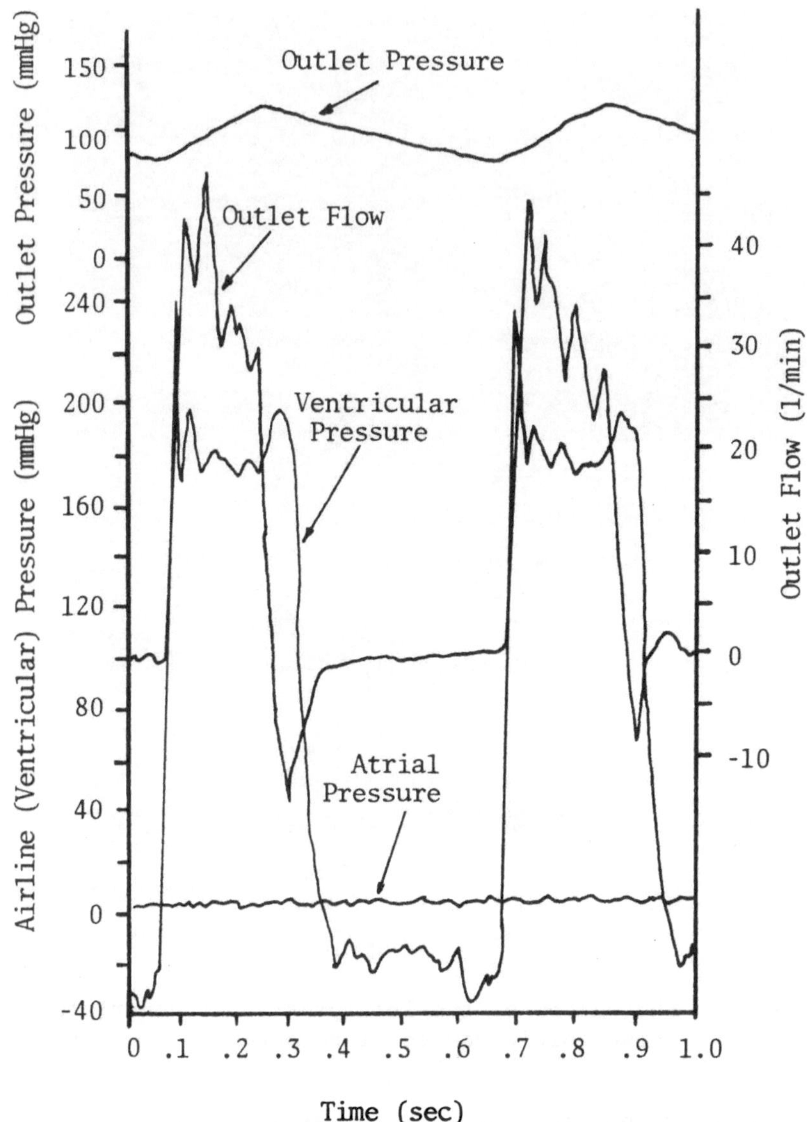

Fig. 3 Sample oscillograph trace during the following drive conditions: Beat rate-100/min, Systolic duration-.200 seconds, Systolic pressure-175mmHg, Diastolic vacuum- -25 mmHg.

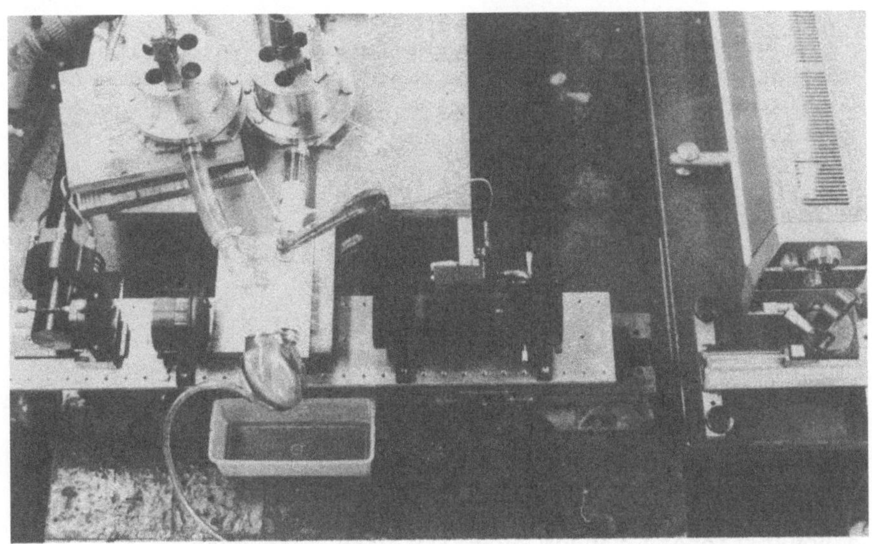

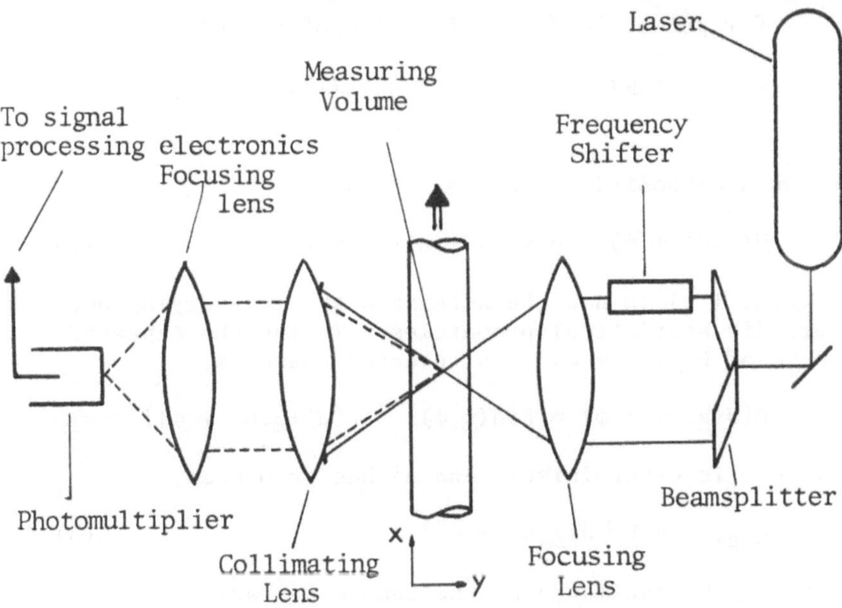

Fig. 4 Laser optics arrangement used for this investigation.

DATA REDUCTION

For separation of mean and turbulent velocities, an unsteady form of Reynolds decomposition is applied. At a point in space, $\underline{x}$ and at time t, the one-dimensional total velocity is:

$$\hat{U}(\underline{x},t) = \overline{U}(\underline{x}) + \tilde{U}(\underline{x},t) + u'(\underline{x},t) \quad (1)$$

where $\hat{U}(\underline{x},t)$ is the total velocity, $\overline{U}(\underline{x})$ the time-averaged mean, $\tilde{U}(\underline{x},t)$ the non-random fluctuation and $u'(\underline{x},t)$ the random fluctuation.

The mean velocity is defined by

$$\overline{u}(\underline{x},t) \equiv \overline{U}(\underline{x}) + \tilde{U}(\underline{x},t)$$

and the turbulent velocity is $u'(\underline{x},t)$. The mean velocity is a deterministic quantity. The turbulent velocity is a nonstationary random quantity with zero mean.

For pulsatile (periodic) flow of period τ,

$$\overline{u}(\underline{x},\eta\tau + \phi) = \overline{u}(\underline{x},\phi), \qquad \eta = 1,2,...$$
$$0 \leq \phi \leq \tau$$

so the decomposition becomes

$$\hat{U}(\underline{x},\eta\tau + \phi) = \overline{u}(\underline{x},\phi) + u'(\underline{x},\eta\tau + \phi) \quad (2)$$

$u'(\underline{x},\eta\tau + \phi)$ is now characterized by time-varying but periodic statistical properties. Taking the expected value of $\hat{U}(\underline{x},\eta\tau + \phi)$ at a selected phase, ϕ,

$$E\{\hat{U}(\underline{x},\eta\tau + \phi\} = E\{\overline{u}(\underline{x},\phi)\} + E\{u'(\underline{x},\eta\tau + \phi)\}$$

Since $\overline{u}$ is deterministic and u' has zero mean,

$$\overline{u}(\underline{x},\phi) = E\{\hat{U}(\underline{x},\eta\tau + \phi)\} \quad (3)$$

This can be estimated by the sample average,

$$\hat{u}(x,\phi) = \frac{1}{N} \sum_{i=1}^{N} \hat{U}(\underline{x}, i\tau + \phi) \qquad (4)$$

where N = number of cycles in the average. We also wish to determine the root-mean-square value of the turbulent velocity (i.e., the turbulent intensity). From Eq. (2),

$$u'_{RMS}(\underline{x},\phi) = \sqrt{E\{[u'(\underline{x},\eta\tau + \phi)]^2\}} = \sqrt{E\{[\hat{U}(\underline{x},\eta\tau + \phi) - \overline{u}(\underline{x},\phi)]^2\}}$$

$$u'_{RMS}(\underline{x},\phi) = \Big[E\{[\hat{U}(\underline{x},\eta\tau + \phi)]^2 - 2\hat{U}(\underline{x},\eta\tau + \phi)\overline{u}(\underline{x},\phi) + [\overline{u}(\underline{x},\phi)]^2\}\Big]^{1/2}$$

Expanding and removing deterministic terms from the expected value,

$$u'_{RMS}(\underline{x},\phi) = \Big[E\{[\hat{U}(\underline{x},\eta\tau + \phi)]^2\} - 2\overline{u}(\underline{x},\phi)E\{\hat{U}(\underline{x},\eta\tau+\phi)\} + [\overline{u}(\underline{x},\phi)]^2\Big]^{1/2}$$

Substituting Eq. (3),

$$u'_{RMS}(\underline{x},\phi) = \sqrt{E\{[\hat{U}(\underline{x},\eta\tau + \phi)]^2\} - [\overline{u}(\underline{x},\phi)]^2}$$

This is simply the standard deviation of $\hat{U}$ at phase ϕ, and can be estimated from a finite sample by

$$u'_{RMS}(\underline{x},\phi) = \sqrt{\frac{1}{N-1}\{\sum_{i=1}^{N} [\hat{U}(\underline{x}, i\tau + \phi)]^2 - N\hat{\overline{u}}(\underline{x},\phi)\}} \qquad (5)$$

The total velocity signal is collected from the LDA signal processor by a Digital Equipment Corporation MINC-11 computer through an analog-to-digital converter. Successive cycles are synchronized, via a Schmitt trigger which is part of the computer's analog input system, by the pressure signal from a transducer which measures pneumatic airline pressure. Cycles are defined to begin at the initiation of systole, characterized by a sharp slope on the airline.

The computer program, written in the Basic language, calculates mean velocity from turbulent intensity according to equations (4) and (5). The resultant records of mean velocity and turbulent intensity versus time, at each point in space of interest, are used to construct profiles of both quantities vs. position. At the 300 points acquired per 0.6 sec. pump cycle, the program processes velocity data in approximately 1/10 real time, i.e., data are acquired during every 10th pump cycle and calculations performed during the intervening 9 cycles. A recently installed Fortran compiler, assembler and real-time operating system now allows more rapid processing.

The number of cycles to be used in the estimation of the mean velocity and turbulent intensity is determined empirically by determining the minimum number of cycles beyond which no significant changes in the calculated values are observed. Fifty cycles were required for the flow field in this study. Seventy-five cycles were used for all data presented here.

RESULTS

Time-dependent velocity profiles were derived for t = 0.01 - 0.06 seconds into the cycle in increments of 0.02 seconds. However, only a small portion of this data will be presented in order to represent a typical evolution of mean and RMS velocities during the distinct phases of a cardiac cycle. Time zero represents the onset of systole.

Figures 5-9 give the mean and RMS velocity curves and flow visualization photographs at X = 1.4 and 3.0 cm downstream. Times into the cycle are 0.03 sec (early systole), 0.11 sec (acceleration), 0.23 sec (end of deceleration), 0.25 sec (early diastole) and 0.39 sec (diastole). All data points were connected with a standard cubic spline curve fit. Curves are drawn within the aortic connector and an approximate opening of the valve disc is shown in order to give a physical feel between the time into the cycle, opening of the disc, and mean velocity magnitude. The physical location to the right of the opened disc (where greater velocities are found) will be referred to as the major orifice, and that to the left, the minor orifice.

At 0.03 seconds into the cycle (Fig. 5) it can be seen that the profiles at both locations are characterized by plug-type flow. This implies that the valve disc has not yet opened to the point where it would have much of an effect on the flow pattern. The early makings of a jet can be seen, however, near the wall at X = 1.4 cm. The effects of the jet have not propagated downstream as yet. The RMS profiles associated with each mean velocity curve gives a good indication of the "turbulence production". As expected, the low values found at X = 3.0 cm indicates a lack of turbulence this early in the cycle. A small peak is seen at the near valve location. These fluctuations are found just distal to the valve disc as it begins to open.

By t = 0.11 sec (Fig. 6) the disc is fully opened and has significantly affected both the near valve flow field (1.4 cm) as well as the one further downstream (3.0 cm). Jetting has occurred near the major orifice wall with velocities on the order of 350 cm/sec. Note the way that the velocity peak in the minor orifice at 1.4 cm smoothes out at the more distant axial location. Also note the low velocity magnitude at 1.4 cm ($\sim$ 25 cm/sec) where the fluid separates from the distal tip of the disc as opposed to that found at the same location further downstream. Both these factors are due to the high inertia of the accelerating fluid near the valve site and create a significant shear layer separating the two major flow areas. The RMS profiles in Fig. 6 show high fluctuations (on the order of 140 cm/sec) due to this shear layer. This is as expected since turbulence production is normally proportional to the mean velocity gradient. RMS fluctuations are also higher in the upstream minor orifice region than at 3.0 cm. It is reasonable to assume that there is strong mixing occurring at the entire near-valve traverse.

The profile shapes determined from preliminary steady flow measurements are similar to those found during this

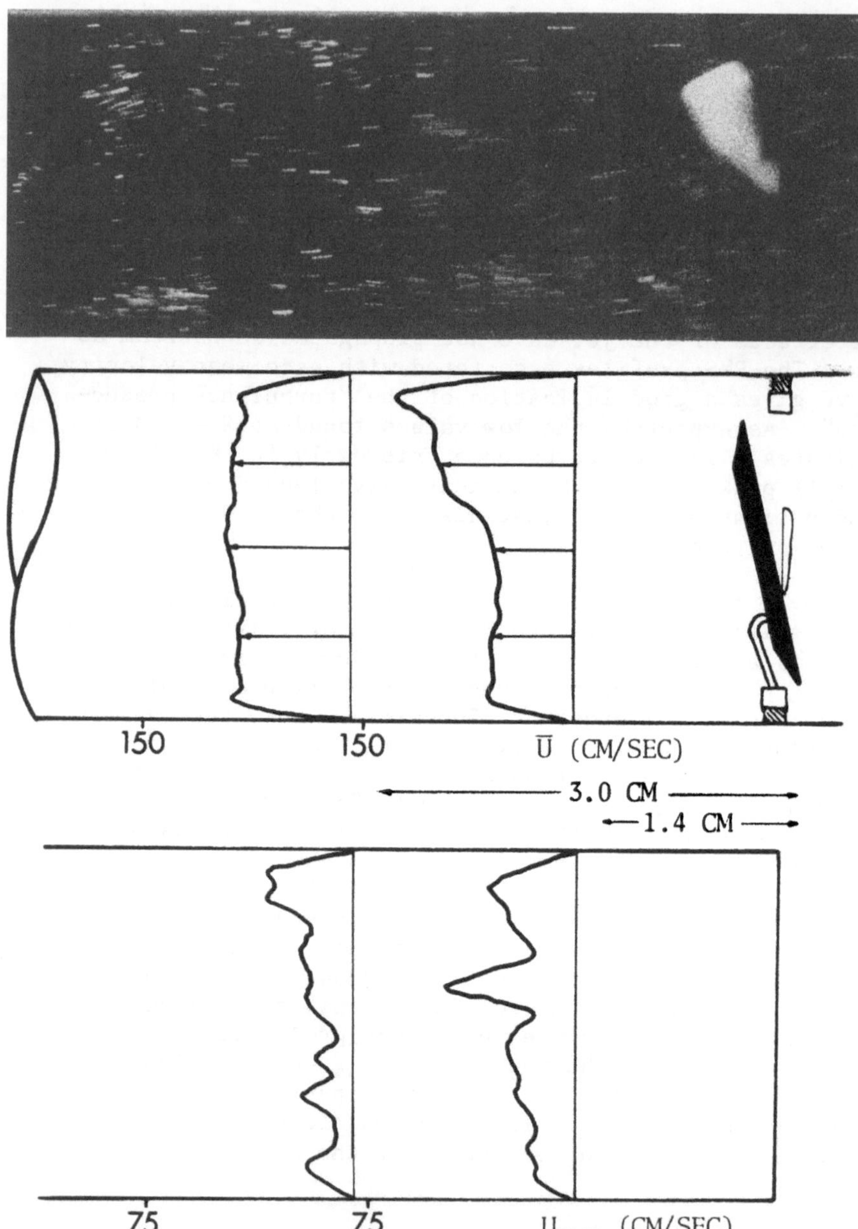

Fig. 5 Flow visualization photograph and corresponding mean and RMS velocity profiles at 0.03 sec into the cycle (early systole).

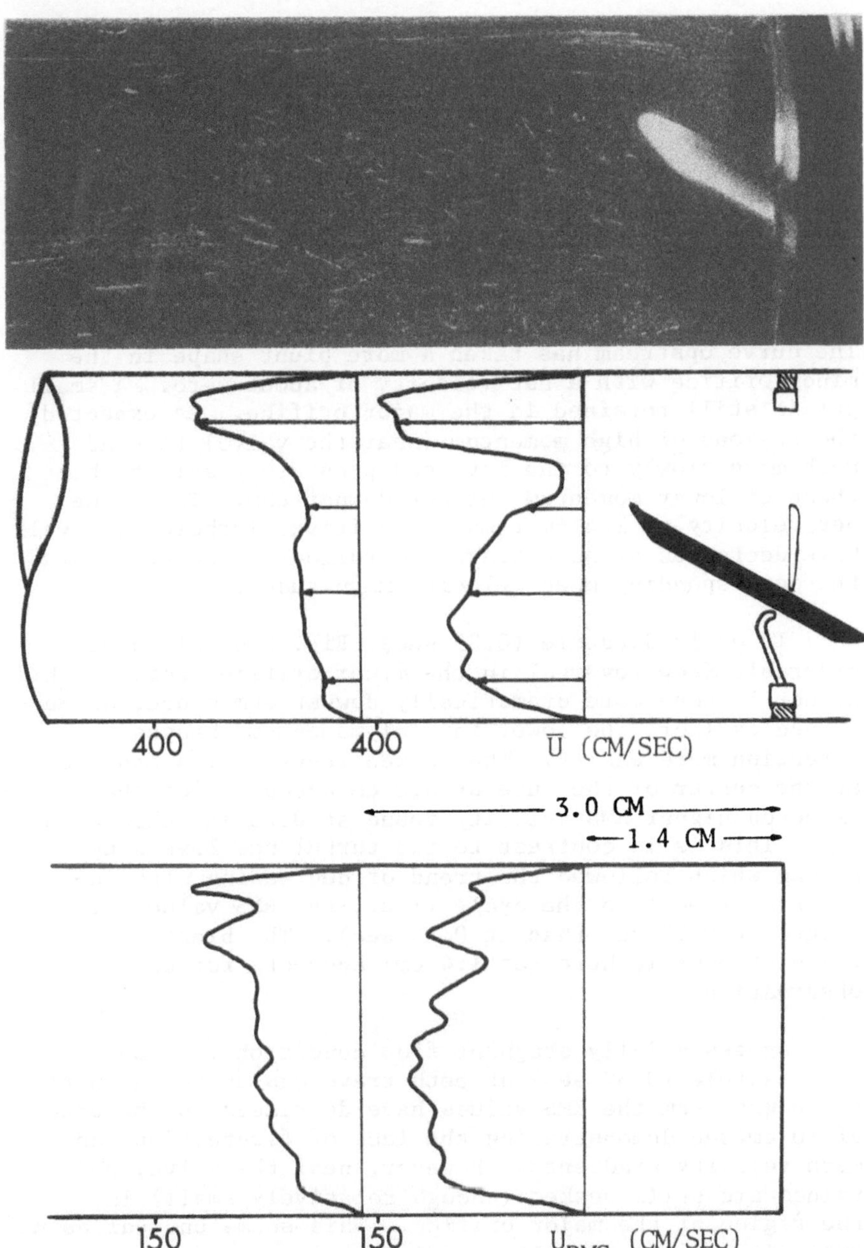

Fig. 6 Flow visualization photograph and corresponding mean and RMS velocity profiles at 0.11 sec into the cycle (acceleration).

acceleration phase of the cycle (0.11 sec). This is not unusual as at this time in the cycle the flow field is inertial in nature and can be assumed to be time-independent.

The velocity profiles at the end of deceleration (0.23 sec), after the valve has just closed, are shown in Fig. 7. At the near-valve site, the magnitudes of both orifice velocities have diminished considerably. Reverse flow is noted in the center of the major orifice at the 1.4 cm traverse, perhaps due to recirculation. The curve upstream has taken a more blunt shape in the minor orifice with a net velocity of about zero. A small jet is still retained in the major orifice. As expected, the regions of high momentum (near the valve) respond much more slowly to the reversed pressure gradient than those of lower momentum further downstream. Thus, the net velocity at 1.4 cm remains positive. Turbulence levels have decreased at both traverses following the trend of the corresponding mean velocity magnitudes.

In early diastole (0.25 sec), Fig. 8 displays considerable flow reversal in the minor orifice region. This effect is seen more dramatically downstream since, as mentioned earlier, the lower initial momentum changes direction more easily. The peaked reverse flow profile at the center of the tube at 3.0 cm accounts for the unexpected higher RMS velocity found at 0.25 sec than 0.23 sec. This is in contrast to the turbulence levels upstream which followed the trend of decreasing with increasing time into the cycle (i.e., the RMS values are higher at 0.23 sec than at 0.25 sec). The blunt mean velocity profile here (at 1.4 cm) accounts for this observation.

An essentially stagnant flow condition is found in mid-diastole (0.39 sec) at both traverses as shown in Fig. 9. Downstream the RMS values have decreased to the order of 10 cm/sec demonstrating the lack of fluctuations and mean velocity gradients. However, near the valve, RMS values are still peaked (though relatively small) in the region of the major orifice. This seems unusual as no appreciable mean velocity gradients are seen during this time. It is possibly due to the remaining effects of a wake produced in the major orifice as the valve closed.

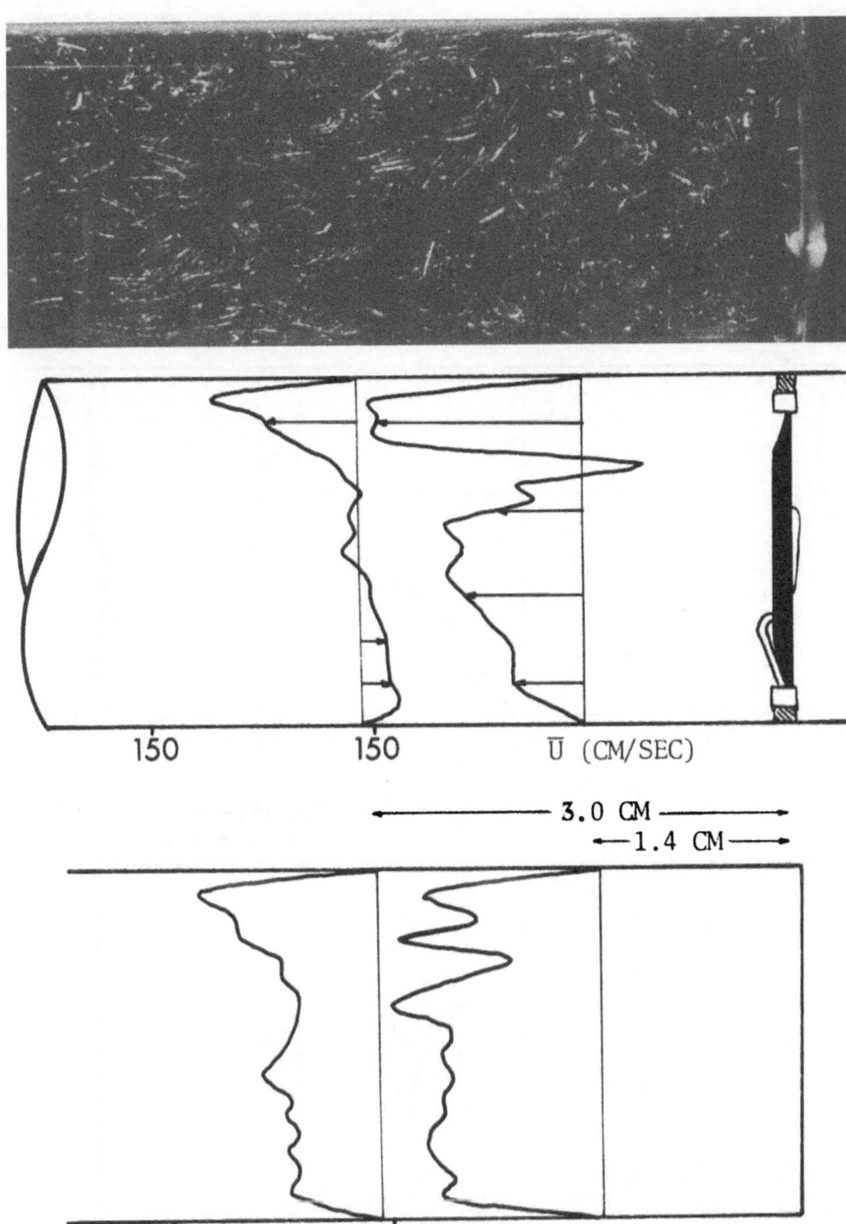

Fig. 7 Flow visualization photograph and corresponding mean and RMS velocity profiles at 0.23 sec into the cycle (end of deceleration).

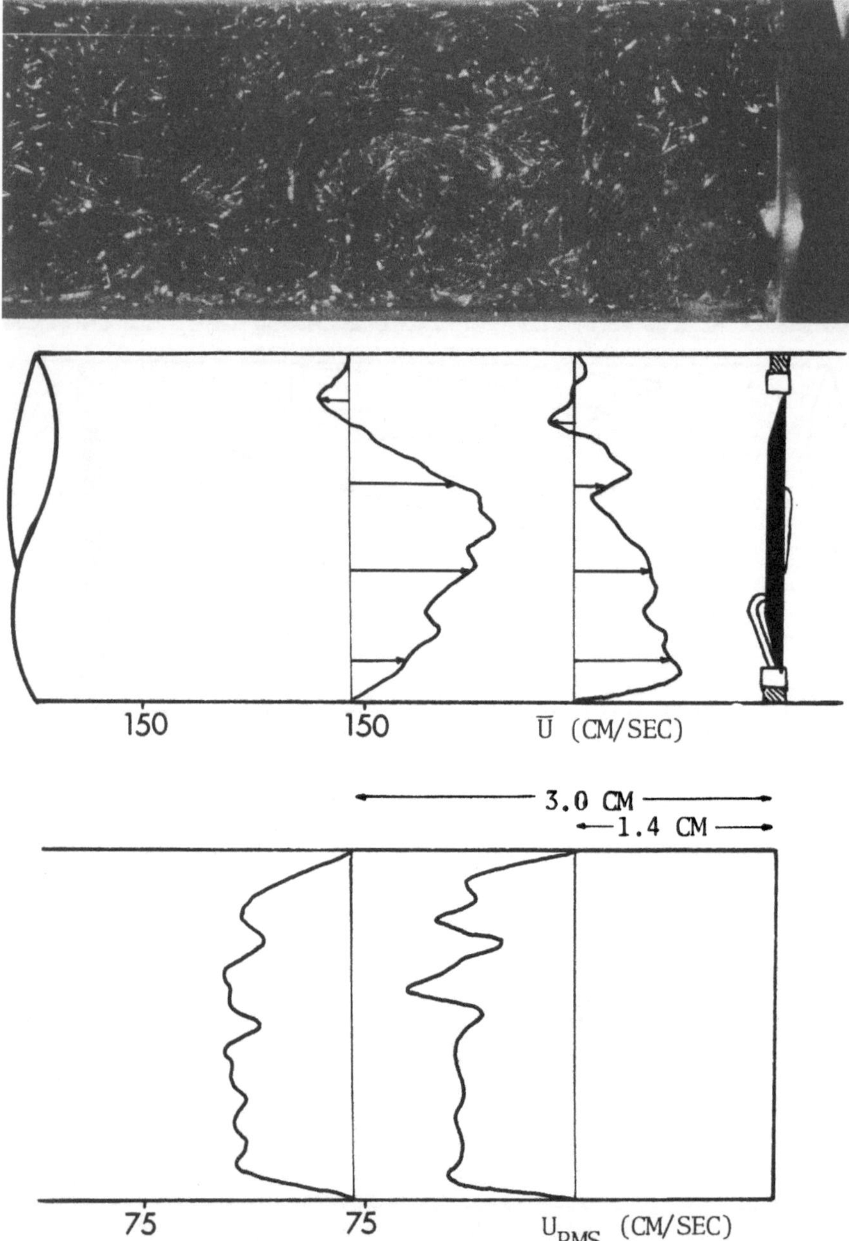

Fig. 8 Flow visualization photograph and corresponding mean and RMS velocity profiles at 0.25 sec into the cycle (early diastole).

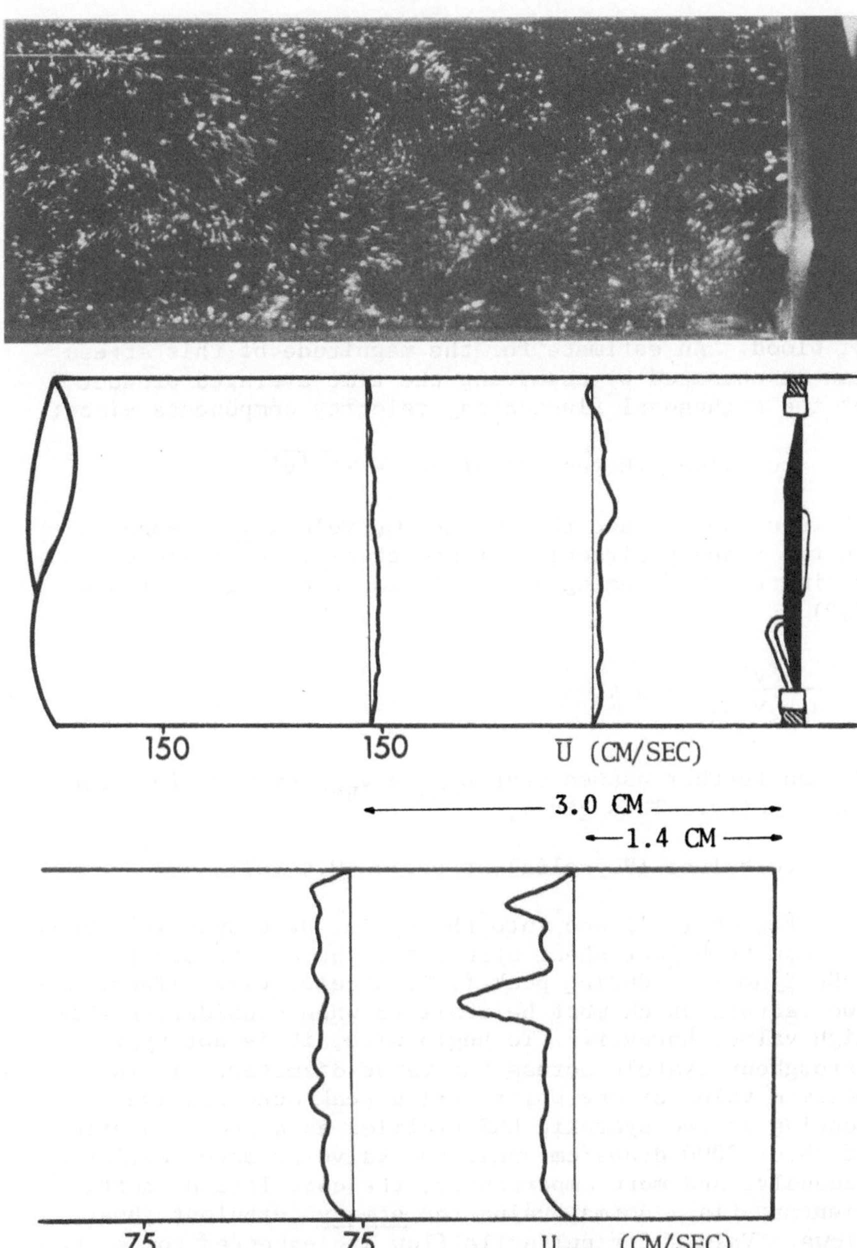

Fig. 9 Flow visualization photograph and corresponding mean and RMS velocity profiles at 0.39 sec into the cycle (diastole).

DISCUSSION

The data give information on the following:

1) turbulent shear stress levels during peak systole,
2) recirculation or regions of stasis distal to the valve,
3) a comparison of the above characteristics for prosthetic valves.

Turbulent shear stresses are of interest because of their implications in damage to the corpuscular elements of blood. An estimate for the magnitude of this stress can be obtained by observing the time-averaged product of the orthogonal fluctuating velocity components since:

$$\text{turbulent (Reynolds) stress} = -\rho\overline{u'v'}$$

where u' and v' are the fluctuating velocity components in the x and y directions respectively. $\overline{u'v'}$ can be estimated by assuming a correlation coefficient of 0.4 (23) or

$$\frac{\overline{u'v'}}{u_{RMS}v_{RMS}} = 0.4$$

We can further assume that $u_{RMS} \simeq v_{RMS}$ by two-dimensional homogeneity. This gives,

$$\text{turbulent (Reynolds) stress} = -0.4\ (\rho)(u_{RMS})^2$$

Fig. 6 (0.11 sec into the cycle, peak systole) shows maximum turbulent shear stresses to be of the order of 6500 dynes/cm^2 during peak fluid acceleration. There are two factors which must be realized when considering this high value, however. To begin with, it is not typical throughout systole across the valve diameter. It is a maximum value at one point during peak acceleration. Looking at the systolic RMS profiles as a whole, a value of about 2000 dynes/cm^2 near the valve is more realistic. Secondly, and more importantly, the correlation coefficient used is a normal value for <u>steady</u> turbulent shear flows. Values for pulsatile flow are expected to be much less than unity.

A similar investigation (10) was carried out in our laboratory using a Bjork-Shiley straight disc #27 aortic prosthesis (22 mm orifice diameter). The aortic connector used with the straight disc had an inner diameter of 1.10" (as opposed to 1.00" used for this study) and the systolic

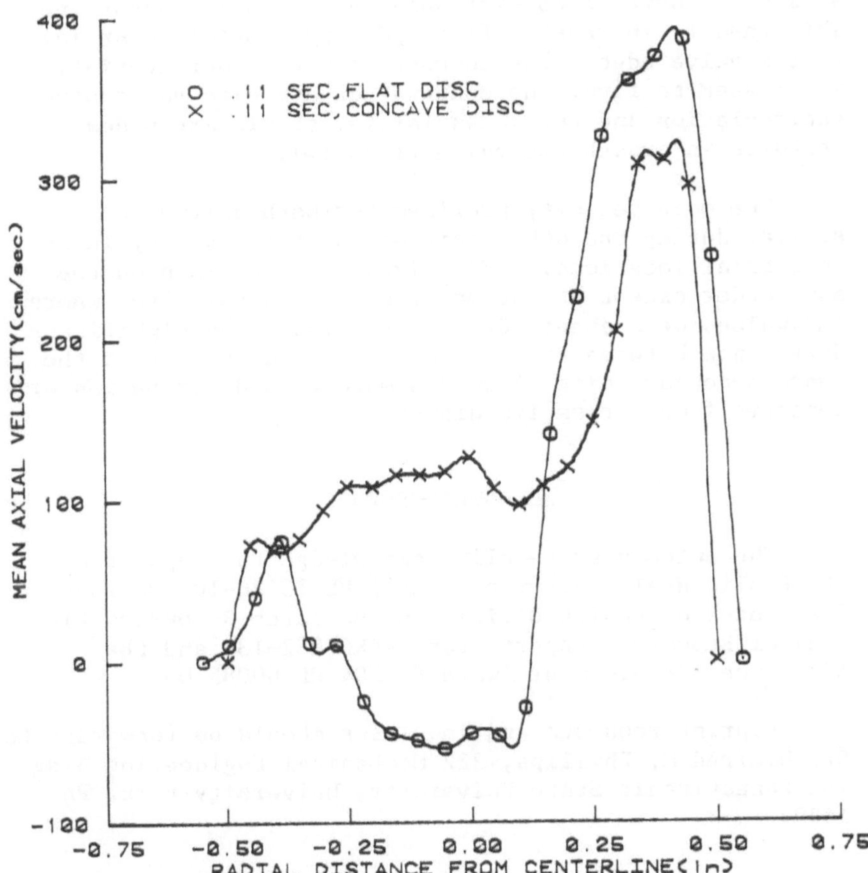

Fig. 10 A Comparison of the Mean Velocity Profiles for the Bjork-Shiley flat vs. convexo-concave disc valve at 0.11 sec into the cycle.

drive pressure was 200 mmHg (as opposed to 175 mmHg). Figure 10 shows a comparison of the mean velocity profiles of the two disc valves at 0.11 seconds into the cycle during peak fluid acceleration and 3.0 cm downstream. It can be seen that the straight disc exhibits a wider jet of higher magnitude (due probably to the higher systolic drive pressure). More importantly, however, is the difference in net positive flow in the minor orifice of this valve as compared to the convexo-concave disc valve. It would seem that there is considerable recirculation occurring. This results in reverse flow and a pronounced shear layer at the valve edge. The absence of these characteristics would seem to favor the convexo-concave disc valve since recirculation and the potential for stasis are undesirable features in prosthetic valve flows (5).

The mean velocity profiles for both valves were similar during the other portions of the heart cycle at both axial locations. RMS values also remained on the same order except at the major orifice shear layer where the values were about 40% higher. Thus, the significant difference between the two valves seemed to be that the convexo-concave disc allowed a more even distribution of positive flow across its diameter.

ACKNOWLEDGEMENTS

The authors gratefully acknowledge the support of the Public Health Service Grant # HL 13426-10, National Institutes of Health Division of Research Resources Biomedical Research Support Grant #RR07082-13, and the NIH Career Development Award #5-K04 HL 00085-05.

Reprint requests and inquiries should be forwarded to Dr. Winfred M. Phillips, 312 Mechanical Engineering Bldg., The Pennsylvania State University, University Park, PA 16802.

REFERENCES

1. Adams, P.M., "A Computer Graphics Simulation Technique and Related Results for Evaluating Blood Flow Characteristics through Prosthetic Heart Valves," ISA BM Paper No. 75328, 1975.

2. Adrian, R.M. and Fingerson, L.M., "Laser Anemometry... Theory, Applications, and Techniques, TSI Short Course and Workshop, Boston, 1977.

3. Agrawal, Y.C., "Laser Velocimeter Study of Entrance Flows in Curved Pipes," University of California, Berkeley, Rept. No. FM-75-1, Jan. 1975.

4. Bellhouse, B.J., "Velocity and Pressure Distributions in the Aortic Valve," J. of Fluid Mechanics 37, pt. 3, 1969.

5. Bjork, Viking O., "The Improved Bjork-Shiley Tilting Disc Valve Prosthesis," presented at Pan Pacific Surgical Convention, April 7, 1978.

6. Figliola, R.S. and Mueller, T.J., "In Vitro Measurements of Fluid Stresses in the Vicinity of a Disc-Type Prosthetic Heart Valve," Proc. 29th ACEMB Conf., Boston, 1976.

7. Furkay, S.S., "Fluid Dynamics of the Bjork-Shiley Aortic Valve Prosthesis and PSU Prosthetic Ventricle," M.S. Thesis, The Pennsylvania State University, August 1979.

8. George, W.K. and Lumley, J.L., "The Laser-Doppler Anemometer and Its Application to the Measurement of Turbulence," J. of Fluid Mechanics, Vol. 60, pt. 2,1973.

9. Greenfield, H., Au, A., Kelsey, S., and Kolff, W., "Simulation of Assumed Detriments to Prosthetic Heart Function," Trans. American Soc. for Art. Int. Organs, Vol. 20-B, 1974.

10. Hussain, A.K.M.F., "Mechanics of Pulsatile Flows of Relevance to the Cardiovascular System," from Cardiovascular Flow Dynamics and Measurements, N.H.C. Hwang and N.A. Normann, editors, University Park Press, Baltimore, 1977.

11. Kreid, D.K. and Goldstein, R.J., "Measurement of Velocity Profiles in Simulated Blood by the Laser-Doppler Technique," Proc. Symposium on Flow, Paper No. 4-2-95, Pittsburgh, May 10-12, 1971.

12. Lenker, J.A., "Flow Studies in Artificial Hearts and LVAD: An Application of Flow Visualization Analysis," Ph.D. Dissertation, The Pennsylvania State University, 1978.

13. Ly, D.P. and Bousquet, A., "Determination of Velocity Profiles by Laser-Doppler Anemometry in Pulsating Laminar Flow of Non-Newtonian Fluids," Proc. LDA-75 Symposium, Technical University ofDenmark, Copenhagen, 1975.

14. Modi, V.J. and Aminzadeh, M., "Fluid Mechanics of an Aortic Valve Implant," Proc. ASME Biomechanics Symp., 1977.

15. Peronneau, P.A., Pellet, M.M., Xhaard, M.C. and Hingalis, J.R., "Pulsed Doppler Ultrasonic Blood Flowmeter Real-Time Instantaneous Velocity Profiles," from Proc. Symp. on Flow, Theme IV: Biological Flows, Paper No. 4-2-146, Pittsburgh, May 10-12, 1971.

16. Peskin, C.S., "Flow Patterns Around Heart Valves: A Numerical Method," J. of Computational Physics, Vol. 10, 1972.

17. Phillips, W.M., Lenker, J.A., Brighton, J.A. and Pierce, W.S., "Flow Visualization Methods for In Vitro Cardiovascular Flow Studies," Proc. 29th Conf. on Eng. in Med. and Biol., Boston, 1976.

18. Pierce, W.S., "Development and Evaluation of Left Ventricular Assist and Artificial Heart," NIH Contractor's Report N01-HV-3-2966, June 1974.

19. Rosenberg, G., "A Mock Circulatory System for In Vitro Studies of Artificial Hearts," M.S. Thesis, The Pennsylvania State University, 1972.

20. Sabbah, H.N. and Stein, P.D., "Turbulent Blood Flow in Humans," Circulation Research, 38, No. 6, 1976.

21. Saklad, E. and Moskowitz, G., "Plane Flow Through a Prosthetic Aortic Valve," Proc. 25th Conf. on Eng. in Med. and Biol. Bal Harbour, Florida, 1972.

22. Swope, R.D. and Falsetti, H.L., "Velocity Profiles in Prosthetic Heart Valves under Steady Flow Conditions," Proc. 29th Conf. on Eng. in Med. and Biol., Boston, 1976.

23. Tennekes, H. and Lumley, J.L., A First Course in Turbulence, MIT Press, Cambridge, Mass., 1972.

24. Underwood, F.N. and Mueller, T.J., "Numerical Study of the Steady Axisymmetric Flow through a Disc-Type Prosthetic Heart Valve in a Constant Diameter Chamber," J. Biomechanical Eng., Vol. 99 (2), 1977.

25. Yoganathan, A.P., Corcoran, W.H., Harrison, E.C. and Carl, J.R., "The Bjork-Shiley Aortic Prosthesis: Flow Characteristics, Thrombus Formation, and Tissue Overgrowth," Circulation 3 (1), 1978.

THE STARR-EDWARDS AORTIC BALL VALVE: FLOW CHARACTERISTICS, THROMBUS FORMATION AND TISSUE OVERGROWTH

A.P. Yoganathan**, H.H. Reamer*, W.H. Corcoran*
E.C. Harrison+, I.A. Shulman+ & W. Parnassus+

**School of Chemical Engineering, Georgia Tech, Atlanta, GA. 30332 *Chemical Engineering Laboratory, Cal Tech, Pasadena, CA. 91125 +Cardiology Section, USC-LA County Medical Center, Los Angeles, CA. 90033

ABSTRACT

The Starr-Edwards ball valve has been one of the more commonly used aortic valve prostheses. In the study reported here, *in vitro* velocity, shear-stress and pressure drop measurements were made under steady-flow conditions and used to interpret some of the failure modes of this prosthesis as observed at autopsy. The findings show that some failure modes can be explained by the nature of the values for velocities and shears in the near vicinity of the valve.

Our results indicate that the Starr-Edwards ball valve has major fluid dynamic drawbacks such as: (a) relatively large pressure drop (17.3 to 31.0 mm Hg at a flow rate of 417 cm^3/sec), (b) hydrodynamically unstable poppet, (c) regions of flow separation at the base of each of the three struts, (d) region of flow stagnation at the apex of the cage (~ 7 to 15 mm in diameter) (e) large wall-shear stresses (~500-2000 dynes/cm^2) and bulk turbulent shear stresses (on the order of 100-5000 dynes/cm^2) in the immediate downstream vicinity of the valve, and (f) large shear stresses adjacent to the poppet surface and struts (on the order of 10^2 - 10^3 dynes/cm^2).

The observed stagnation zone could encourage thrombus formation on the apex of the cage, while the observed regions of flow separation could lead to thrombus formation and tissue overgrowth at the base and upwards along the struts. The observed wall shear could lead to damage of endothelial tissue in the proximal ascending aorta, to hemolysis, and to thrombus formation. In addition, the elevated shears adjacent to the struts and the surface of the poppet could lead to increased hemolysis with those Starr-Edwards ball valves having cloth covered struts.

Examinations have been made at the USC-LA County Medical Center of 13 Starr-Edwards aortic ball valves recovered during autopsy. Thrombus formation and tissue overgrowth were observed at various locations on the recovered valves. The locations of thrombus formation and tissue overgrowth correlate well with those predicted by the *in vitro* fluid dynamic data. In addition, endothelial damage and tissue proliferation in the proximal ascending aorta were observed in some cases. Similar pathologic findings have also been observed by other investigators.

INTRODUCTION

Of the nearly fifty different designs of prosthetic heart valves introduced during the past two decades (since 1960) many have been discarded, and of those remaining usually several modifications have been made. One valve that has been continuously used over these two decades is the Starr-Edward aortic ball valve prosthesis. The Starr-Edwards aortic ball valve has been in clinical use since about September 1961. The valve since its initial use has undergone many modifications (a word not always synonymus with improvements). The models of the Starr-Edwards ball valves most commonly used at present are #1260, 2400 and 2320. A brief description of each of these models is given below.

The Starr-Edwards 1260 valves are closed caged silastic ball valves. They are comprised of a polished Stellite alloy 21 cage with a combination of Teflon and polypropylene cloth sewing ring. The cloth on the sewing ring extends to the orifice, leaving no exposed metal on the in-flow face of the valve. The ball is made of silicone rubber and contains 2 percent-by-weight barium sulfate for radiopacity. The radiopacity aids visualizaiton of ball motion on cinefluoroscopy. The 1260 model was first made generally available in July 1969 as a modification of the model 1200 aortic prosthesis.

The Starr-Edwards Model 2320 closed caged ball valve prosthesis is comprised of a Stellite alloy 21 cage with completely cloth covered struts and a composite orifice consisting of metal and cloth to form the poppet seating surface. The metallic stops project through the knitted orifice at regular intervals. The cloth covering which is made of Teflon and polypropylene is intended to promote tissue invasion and encapsulation of the valve cage and its orifice. The ball is hollow and made of Stellite alloy 21. It has potentially greater resistance to biodegradation as compared with silicone rubber. This model has been on the market since 1968.

The Starr-Edwards 2400 closed caged ball valve is known as a composite track valve prosthesis. It is comprised of a partially cloth covered Stellite allow 21 and a hollow poppet made of the same material. The cage legs are covered with a porous knit polypropylene cloth except for an exposed metal track on the inner aspect which protects the strut cloth and metal stops which protrude through the cloth at regular intervals and protect the orifice cloth from poppet impact. All other surfaces which do not come into contact with the poppet are completely cloth covered.

The Starr-Edwards aortic ball valves have over the past 19 years had generally adequate but not the best hemodynamic characteristics. During this time the <u>in vitro</u> fluid dynamic behavior of the design has not been satisfactorily investigated. Important missing information are the velocity and shear-stress (laminar and turbulent) profiles in the vicinity of the valve, and the hydrodynamic stability of the valve poppet. Recently, Yoganathan (1978) and Figliola (1979) have made detailed <u>in vitro</u> fluid dynamic studies on the Starr-Edwards aortic ball valves.

The present article reports two studies: (1) <u>in vitro</u> measurements of velocity, shear stress and pressure drop in the near vicinity of normally functioning Starr-Edwards aortic ball valves, and (2) clinical pathology findings of recovered Starr-Edwards aortic ball prostheses. Correlation of these studies is unique (Yoganathan et al., 1978) and very useful in understanding <u>why</u> and <u>where</u> thrombus formation and tissue overgrowth occur.

EXPERIMENTAL APPARATUS AND METHODS

Laser-Doppler Anemometer (LDA)

Velosity measurements were made using a laser-Doppler anemometer in the dual-beam, forward and back-scatter modes. Figure 1 shows a block diagram of the anemometer. The laser-Doppler system consists of a 15 mW He-Ne laser (Spectra Physics 124A), a beam splitter and modulator section (DISA 55L83), a back-scatter section (DISA 55L86), a beam expander section (DISA 55L77), a photomultiplier tube (DISA 55L12), a LDA control unit (DISA 55170) and a frequency tracker anemometer used in this study and its operation have been published elsewhere by one of the authors (Yoganathan, 1978., Yoganathan et al., 1979a). The frequency response of the LDA ststem for the present experimental measurements was at least of the order of 5 X 10^4 to 10^6 Hz. Turbulence intensity levels were measured using a trus RMS volt-meter (HP-3400A).

Flow Apparatus

The **flow** apparatus consists of the following sections: (a) immersible centrifugal pump (little Giant), (b) entrance section, (c) flow channel, (d) rotameter (Brooks 10-1110),and (e) needle valve. The centrifugal pump is immersed in a bucket containing the liquid used in the experiment. The outlet of the pump is connected to the entrance section by a 1675 mm length of 25.4 mm i.d. rubber hosing. The entrance section consists of a 1370 mm length of 25.4 mm i.d. Lucite tubing. This Lucite tube is coupled to the flow channel and held in place by a Lucite collar. The entrance section ensures that the flow entering the flow channel has reached a steady state (i.e., entrance effects are not present).

The Lucite flow channel consists of three sections as shown in Figure 2, 'A' is the inlet section and represents the outflow tract from the left ventricle into the aorta. 'B'is the heart valve chamber and simulates the root of the aorta and the sinuses of Valsalva. The dimensions of chamber 'B' were obtained from fluoroscopic movies made of patients with aortic prostheses, and whose ascending aortas had mean internal diameters of 25.4 mm. Dimensions are shown in Figure 2. The heart-valve chamber is enclosed by a square Lucite pot which contains glycerine because Lucite and glycerine have almost the same index of refraction, and the

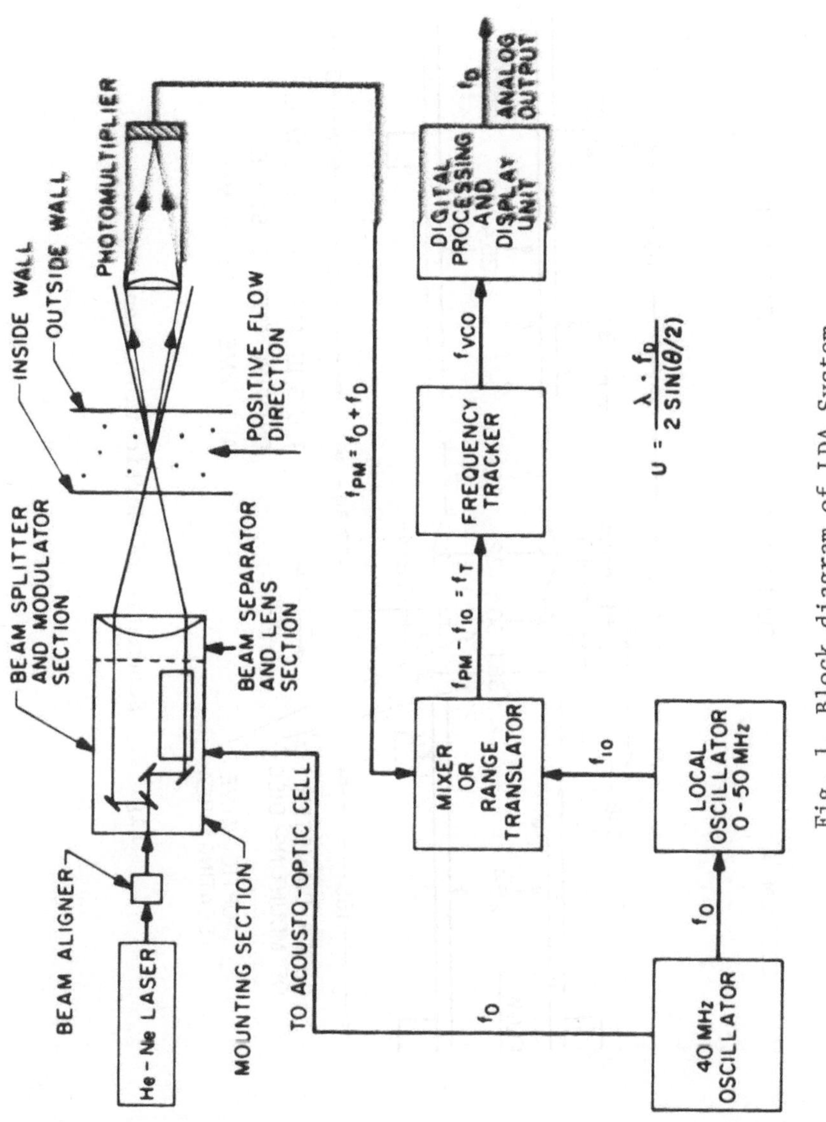

Fig. 1. Block diagram of LDA System in forward-scatter mode.

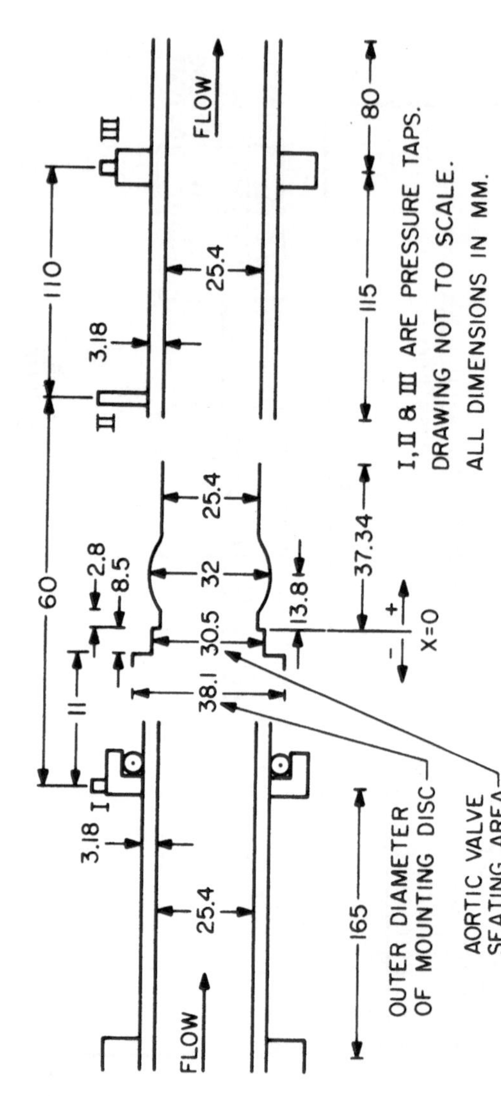

Fig. 2. Schematic of aortic flow channel.

pot reduces the effects of optical distortion when making velocity measurements in this section. 'C' is the outlet section and represents the ascending aorta. Even though this section is 150 mm in length, velocity measurements are made only in the first 100 mm. Pressure taps I, II, and III on the inlet and outlet sections are used for monitoring pressure drops in the flow channel. The entrance section and the flow channel are always maintained in the same horizontal plane. The downstream end of the outlet section is attached to rubber tubing with an i.d. of 25.4 mm, and the liquid is returned to the bucket through a rotameter and a needle valve. The rotameter had a 100 per cent reading of 417 cm^3/sec.

The aortic prostheses are mounted on metallic circular discs that have been carefully machined, and the sewing rings of the valves fit the orifices for the respective discs. The outer dimensions of all the discs are identical (38. 1 mm) and fit into the heart-valve chamber. The aortic valves are placed in the heart-valve chamber such that the front end of their sewing rings coincide with the X = 0 position as shown in Figure 2.

Liquid used in the velocity measurement experiments was about 6 per cent by-weight Pluracol Polyol V-10 (Wyandotte Chemicals) in water. At this low concentration the solution is Newtonian in nature. The solution contained a sprinkling of corn starch particles which were 10-12 μm in diameter and which were suspended in the liquid so as to scatter light. The concentration of the cornstarch in the liquid was very small. The experiments were conducted at a temperature of about 32°C. At 32°C the Polyol solution has a viscosity of about 0.035 dyne sec/cm^2, a density of 1.01 g/cm^3 and a refractive index of 1.33. The physical properties of the solution were checked every 2 to 3 days.

Pressures were measured with either Statham physiological pressure transducers (P23De and P23 Db) which were connected to a Honeywell 218-1 bridge amplifier or a water manometer system. Pressures were measured at taps I, II and III (see Fig. 2). The liquids used in the pressure drop studies were saline at 22°C and a 6-per-cent-by-weight Pluracol Polyol V-10 solution at 32°C. Saline at 22°C had a viscosity of 0.01 dyne sec/cm^2, and the Polyol solution had a viscosity of 0.035 dyne sec/cm^2 at 32°C.

The steady flow velocity and pressure drop measurements were made at flow rates in the range of 167 to 417 cm^3/sec. These steady flow rates correspond to the peak systolic flow rate for cardiac outputs between about 2.0 and 5.5 l/min.

The pulsatile flow pressure drop experiments were conducted in the Caltech pulse duplicator system. A detailed description of the pulse duplicator has been published elsewhere (Yoganathan, 1978). The flow channel used in the steady flow experiments was also used in the experiments on pulsatile flow. In all of the pulsatile experiments a pulse or heart rate of 70 beats/minute and a systolic ejection time period (t_s) of 300 msec was used. The aortic pressure was maintained at ~120/80 mm Hg. Two different types of experiments were performed under pulsatile flow. (1) Peak pressure drop and flow measurements during systole. (2) Mean pressure drop and flow measurements during systole. Pressure measurements were made with the Statham pressure transducers, while the flow rate was monitored by a Statham 2202 electromagnetic flowmeter and a cannulating type flow probe having an i.d. of 25 mm. The aortic and ventricular pressure wave signals were not filtered in the Honeywell bridge amplifier.

Details of the experimental techniques for the velocity and pressure drop measurements have been recently published (Yoganathan, 1978; Yoganathan et al., 1979b; Yoganathan et al., 1979c).

The Starr-Edwards aoritc ball valves used in the study are listed in Table 1.

Table 1: Starr-Edwards Aortic Ball Valves Studied

Name ot valve	Sewing ring diameter (mm)	Primary orifice diameter (mm)
Starr-Edwards 1260-12A	27	16.62
Starr-Edwards 1260-10A	24	15.47
Starr-Edwards 2320-10A	24	15.62
Starr-Edwards 2400-10A	24	15.75

RESULTS AND DISCUSSION

In Vitro Pressure Drop Studies

Steady flow pressure drop measurements were obtained across the Starr-Edwards 1260-12A, 1260-10A, 2400-10A and 2320-10A ball valves. Of all the prosthetic aortic heart valves studied (Yoganathan et al., 1979c) the pressure drop results of the four Starr-Edwards ball valves were the hardest to reproduce. Under steady flow conditions at times the results obtained were observed to differ by almost a factor of two. It was observed from the very beginning that the poppets of these valves became hydrodynamically unstable at a flow rate of about 133 cm^3/sec for both test liquids. The instability of the poppet caused it to oscillate and rotate at different positions in the cage. Such hydrodynamic instability of the poppets of the Starr-Edwards caged ball valves has also been observed by other investigators (Weiting, 1969; Figliola, 1979; Davey et al., 1966). These investigators, however, did not realize or notice that the ball could oscillate and rotate at different positions in the cage.

During our study it was observed that the ball could and would oscillate at three different positions, namely, top (apex) of the cage, middle of the cage, and bottom (near seat) of the cage. Depending on the position of oscillation of the ball of a given Starr-Edwards valve, the pressure drops obtained were observed to differ by as much as a factor of two as illustrated by Figures 3-6. In these figures results are shown for the Starr-Edwards 1260-12A ball valve, but the same trends are observed with the other three valves tested. The pressure drops shown in Figures 3-6 were obtained between taps I and II (Δp_1). For example, with the Starr-Edwards 1260-12A valve as the position of the ball moved from the apex to the back of the cage the pressure drop across the valve at a steady flow rate of 417 cm^3/sec increased from 17.3 to 31.0 mm Hg.

During a given experiment the ball would oscillate at any three of the given locations, and it could change its position of oscillation during the middle of the experiment. It was, however, possible to maintain the position of the ball at a given location during an experiment with a great deal of care. This was accomplished by squeezing by hand the rubber hosing downstream and upstream of the flow

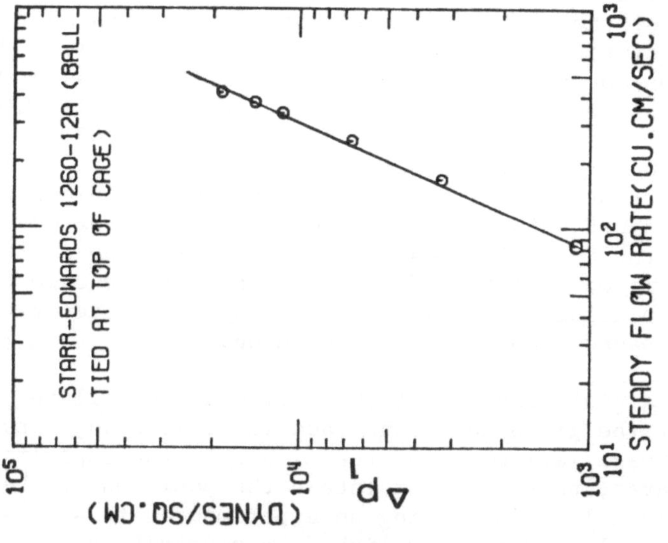

Fig. 3. Steady flow pressure drop across Starr-Edwards ball valve.

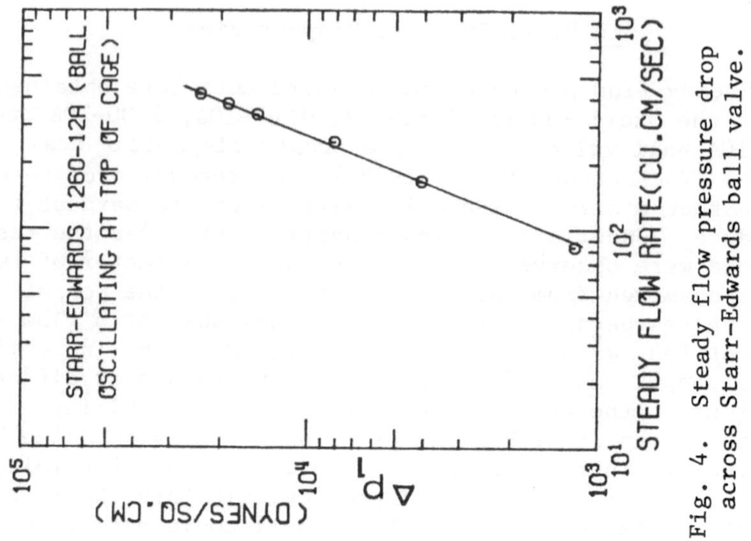

Fig. 4. Steady flow pressure drop across Starr-Edwards ball valve.

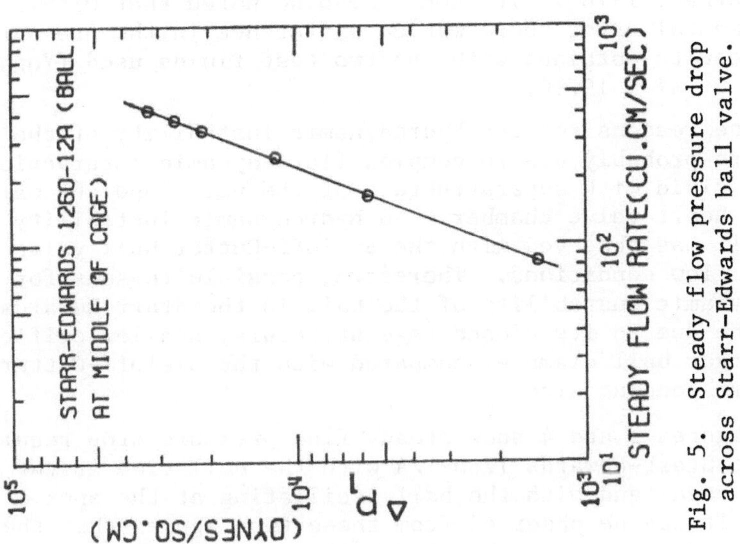

Fig. 5. Steady flow pressure drop across Starr-Edwards ball valve.

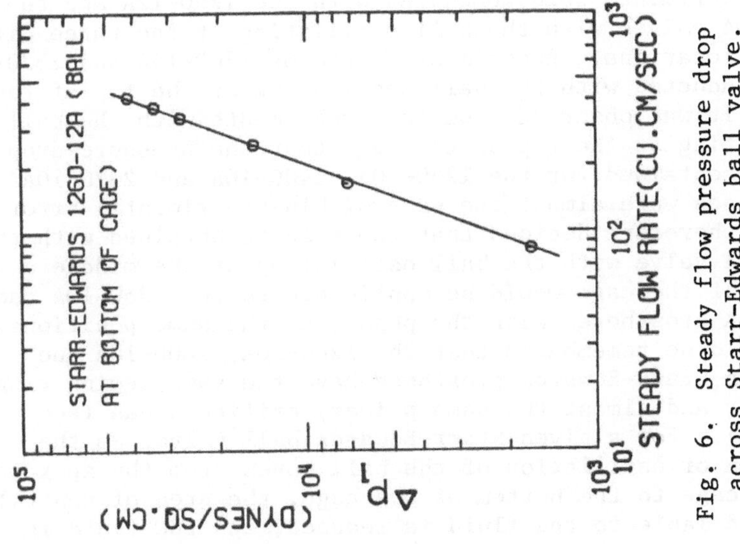

Fig. 6. Steady flow pressure drop across Starr-Edwards ball valve.

channel, and also by increasing or decreasing the flow rate in the flow system in a gradual manner. Steady flow pressure drop experiments were conducted with the 1260-12A and the 2400-10A valves with the ball oscillating at the three different locations. Experiments with the 1260-10A and 2320-10A were conducted with the ball oscillating at the top of the cage. It was observed from the experiments with the ball oscillating at the top of the cage that the pressure-drop results obtained for the 1260-10A, 2400-10A and 2320-10A prostheses were almost the same within experimental error. It was therefore decided that the results obtained with the 2400-10A valve with the ball oscillating at the middle and bottom of the cage would be applicable to the 1260-10A and 2320-10A prostheses with the poppet at the same positions. It should be remembered that the 1260-10A, 2400-10A and 2320-10A Starr-Edwards prosthese have the same sewing ring diameter and almost the same primary orifice areas (see Table 1). For a given Starr-Edwards ball valve, as the position of oscillation of the ball moves from the apex (top) of the cage to the bottom of the cage, the area of central flow available to the fluid is reduced, and the fluid is forced to flow peripherally around the ball in a shorter distance form the orifice, leading to larger pressure drops. Detailed results of this study can be found elsewhere (Yoganathan, 1978). It should also be noted that within experimental error there was no difference in the pressure drop results obtained with the two test fluids used (Yoganathan et al., 1979C).

The reasons for the hydrodynamic instability of the ball are probably due to complex fluid dynamic interactions of the fluid with superstructure of the valve and the walls of the heart valve chamber. No hydrodynamic instability of the ball was observed with the Smeloff-Cutter ball valves under steady flow conditions. Therefore, possible reasons for the hydrodynamic instability of the ball in the Starr-Edwards valve could be due to its closed cage structure, smaller orifice area and larger ball diameter compared with the Smeloff-Cutter valve of corresponding size.

Figures 3 and 4 show steady flow pressure drop results for the Starr-Edwards 1260-12A with the ball tied at the apex of the cage, and with the ball oscillating at the apex of the cage. It can be observed from these two figures that the pressure drops across the tied ball are significantly smaller than those measured across the oscillating ball. This observation is not unusual and has been observed in flow past stationary and oscillating circular cylinders. When the sphere or ball oscillates, the width of the wake downstream of the ball increases. This leads to increased pressure drag and therefore to a larger pressure drop across the ball. If the poppets of the Starr-Edwards ball valves were hydrodynamically stable, the pressure drops across these valves under steady flow would be very similar to those obtained with the balls tied at the apecies of their respective cages. For example, if this were true with the Starr-Edwards 1260-12A valve the pressure drops across this valve would be less than those measured across the Smeloff-Cutter A5 valve. At a steady flow rate of 417 cm^3/sec the pressure drop(Δp_1) across the tied Starr-Edwards 1260-12A valve is 13.8 mm Hg, as against 17.1 mm Hg across the Smeloff-Cutter A5 valve (cf.Δp_1 across 1260-12A ball oscillating at apex of cage 17.3 mm Hg).

The pulsatile flow pressure drop experiments strongly suggest that the poppets of the Starr-Edwards aortic ball valves stay at the middle of their cages during the major part of systole instead of staying at the top (apex) of the cages as was envisioned when originally designed. Recent photographic studies in our laboratory confirm these findings. Figure 7 shows the mean systolic pressure drops (across taps I and II) across the Starr-Edwards 1260-12A valve. During the opening phase of systole the ball strikes the apex of the cage and then bounces back to the middle of the cage, where it seems to attain an equilibrium position. The observations made with the Starr-Edwards aortic ball valves in this study under pulsatile flow conditions may not occur in every *in vitro* test performed, or in every patient who has such a valve. The above discussions indicate the importance of having a hydrodynamically stable poppet.

Hamby et al., (1974) in an excellent clinical study demonstrated the hydrodynamic instability of the Starr-Edwards aortic ball valves in 41 patients. The study combined cinefluoroscopy, phonocardiography and hemodynamic measurements.

In Vitro Velocity Studies

Two sets of velocity measurement experiments were performed with the Starr-Edwards 1260-12A ball vlave. The first set was made with the ball tied at the apex of the cage with sewing thread. A second series of experiments were conducted

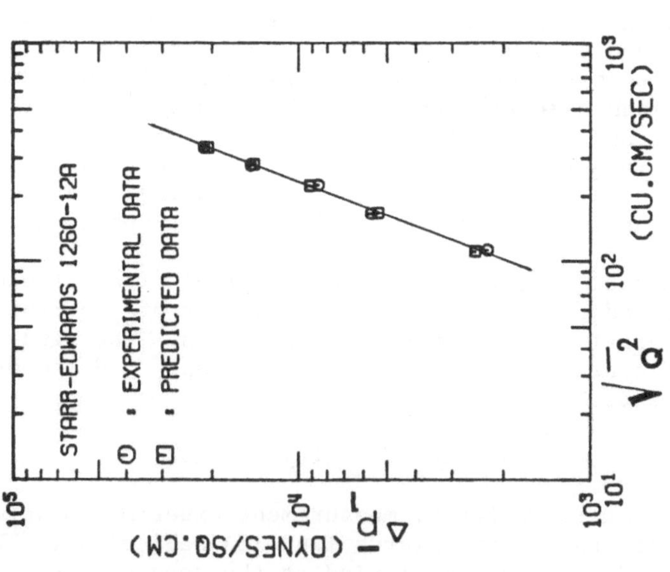

Fig. 7.1 Mean systolic pressure drop across Starr-Edwards ball valve.

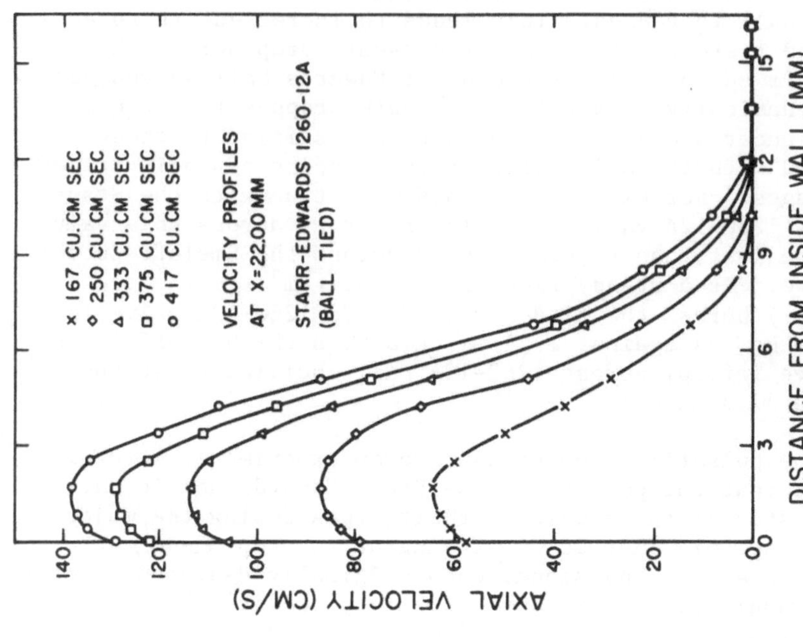

Fig. 9. Velocity 22 mm downstream of Starr-Edwards ball valve (ball tied)

with the ball free and oscillating at the apex of the cage.

Figures 8, 9, 10 and 11 show the axial velocity profiles obtained 22 mm and 42 mm downstream of the front end of the valve sewing ring with the ball tied and untied, respectively. Unless otherwise stated, all axial velocity profiles shown were obtained in the horizontal plane through the center of the flow channel. In the velocity profiles plots the data points which appear to lie on the vertical axes correspond to velocity measurements made at distances 0.034 mm from the flow channel walls. Figures 8 and 10 show velocity profiles made in sinus region (section 'B' of flow channel) of the flow channel. In these two figures the velocity profiles are drawn from the inside wall to about the center line of the channel in order to highlight the measurements in this region. In Figures 9 and 11 the velocity profiles are drawn from the inside wall to the outside wall.

The location X=22 mm is about 2 mm downstream from the apex of the cage. The fact that the ball was tied is illustrated by the profiles in Figure 8, which show an area of zero flow in the very near vicinity of the stationary ball. Figures 10-12 show velocity profiles obtained downstream of the untied (ball oscillating at apex of cage) ball. At X=42.0 and 55.0 mm (Figs. 11 and 12) the velocity profiles obtained at the flow rates 167 and 250 cm^3/sec are quite different in shape from those obtained at 333,375 and 417 cm^3/sec. The reason for the change in shape of the profiles is probably associated with some change in the nature of the oscillations and rotations of the ball at the apex of the cage. In the near vicinity of the valve ($X \leq 55$ mm), there are differences in the profiles obtained with the ball tied and untied. The area of stagnation at X = 22 mm is smaller in the case of the untied ball and is confined more towards the apex of the cage (see Fig. 10), and not the entire area occupied by the top surface of the ball as was observed with the tied ball (Fig. 8). It should be noted in Figures 8 and 10 that the velocities in the stagnant zone are plotted as zero velocity. In the experiments, however, the velocities in both stagnation zones oscillated between ± 2.0 cm/sec. Velocity measurements at X = 55.0 mm and 67.4 mm tend to indicate that the profiles of the untied ball tend to return to turbulent pipe flow profiles somewhat faster than the profiles of the tied ball. The velocity profiles of the tied ball created a larger velocity defect (i.e., dip in the center of the velocity profile) phenomenan at X = 55.0

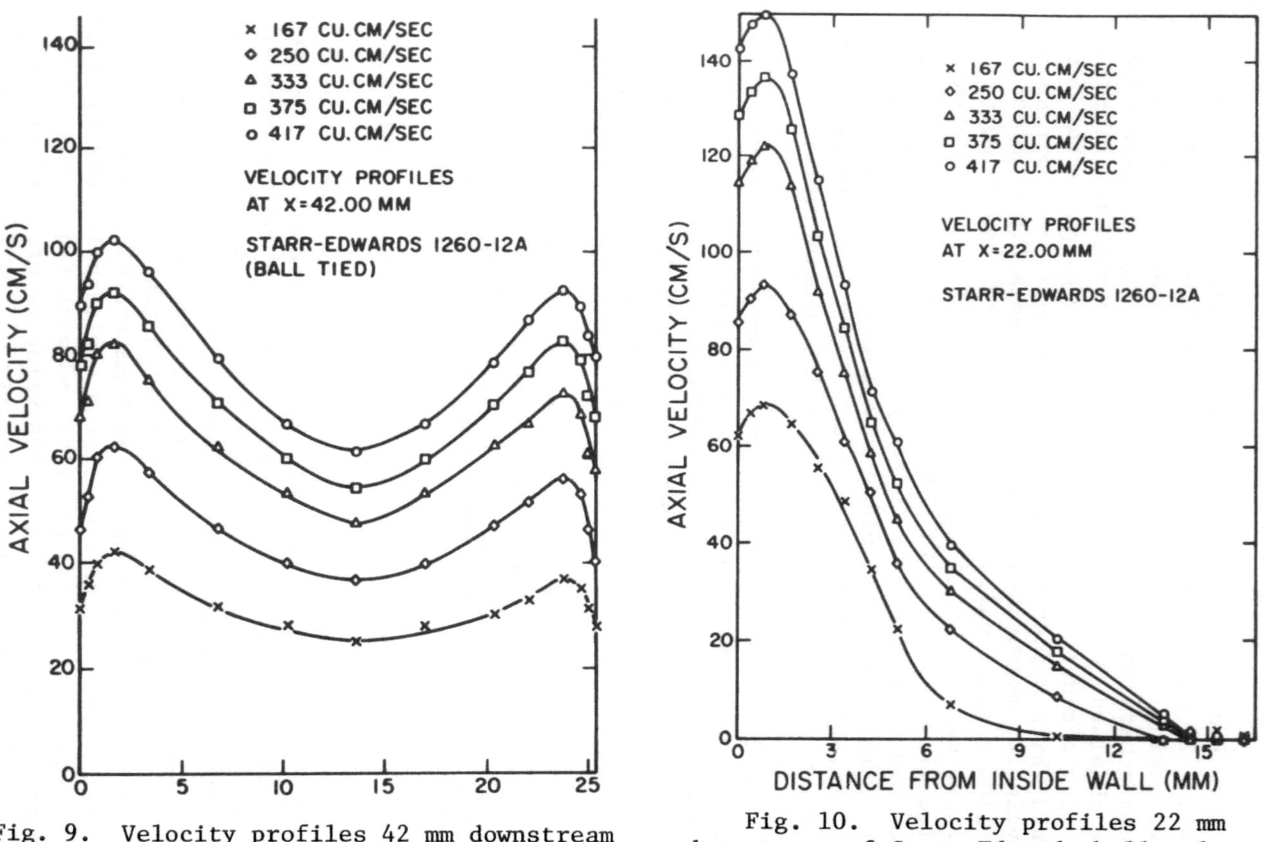

Fig. 9. Velocity profiles 42 mm downstream of Starr-Edwards ball valve (ball tied).

Fig. 10. Velocity profiles 22 mm downstream of Starr-Edwards ball valve.

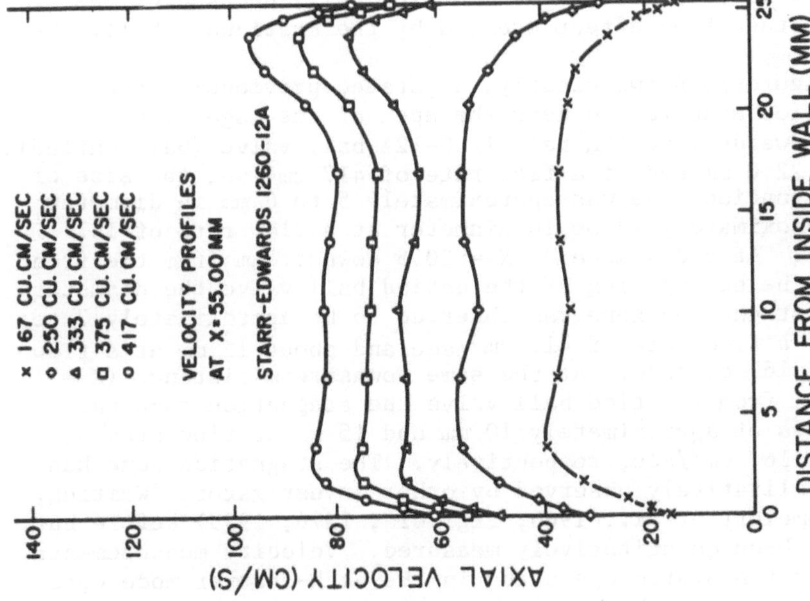

Fig. 12. Velocity profiles 55 mm downstream of Starr-Edwards ball valve.

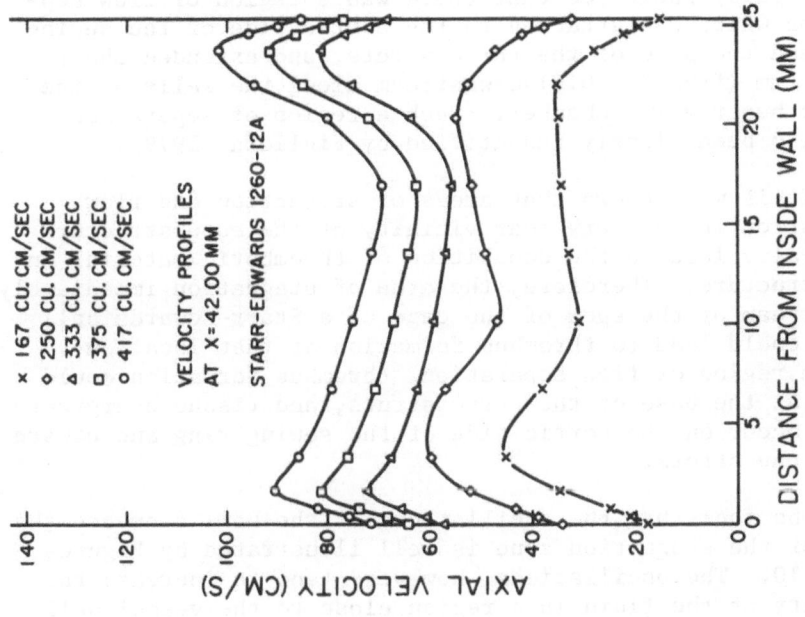

Fig. 11. Velocity profiles 42 mm downstream of Starr-Edwards ball valve.

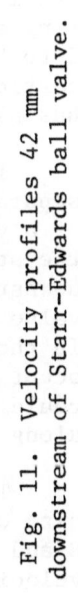

and 67.4 mm compared with the untied ball, because of the larger stagnation effect created by the stationary ball.

Figure 10 shows clearly, as stated previously, the stagnation zone formed near the apex of the cage of the Starr-Edwards 1260-12A ball 1260-12A ball valve (ball untied). At X = 22.0 mm and at a flow rate of 417 cm^3/sec the size of the stagnation zone was approximately 5 to 6 mm in diameter and approximately 12 mm in diameter at a flow rate of 167 cm^3/sec. At a distance of X = 20.5 downstream from the front end of the sewing ring of the untied ball valve the diameter of the stagnation zone was observed to be approximately 7 to 8 mm at a flow rate of 417 cm^3/sec and about 12 mm at a flow rate of 167 cm^3/sec. At the same downstream distance (X = 20.5 mm) from the tied ball valve the stagnation zone had diameters of approximately 10 mm and 15 mm at flowrates of 417 and 167 cm^3/sec, respectively. The stagnation zone has been qualitatively observed by other investigators (Weiting, 1969; Smeloff et al., 1966; Figliola, 1976; 1979) before but has not been quantitatively measured. Velocity measurements with the LDA system operating in the back-scatter mode were made adjacent to the aortic side of the valve sewing ring and the base of the three struts. These measurements (with untied ball) indicated that there was a region of flow separation which was attached to the aortic side of the sewing ring and the base of the three struts, and extended about 2 to 5 mm (from X = 0.0) downstream along the walls of the aortic heart valve chamber. Such a region of separation has also been clearly indentified by Figliola (1979).

It is well known that areas of stagnation and flow separation in the very near vicinity of the superstructure of a valve lead to the deposition of thrombotic material on the structure. Therefore, the area of stagnation immediately downstream of the apex of the cage of a Starr-Edwards ball valve could lead to thrombus formation at that location. In the region of flow separation, thrombus formation could occur at the base of the three struts, and tissue overgrowth could occur on the aortic side of the sewing ring and upward along the struts.

The fact that the oscillations of the ball decrease the size of the stagnation zone is well illustrated by Figures 8 and 10. The oscillations, however, tend to increase the velocity of the fluid in a region close to the vessel wall (distance from inside wall $\leq$ 3.5 mm) as illustrated by Figures 8 and 10. For example, with the ball oscillating at the apex of the cage, at X = 22 mm the maximum velocity very near the inside wall was about 145 cm/sec at a flow rate of 417 cm^3/sec; with the ball oscillating at the back of the cage, however, the velocity measured was about 190 cm/sec. As the velocities along the vessel wall in the sinus region increase, the wall shears in the very near vicinity of the valve also increase. Also the increased oscillations of the ball lead to increased turbulence levels and turbulent shear stresses both in the annular flow region between the channel wall and the poppet, and the region immediately downstream of the apex of the cage. Therefore, even though the oscillations of the ball lead to a smaller area of stagnation at the apex of the cage, the speed of the oscillations tend to increase the shear levels (wall and turbulent) in the very near downstream vicinity of the valve. It should be noted that the speed of the oscillations increase as the ball moves from the apex to the back of the cage.

Velocity measurements were made in the annular region between the inside wall and the surface of the ball for the case where the ball was oscillating at the apex of the cage. The LDA system had to be operated in the back-scatter mode and the poppet of the 1260-12A valve had to be painted black for these measurements. In this annular region very high jet like velocities on the order of 280 cm/sec and 90 cm/sec were observed at flow rates of 417 and 167 cm^3/sec, respectively. It was also possible to measure velocities very close to the poppet surface (0.17 mm from the surface). At this distance from the poppet surface velocities on the order of 250 cm/sec and 100 cm/sec were measured at steady flow rates of 417 and 167 cm^3/sec, respectively.

The wall shear stresses measured in the immediate downstream vicinity of the 1260-12A valve are listed in Table 2. As indicated previously these results show that the wall shears in the sinus region ($0 \leq X \leq 37$ mm) are larger for the case of the untied ball. With the back-scatter studies it was possible to measure the shear stresses on the poppet surface for the case of the untied ball. Poppet wall shears on the order of 2500-2800 dynes/cm^2 and 900-1000 dynes/cm^2 were measured at flow rates of 417 and 167 cm^3/sec, respectively. These results are in good agreement with those obtained by Figliola (1979).

Table 2. Experimentally Measured Wall Shear Stresses

Name of valve	X* (mm)	Flow rate (cm^3/sec)	τ_{wi} (dynes/cm^2)	τ_{wo} (dynes/cm^2)
Starr-Edwards 1260-12A	22	167	676	659
(ball tied)	22	417	1510	1501
	32	167	650	636
	32	417	1432	1401
	118	167	173	102
	118	417	594	533
Starr-Edwards 1260-12A	22	167	726	750
(ball tied)	22	417	1666	1751
	32	167	685	706
	32	417	1598	1669
	118	167	142	109
	118	417	584	624

* $0 \leq X \leq 37$ mm is the sinus region

τ_{wi} = shear stress on inside wall

τ_{wo} = shear stress on outside wall

Turbulence intensity levels, I_x, (Yoganathan, et al., 1979b) were measured for both the tied and untied ball. These measurements indicated turbulence intensity levels as high as 30% to 50% in both sets of experiments. These high turbulence intensity levels led to large bulk turbulent shear stresses both in the annular region between the flow channel wall and the poppet surface, and in the region immediately downstream of the apex of the valve cage. Preliminary bulk turbulent shear stress measurements in both regions indicate maximum turbulent shears of the order of 2000 to 5500 dynes/cm^2 and 100 to 900 dynes/cm^2 at flow rates of 417 and 167 cm^3/sec, respectively. The measured turbulent shear stress valves are in good agreement with previous estimations made by the authors (Yoganathan et al., 1979b). The values mentioned above apply to experiments conducted with the tied and untied ball. The turbulent shear stresses measured with the tied ball were generally about 80-85% of the values measured with the untied ball (ball oscillating at apex of cage).

The wall shear stresses measured in the near vicinity of the Starr-Edwards aortic ball valve could damage and erode off the endothelial lining of the vessel wall immediately downstream of the valve. Fry (1968; 1969) in his studies observed that endothelial cells could be damaged by wall shear stresses of about 400 dynes/ cm^2 and could be eroded off by wall shears of about 950 dynes/cm^2. Blackshear and his co-workers (1972 a,b) suggest that the shear stresses required in the bulk of the flow to damage red blood cells are about 40,000 dynes/cm^2. Nevaril and his co-workers (1969) contend, however, that this value could be as low as 1500 dynes/cm^2. *In vitro* experiments (Hellums and Brown, 1977; Hung et al., 1976; Ramstack et al., 1979) have also shown that platelets could be damaged by shear stresses of the order of 100-500 dynes/cm^2. A formed element such as a red blood cell which adheres to the vessel wall or to a foreign surface may be damaged by shear stresses of the order of 10-10^2 dynes/cm^2 (Blackshear, 1972a; 1972b, Mohandas et al., 1974). It is therefore quite clear that the wall shear stresses caused by the Starr-Edwards aortic ball valve are capable of damaging the endothelial lining of the ascending aorta and the formed elements of blood such as red cells which may adhere to the vessel wall.

The bulk turbulent shear stresses in the immediate downstream vicinity of the Starr-Edwards ball valve are also large enough that they could cause either sub-lethal or lethal damage to the formed elements of blood in the bulk fluid. Sub-lethal damage could morphologically damage the formed element. The shear stresses immediately adjacent to the valve cage (in the annular region) are large enough to lethally damage any formed elements of blood whicn may adhere to the valve cage. It has been clinically observed (Roberts 1976; Santinga et.al., 1973; Eyster et al., 1971; Lefemine et al., 1974; Santinga and Kirsh, 1972) that the Starr-Edwards ball valves with cloth covered struts cause more hemolysis compared with the non-cloth covered Starr-Edwards ball valves. A very probable and logical explanation for this observation is that the porous cloth covering which is rough provides an ideal foreign surface for the adhesion of the red cells as they flow past the valve struts. Once adhered, the red cells undergo shear stresses on the order of 10^2-10^3 dynes/cm^2 which lead to their destruction and cause hemolysis. Damage caused to the red cells, platelets and the endothelial lining of the ascending aorta could in turn all lead to thrombus formation.

Clinical Pathology Results

Examinations were made at the USC - Los Angeles County Medical Center of 13 Starr-Edwards aortic ball valve prostheses recovered during autopsy. Thrombus formation and tissue overgrowth were observed at various locations on the recovered as illustrated schematically by Figure 13. Figures 14-16 show examples of some of the recovered ball valves. Varying amounts of thrombus formation were observed predominantly at the base of the struts and the apex of the cage, while varying degrees of tissue overgrowth were predominantly observed on the aortic side of the sewing ring and on the struts of the cloth covered valves. In some cases the thrombus had grown along the struts, probably starting either at the apex or the base of the struts. In addition, examination of the vessel walls immediately downstream of the valve sewing rings indicated varying degrees of endothelial damage and fibrous tissue proliferation in some of the specimens. Detailed pathologic studies on the recovered valves are in progress at the present time.

Clinical and patholigic studies on recovered Starr-Edwards aortic ball valves by other investigators have also identified: (a) thrombus formation on the apex of the cage, at the base of the three struts and occasionally along the

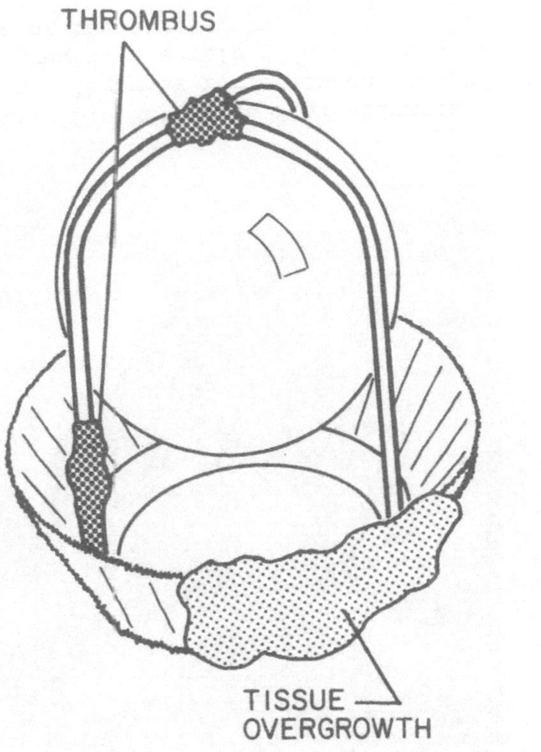

Fig. 13. Schematic example of thrombus formation and tissue overgrowth on a Starr-Edwards aortic ball valve.

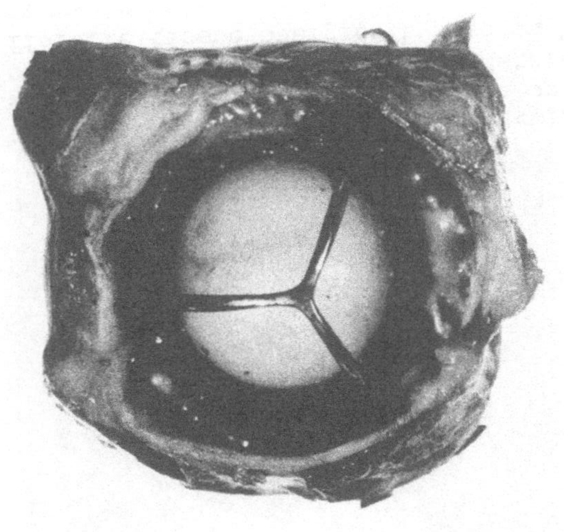

Fig. 14. Recovered Starr-Edwards 1260 aortic valve. Thrombus formation at base of the three struts and tissue overgrowth on aortic side of sewing ring.

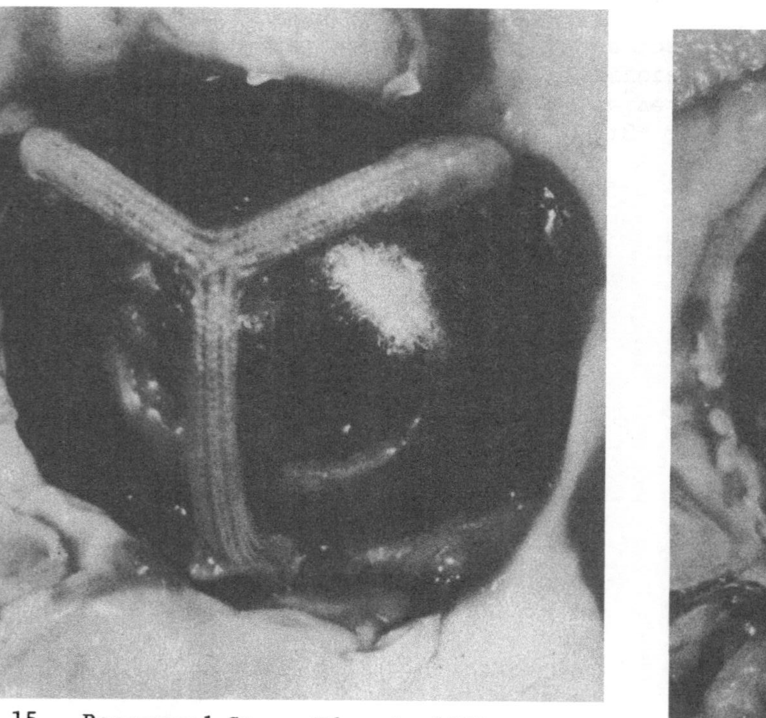

Fig. 15. Recovered Starr-Edwards 2320 aortic valve. Thrombus formation on apex of cage; tissue overgrowth on aortic side of sewing ring and along base of the struts.

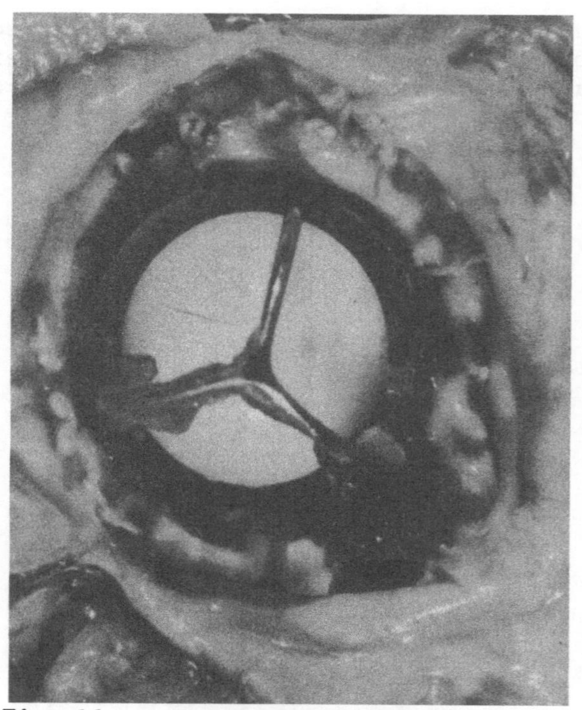

Fig. 16. Recovered Starr-Edwards 1260 aortic valve. Thrombus formation on apex of cage and at the base of the struts; fibrotic tissue overgrowth also on base of two struts.

struts, (b) tissue overgrowth on the aortic side of the sewing ring and on the struts of the cloth covered models, (c) endothelial damage and tissue proliferation of the proximal ascending aorta, and (d) increased hemolysis with the cloth covered strut models (Roberts, 1976; Smithwick et al., 1975; Roberts and Morrow, 1967a,b; Crawford et al., 1973; Stein, et al., 1976; Davilla et al., 1966; Lamberti et al., 1977).

CONCLUSIONS

We conclude that the presence and locations of thrombus formation and tissue overgrowth, damage to the endothelial lining of the proximal ascending aorta, increased hemolysis of the cloth covered (struts) models, and hydrodynamically unstable poppet observed with the Starr-Edwards aortic ball valves are inherent to their fluid mechanical designs. All prosthetic aortic heart valves are liable to thrombus formation, tissue overgrowth etc. Therefore, the correlative studies described in this article should be extended to other heart valve prostheses. Such studies would be helpful in improving future designs of prosthetic heart valves.

ACKNOWLEDGEMENTS

Funds for the support of the work described here were provided by the Donald E. Baxter Foundation, Children's Heart Foundation of Southern California, American Heart Association-Los Angeles Chapter and the Georgia Tech Biomedical Engineering Research Support Grant(#5 S07 RR07024-14). The valves tested were provided by Edwards Laboratories, Santa Ana, California.

REFERENCES

Blackshear, P.L. (1972a) Hemolysis at prosthetic surfaces. Chemistry of Biosurfaces (Edited by Hair, M.L.) Vol. 2. pp. 523-561. Marcel Dekker, New York.

Blackshear, P.L. (1972b) Mechanical hemolysis in flowing blood. Biomechanics - Its Foundations and Objectives (Edited by Fung, Y.C.,Perrone, N. and Anliker, M), pp. 501-528. Prentice - Hall, Englewood Cliffs.

Crawford, F.A., Sethi, G.K. Scott, S.M., and Takaro, T. (1973) Systemic emboli due to cloth wear in a Starr-Edwards model 2320 aortic prosthesis. Ann Thorac. Surg 16, 614-619.

Davey, T.B., Kaufaman, B., and Smeloff, E.A (1966) Pulsatile flow studies of prosthetic heart valves. J. Thorac, Cardiovasc. Surg. 51, 264-267.

Davila, J.C., Palmer, T.E., Sethi, R.S., DeLaurentis, D.A., Enriquez, F., Rincorn,N., and Lautsch, E.V. (1966) The problem of thrombosis in artificial cardiac valves. Heart Substitutes: Mechanical and Transplant(Edited by Brest, A.N.), pp.25-36. Charles C. Thomas, Springfield.

Eyster, E., Rothchild, J., and Mychajliw, O. (1971) Chronic intravascular hemolysis after aortic valve replacement. Circulation XLIV 657-665.

Figliola, R.S. (1976). A study of the hemolytic potential of prosthetic heart valve flows based on local in vitro stress measurements. M.S. thesis, University of Notre Dame.

Figliola, R.S. (1979) In Vitro Velocity and Shear Stress Measurements in the Vicintiy of Prosthetic Heart Valves Using Laser Doppler and Hot-Film Anemometry. Ph.D. thesis, University of Notre Dame.

Fry, D.L. (1968) Acute vascular endothelial changes associated with increased blood velocity gradients. Circ. Res. 22, 165-197.

Fry, D.L. (1969) Certain histological and chemical responses of the vascular interface to acutely induced mechanical stress in the aorta of the dog. Circ. Res. 24, 93-108.

Hamby, R.I., Lee, R.L., Aintablian, A., Wisoff, B.G., and Hartstein, M.L. (1974) Cinefluorographic study of the aortic ball-cage prosthetic valve. Am.J. Cardiol. 34, 276-283.

Hellums, J.D., and Brown III, C.H. (1977) Blood cell damage by mechanical forces. Cardiovascular Fluid Dynamics (Edited by Hwang, N.H.C. and Normann, N.A.),pp. 799-823. University Park Press, Baltimore.

Hung, T.C., Hochmuth, R.M., Joist, J.H., and Sutera, S.P. (1976) Shear-induced aggregation and lysis of platelets. Trans. Am. Soc. Artis. Intern. Organs 22, 285-290.

Lamberti, J.J., Das Gupta, D.D., Falicov, R., and Anagnostopoulos, C.E. (1977) An unusual form of late stenosis after aortic valve replacement with cloth-covered Starr-Edwards and prosthesis. Chest 71, 680-684.

Lefemine, A.A., Miller, M., and Pinder, G.C. (1974) J. Thorac. Cardiovasc. Surg. 67, 857-862.

Mohandas, N., Hochmuth, R.M., and Spaeth, E.E. (1974) Adhesion of red cells to foreign surfaces in the presence of flow. J. Biomech. Mat. Res. 8, 119-136.

Nevaril, C.G., Hellums, J.D., Alfey Jr., C.P. and Lynch, E.C. (1969) Physical effects in red blood cell trauma. A. I. ChE. J. 15, 707-711.

Ramstack, J.M. Zuckerman, L., and Mockros, L.F. (1979) Shear induced activation of platelets. J. Biomech. 12, 113-125 (1979).

Roberts, W.C., and Morrow, A.G. (1979) Anatomic studies of hearts containing caged-ball prosthetic valves. Johns Hopkins Med. J. 121, 271-295.

Roberts, W.C., and Morrow, A.G. (1967b) Late post-operative pathological findings after cardiac valve replacement. Circ. Suppl. (1) 35-36,48-62.

Roberts, W.C. (1976) Choosing a substitute cardiac valve: type, size, surgeon. Am. J. Cardiol. 38, 633-644.

Santinga, J.T., and Kirsh, M.M. (1972) Hemolytic anemia in series 2300 and 2310 Starr-Edwards prosthetic valves. Ann Thorac. Surg. 14, 539-544.

Santinga, J.T., Kirsh, N.N., and Batsakis, J.T. (1973) Hemolysis in different series of the Starr-Edwards aortic valve prostheses. Chest 63, 905-908.

Smeloff, E.A., Huntley, A.C., Davey, T.B., Kaufman, B., and Gerbode, F. (1966) Camparative study of prosthetic heart valves. J. Thorac. Cardiovasc. Surg. 52, 841-846.

Smithwick III, W., Kouchoukos, N.T., Karp, R.B., Pacifico, :D., and Kirklin, J.W. (1975) Late stenosis of Starr-Edwards cloth-covered prostheses. 20, 249-255.

Stein, D.W., Rahimtoola, S.H., Kloster, F.E., Selden, R., and Starr, A. (1976). J. Thorac. Cardiovasc. Surg. 71, 680-684.

Weiting, D.W. (1969) Dynamic Flow Characteristics of Heart Valves. Ph.D. thesis, University of Texas, Austin.

Yoganathan, A.P. (1978) Cardiovascular Fluid Mechanics. Ph.D. thesis, California Institute of Technology, Passadena.

Yoganathan, A.P., Corcoran, W.H., Carl, J.R., and Harrison, E.C. (1978). The Bjork-Shiley aortic prosthesis: flow characteristics, thrombus formation and tissue overgrowth. Circulation 58, 70-76.

Yoganathan, A.P. Reamer, H.H. Corcoran, W.H., and Harrison, E.C. (1979a) A laser-Doppler anemometer to study velocity fields in the vicinity of prosthetic heart valves. Med. Biol. Engng. Computing 17, 38-44.

Yoganathan, A.P., Corcoran, W.H., and Harrison, E.C. (1979b) In vitro velocity measurements in the vicinity of aortic prostheses. J. Biomech. 12, 135-152.

Yoganathan, A.P., Corcoran, W.H., and Harrison, E.C. (1979c) Pressure drops across prostletic aortic heart vlaves under steady and pulsatile flow - in vitro measurements. J. Biomech. 12, 153-164.

IN VITRO FLUID DYNAMICS OF ST. JUDE, IONESCU-SHILEY AND CARPENTIER-EDWARDS AORTIC HEART VALVE PROSTHESES

A.P. Yoganathan*, W.H. Corcoran**, E.C. Harrison+ and A. Chaux++

*School of Chemical Engineering, Georgia Institute of Technology, Atlanta, Ga 30332
**Chemical Engineering Laboratory, California Institute of Technology, Pasadena, CA 91125
+Cardiology Section, USC-LA County Medical Center, Los Angeles, CA 90033
++Dept. of Thoracic Surgery, Cedars-Sinai Medical Center, Los Angeles, CA 90048

ABSTRACT

In the study reported here the *in vitro* fluid dynamic characteristics of the St. Jude (mechanical bi-leaflet), Carpentier-Edwards (porcine) and Ionescu-Shiley (calf pericardial) aortic valve prostheses were investigated. The experiments conducted were (a) pressure drop measurements, (b) preliminary photography of the opening of the tissue valve leaflets, and (c)velocity and shear stress measurements. The pressure drop, velocity and shear stress measurements were conducted under steady flow conditions, while the preliminary photography studies were conducted under steady and pulsatile flow conditions. The pressure drop results indicated that the St. Jude and Hall-Kaster valves have the lowest pressure drops compared to any of the other valves used clinically at present. The two bioprostheses had larger pressure drops than would be expected for their basic designs. The smaller sizes of the Carpentier-Edwards valve had excessively large pressure-drops. The photographs of the opening of the valve leaflets indicated that the two bioprostheses do not open as ideally as the natural aortic valve. It was also observed that the Ionescu-Shiley aortic valves opened more symmetrically and with reproducability than the corresponding Carpentier-Edwards valves.

Detailed velocity and shear stress measurements were made with a laser-Doppler anemometer system. These measurements indicated that the flow that emerged from the leaflets of both types of tissue valves was jet-like and could lead to turbulent shear stresses on the order of 1000-3000 dynes/cm^2. Such turbulent shear stresses could be harmful to blood components. The jet type flow could also damage the the wall of the ascending aorta. Velocity measurements in the immediate downstream vicinity of the St. Jude valve showed that the flow field which emerged from the valve was centralized. The velocity measurements also indicated that there was a region of flow separation adjacent to the vessel wall and immediately downstream from the sewing ring. Such a region of flow separation could lead to excessive tissue overgrowth along the aortic side of the sewing ring. All three types of valve designs, however, created relatively low wall shear stresses on the order of 100-600 dynes/cm^2. This result is definitely a positive aspect of these valves when you consider that most of the rigid aortic prostheses we have studied created wall shears on the order of 1000-3000 dynes/cm^2.

INTRODUCTION

Prosthetic heart valves have been in clinical use since 1960. During the past two decades as many as fifty different designs of heart valves have been clinically tested, many of them unsuccessfully. At the present time the tissue bioprostheses and the tilting disc type designs of mechanical valves are the most widely used prosthetic heart valves.

Since July 1976, St. Jude Medical, Inc. has been testing and evaluating a new bi-leaflet prosthetic valve made of pyrolytic carbon. At present this valve is undergoing clinical testing and evaluation at selected medical centers throughout the world, according to FDA regulations. As of November 1979, approximately 1500 implants have been made in the aortic, mitral and/or tricuspid positions.

Tissue bioprostheses have been in clinical use since about 1968. However, it is only during the past three to five years have they gained widespread use throughout the world. Their main advantage compared to the rigid prosthetic heart valves is that they have a lower incidence of thromboembolic complications (Stinson, et al., 1977; Ionescu, et al.,

1977). The tissue bioprostheses clinically used at present do, however, have disadvantages such as: (a) relatively large pressure drops compared to some of the rigid prostheses, especially in the smaller sizes, (b) jet-like flow through the valve leaflets, (c) material fatigue or wear of the valve leaflets, and (d) valve leaflets are subject to changes that affect the natural valves, namely, lipid deposition and calcification. Problems (a) and (b) are studied in this article while (c) and (d) have been studied by Spray and Roberts (1977), Broom (1978), and Ferrans, et al. (1978).

At the present time the Hancock, Carpentier-Edwards and Angell-Shiley porcine valves, and the Ionescu-Shiley calf pericardial valve are the most widely used tissue heart valve prostheses. Approximately 75,000 heart valve prostheses are used annually throughout the world. Of these about 30% are tissue bioprostheses. It is projected that within the next two to four years this percentage will rise to 50%. As of January 1979 approximately 40,000 Hancock valves, 20,000 Carpentier-Edwards valves have been implanted throughout the world.

In the study reported here, the *in vitro* fluid dynamics of the St. Jude, Ionescu-Shiley and Carpentier-Edwards aortic heart valves were investigated. Even though the two bioprostheses (tissue valves) have been on the market since about 1975 a detailed study of their *in vitro* fluid dynamic characteristics has *not* been conducted. Recently, Grabbay, et al.(1979) studied the *in vitro* pressure drop characteristics of the Ionescu-Shiley and Carpentier-Edwards tissue valves in the mitral position. Dellsperger and Wieting (1979) have studied the pressure drop and leakage characteristics of the St. Jude valve. Until now no velocity or shear stress measurements in the immediate downstream vicinity of the St. Jude, Ionescu-Shiley and Carpentier-Edwards aortic valves have been reported in the open literature.

EXPERIMENTAL APPARATUS AND PROCEDURES

Table 1 lists the aortic heart valves used in this study together with their sewing ring and stent or primary orifice diameters.

Table 1

Heart Valves Used in Study

Name of Valve	Sewing Ring Diameter mm	Stent or Primary Orifice Diameter mm
Ionescu-Shiley	27	23.4
Ionescu-Shiley	25	21.4
Ionescu-Shiley	21	17.4
Carpentier-Edwards	27	23.5
Carpentier-Edwards	25	23.1
Carpentier-Edwards	21	17.5
St. Jude	27	22.3
St. Jude	25	20.4

In Vitro Velocity and Pressure Drop Measurements

In vitro velocity and shear stress measurements were made in the immediate downstream vicinity of a #27 (aortic) Ionescu-Shiley, a #27 (aortic) Carpentier-Edwards and a #27 (aortic) St. Jude valve prosthesis. The velocity measurements were made using a dual beam laser-Doppler anemometer (LDA) system in both forward and back-scatter modes. A description of the LDA system and its general principles, and a detailed description of the experimental technique have been published previously (Yoganathan, et al. 1979a and 1979b).

Pressure drop studies were conducted with St. Jude valves having sewing ring diameters of 27 and 25 mm, and Ionescu-Shiley and Carpentier-Edwards valves with sewing ring diameters of 27, 25 and 21 mm. The pressure drop measurements were made with Statham P23 De and P23 Db transducers, and a water manometer system. A detailed description of the pressure drop measuring technique has been published elsewhere (Yoganathan, et al. 1979c).

The velocity and pressure drop studies were conducted under steady flow conditions. The tissue valves were tested in saline solution at 22°C, and the mechanical valves in 6% by weight Polyol V-10 (Wyandotte Chemical) aqueous solution at 32°C. The saline has a viscosity 0.01 dyne sec/cm^2, and the Polyol solution a viscosity of 0.035 dyne sec/cm^2. The experiments were conducted in an aortic flow channel which simulated the root of the ascending aorta, the sinuses of Valsalva and the ascending aorta. Figure 1 shows a schematic of an aortic flow channel which was fairly similar to the one that was used in this study. The aortic heart valve chamber was built to accept a prosthetic aortic heart valve with a maximum sewing ring diameter of 27 mm.

The valves are mounted on circular metallic discs that have been carefully machined, and the sewing rings of the valves fit the orifices for the respective discs. The outer dimensions of all the discs are identical and fit into the heart valve chamber. The aortic valves are placed in the heart-valve chamber such that the front end of their sewing rings coincide with the X = 0 position as show in Figure 1. Table 1 lists the sewing ring (tissue annulus) diameters and the stent orifice diameters of the valves used in the study.

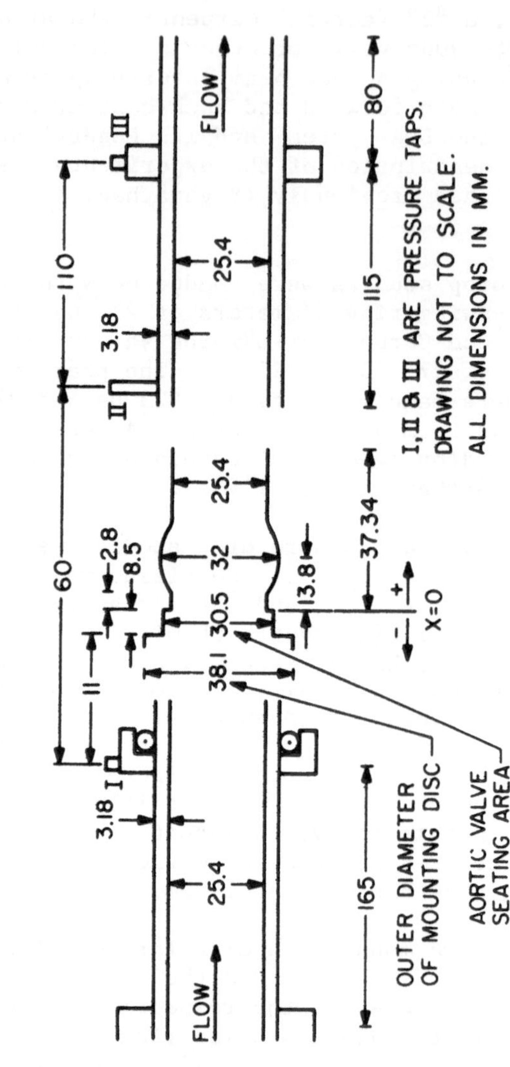

Fig. 1. Schematic of aortic flow channel.

The velocity and pressure drop experiments were conducted over a flow rate range of 167 to 417 cm^3/sec. These steady flow rates correspond to peak systolic flow rates at cardiac outputs of about 2.0 to 5.5 liters/min.

Photography of the Opening of the Valve Leaflets

Photographs of the opening of the valve leaflets were taken under steady and pulsatile flow conditions. The photographs were taken looking down through the downstream section ('C) of the aortic flow channel, with an Asahi-Pentax Electrospotmatic 35 mm camera. A shutter speed of 1/1000 of a second was used with Ektachrome (ASA #160) Tungstan film. Tungstan movie lamps were used to illuminate the prosthetic heart valve.

In the pulsatile flow experiments the camera was synchronized with the timer on the pulse duplicator system. By using an electronic delay circuit unit it was possible to take photographs of the motion of the valve leaflets at any given instant during the heart cycle. The instant the photograph was taken an electronic pulse signal was omitted form the stroboscope connection on the camera body. Figure 2 is a schematic drawing of the photography technique employed under pulsatile flow conditions. The electronic pulse was recorded on a dual beam storage oscilloscope (Tektronix 466) together with either the aortic flow or aortic pressure curve. Subsequently a polaroid photograph of the stored electronic pulse signal and the aortic flow or pressure curve was taken. The pulsatile flow experiments were conducted in the Caltech pulse duplicator system at cardiac outputs of about 4.0 to 6.5 liters/min. Figure 3 is a schematic drawing of the Caltech pulse duplicator system. A detailed description of it can be found elsewhere (Yoganathan, 1978). The steady flow experiments were conducted over a flow rate range of 21 to 417 cm^3/sec. By planimetering the photographs it was possible to calculate the area of opening of the valve leaflets.

RESULTS AND DISCUSSION

Pressure Drop Measurements

Pressure drop measurements made in the immediate downstream vicinity of the various sizes of St. Jude, Ionescu-

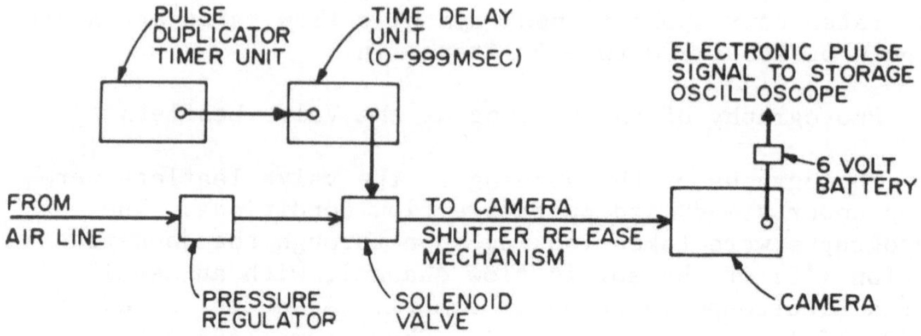

Fig. 2. Schematic of photography system used for pulsatile flow experiments.

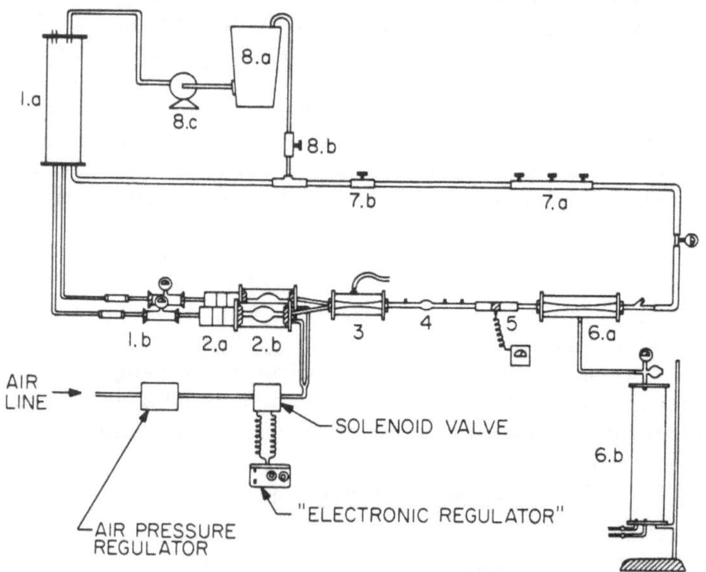

Fig. 3. Schematic of Caltech pulse duplicator system.

Shiley and Carpentier-Edwards valve prostheses are shown in Figures 4-6. The pressure drops shown are those measured between taps I and II (Δp_1) except those indicated by dotted lines. The dotted lines represent pressure drop measurements between taps I and III (Δp_2).

The pressure-drop measurements indicated that the St. Jude valve had a much smaller pressure drop compared to the other designs of prostheses tested. At a steady flow rate of 417 cm^3/sec the #27 St. Jude valve had a pressure drop of about 5.2 mm Hg while the #25 had a pressure drop of about 6.2 mm Hg. The above conclusion was also confirmed by post-operative hemodynamic clinical data obtained by one of the authors (A.C.). The results also indicate quite clearly that the two bioprostheses cause larger pressure drops compared to the low profile mechanical valves (St. Jude, Hall-Kaster, Björk-Shiley) of the corresponding size. The difference in the pressure drops between the tissue valves and the low profile mechanical valves becomes larger as the valve size becomes smaller. The Carpentier-Edwards porcine valve had significantly higher pressure drops compared to the Ionescu-Shiley calf pericardial valve of the corresponding size. As illustrated in Figures 5 and 6 the Δp_2 measurements across the Carpentier-Edwards #25 and #21 valves were higher than the Δp_1 measurements for the Ionescu-Shiley #25 and #21 valves, respectively. For the #27 size valves the Δp_2 results for the Carpentier-Edwards valve and the Δp_1 results for the Ionescu-Shiley valve lay almost on the same curve. The above findings were surprising since it was not expected that the Carpentier-Edwards porcine prostheses would exhibit such large pressure drops. The comparatively large pressure drops observed across the tissue valves can be mainly attributed to: (i) the elastic characteristics of the valve leaflets and (ii) the restricted downstream opening of the leaflets due to structural designs of the valves.

Photography Studies

The preliminary photography studies indicate very clearly that the leaflets of the two bioprostheses do not open as ideally as the leaflets of the natural aortic valve. Table 2 shows the amount of opening of the valve leaflets at various steady flow rates. The results given in Table 2

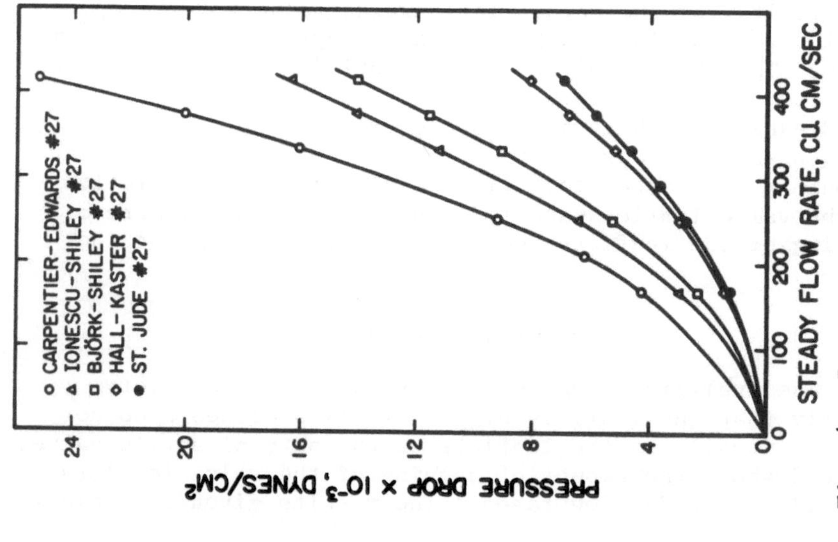

Fig. 4. Pressure drop measurements across size #27 aortic heart valves

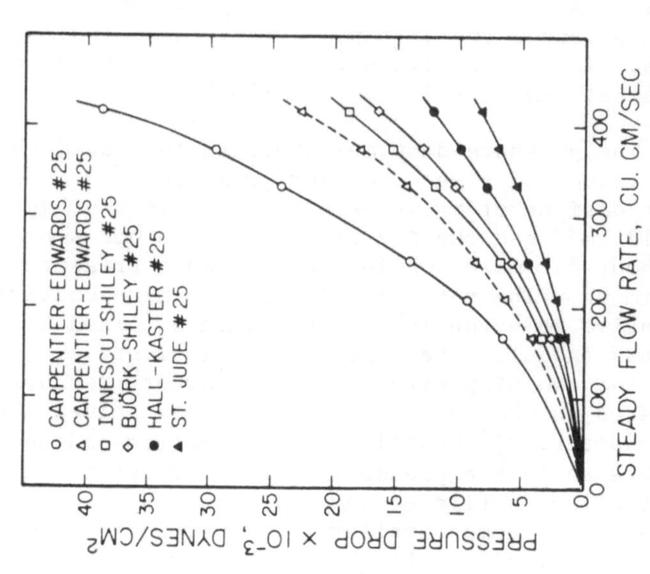

Fig. 5. Pressure drop measurements across size #25 aortic heart valves.

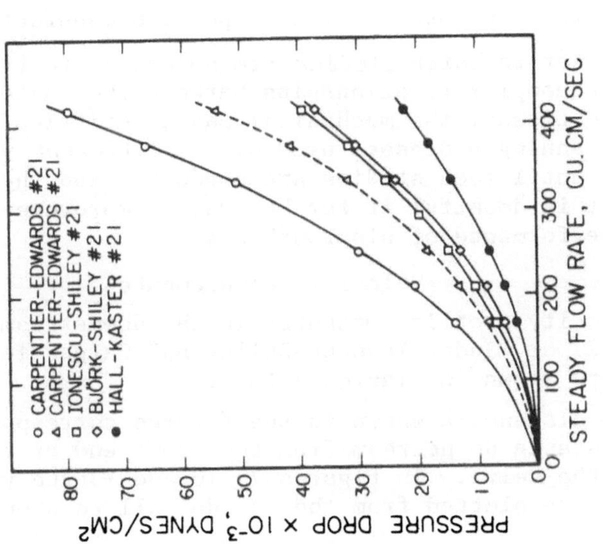

Fig. 6. Pressure drop measurements acros size #21 aortic heart valves.

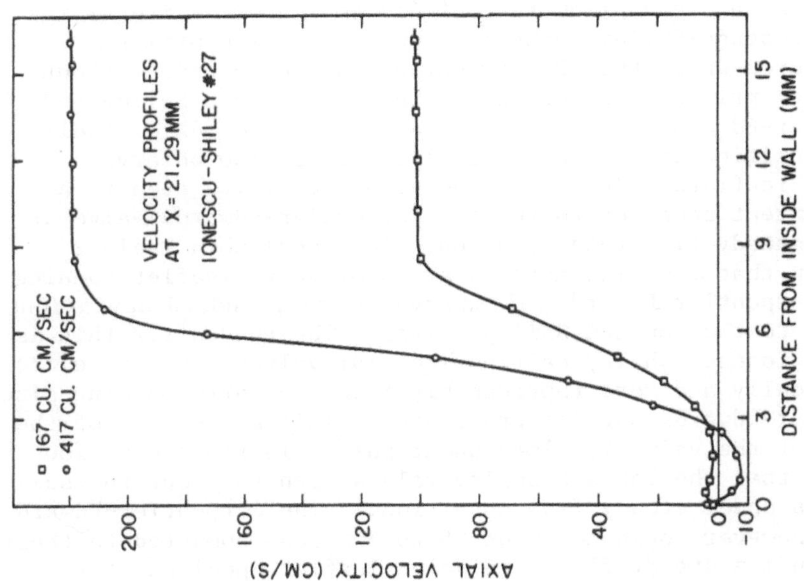

Fig. 7. Velocity profiles 21.29 mm downstream from the Ionescu-Shiley valve.

are the average values for at least four independent measurements. At a steady flow rate of 417 cm^3/sec (25 liters/min) the #27 Ionescu-Shiley opens to about 67% of its primary orifice area while the #27 Carpentier-Edwards opens to about 61% of its primary orifice area. The #25 Ionescu-Shiley and Carpentier-Edwards valves open to about 74% and 54% of their primary orifice areas, respectively. It is also observed that the leaflets of the #25 Ionescu-Shiley valve open to a larger extent compared to the #27 Carpentier-Edwards valve at all the steady flow rates studied. The results in Table 2 also show that the planimetered areas of valve leaflet opening for the Carpentier-Edwards valves have larger standard deviations compared to the Ionescu-Shiley valves. The reason for this is that the Ionescu-Shiley calf pericardial valves open more symmetrically and more reproducibly than the corresponding size Carpentier-Edwards porcine prostheses. Initial results of the opening of the valve leaflets under pulsatile flow conditions indicate that the Ionescu-Shiley valves open to about the same extent as under steady flow conditions. The Carpentier-Edwards valves, however, open to about 25 to 30% less compared to their corresponding steady flow rate valve leaflet openings. It should be noted that these results are preliminary and detailed work in this area is in progress at present. The opening characteristics of the valve leaflets can be attributed to their anatomical and elastic characteristics, as well as the tanning processes used by the respective manufacturers.

Further research studies are necessary to try and understand the complex relationships between the anatomic nature of the valve tissue, the mechanical characteristics of the tissue, and the tanning processes used by the different valve manufacturers. Until such studies are conducted and the results understood, it is doubtful if further major improvements can be made in the performance of bioprostheses.

Velocity Measurements

Velocity profiles measured in the downstream vicinity of the #27 St. Jude, Ionescu-Shiley and Carpentier-Edwards valves are shown in Figures 7 to 15.

The distance X shown in the figures corresponds to the axial distance downstream from the front end of the sewing ring of the valve. In Figures 7, 10 and 13 the velocity profiles are plotted from the inside wall to about the center

Table 2

Extent of Opening of Valve Leaflets Under Steady Flow Conditions

Name of Valve	Stent Orifice Area cm^2	Approximate Primary Orifice Area, cm^2	Planimetered Area of Opening of the Valve Leaflets, cm^2 (Steady Flow Rate, cm^3/sec)				
			(417)	(333)	(167)	(83)	(21)
Ionescu-Shiley #27	4.30	4.15	2.80 ± 0.03	2.68 ± 0.04	2.34 ± 0.03	2.11 ± 0.03	1.51 ± 0.02
Carpentier-Edwards #27	4.34	4.15	2.55 ± 0.28	2.21 ± 0.24	1.77 ± 0.20	1.71 ± 0.24	1.17 ± 0.23
Ionescu-Shiley #25	3.60	3.46	2.59 ± 0.02	2.41 ± 0.04	2.18 ± 0.02	1.95 ± 0.05	1.50 ± 0.02
Carpentier-Edwards #25	4.19	3.46	1.88 ± 0.28	1.55 ± 0.24	1.26 ± 0.20	1.18 ± 0.17	0.93 ± 0.05

of the flow channel in order to accentuate the velocity characteristics immediately downstream of the three prostheses. Those three (Figures 7, 10, 13) sets of profiles were measured in the sinus region (section 'B') of the flow channel. Figures 8, 9, 11, 12, 14 and 15 show velocity profiles which are plotted from the inside wall to the outside wall of the flow channel. All the profiles shown were obtained in the horizontal plane through the center of the flow channel.

Velocity measurements made in the immediate downstream vicinity of the St. Jude valve show that the flow field that emerges from the valve is centralized (see Figs. 13-15). Note that the velocity measurements shown in Figures 13 and 14 were made in the sinus region of the flow channel. The velocity measurements also indicated that there was a region of flow separation adjacent to the vessel wall and immediately downstream from the sewing ring as can be seen in Figure 13. Such a region of flow separation could lead to tissue overgrowth along the aortic side of the sewing ring of the valve. The St. Jude valve did, however, create relatively low wall shear stresses on the order of 50 to 600 dynes/cm^2 in the immediate downstream vicinity of the prosthesis. This result is definitely a positive aspect of the St. Jude valve when you consider that the other rigid prostheses we have studied produced wall shears on the order of 500-3000 dynes/cm^2(Yoganathan et al., 1978). The flow field immediately downstream of the valve was turbulent and let to Reynolds shear stresses on the order of 100 to 600 dynes/cm^2. These bulk turbulent-shear stresses are lower than those created by the other rigid and biological prostheses studied.

Figures 7 and 10 show very distinctly that the flow that emerges from the two bioprostheses is jet like. The profiles shown in these two figures were measured immediately (∿2mm) downstream of the fully open valves. At these locations it was possible to obtain complete velocity profiles (inside wall to outside wall) in the forward scatter mode. The jet type flow that emerged from the two valves led to relatively large bulk turbulent (Reynolds) shear stresses. Initial experimental results indicate turbulent shear stresses on the order of 1000-3000 dynes/cm^2 at a steady flow rate of 417 cm^3/sec. In Figures 7 and 10 the maximum turbulent shears occurred at the distance of about 6.0 to 8.0 mm and 6.5 to 8.5 mm from the inside wall at steady volumetric flow rates of 417 and 167 cm^3/sec, respectively. The Ionescu-Shiley and

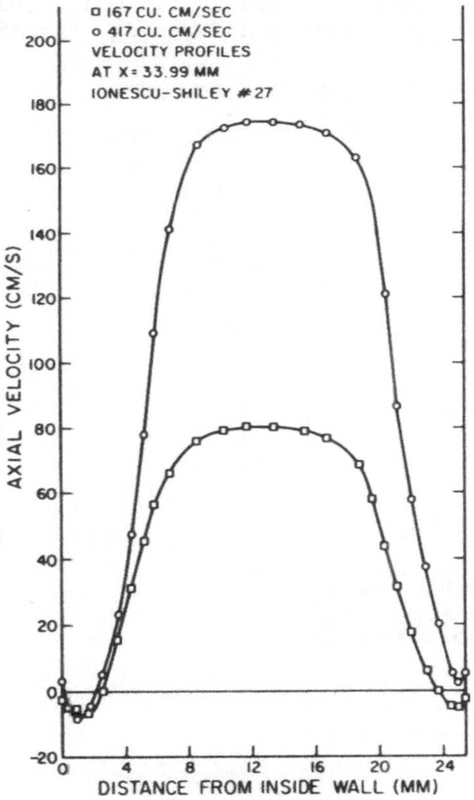

Fig. 8. Velocity profiles 33.99 mm downstream from the Ionescu-Shiley valve.

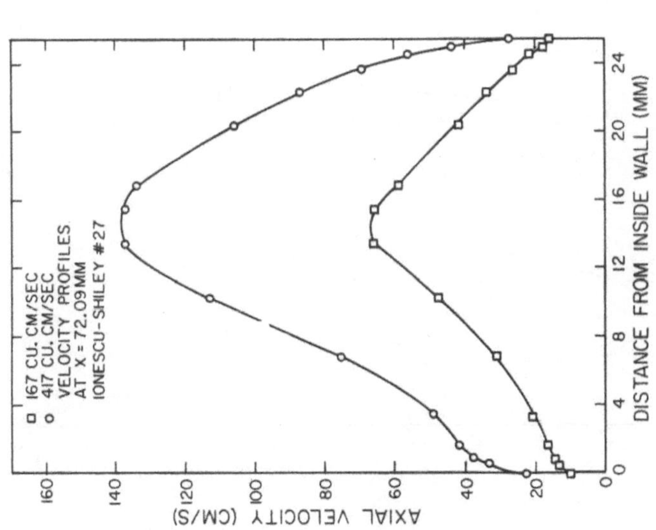

Fig. 9. Velocity profiles 72.09 mm downstream from the Ionescu-Shiley valve.

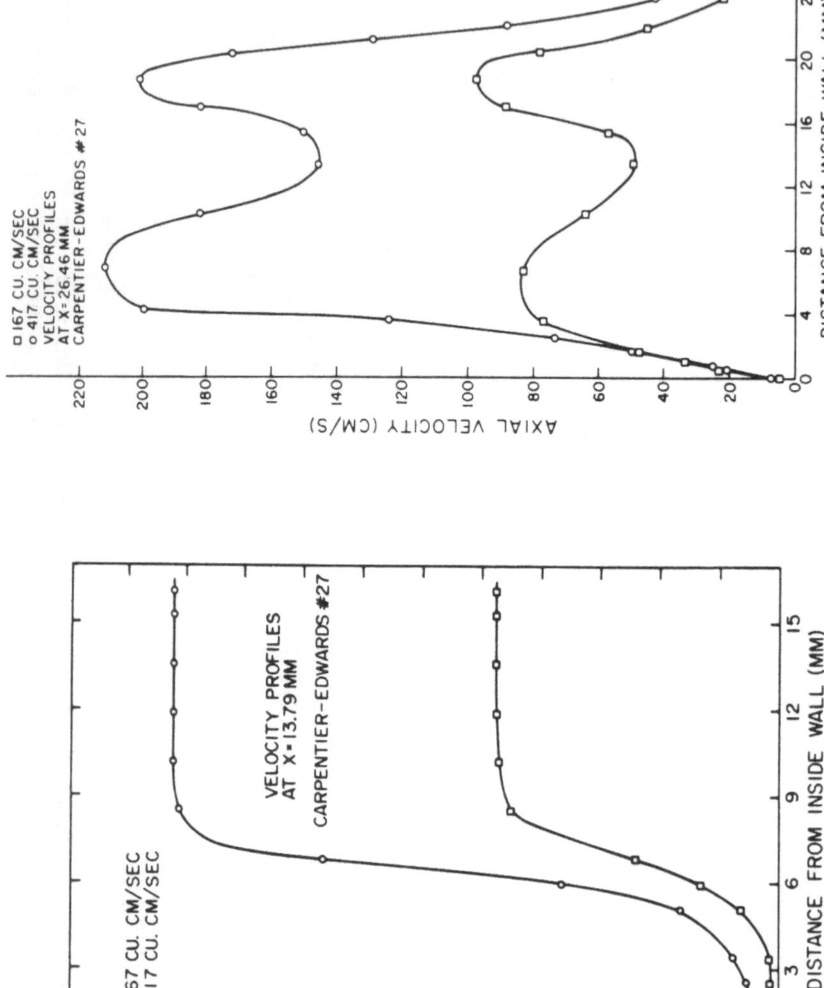

Fig. 10. Velocity profiles 13.79 mm downstream from the Carpentier-Edwards valve.

Fig. 11. Velocity profiles 26.46 mm downstream from the Carpentier-Edwards valve.

Carpentier-Edwards tissue valves created relatively low wall-shears compared to the majority tissue valves created relatively low wall-shears compared to the majority of the mechanical aortic prostheses we have studied (Yoganathan, et al., 1978). The maximum measured wall-shears in the sinus region (section 'B') of the aortic flow channel were 517 and 206 dynes/cm^2 for the Ionescu-Shiley valve at steady flow rates of 417 and 167 cm^3/sec, respectively. The corresponding maximum wall-shears for the Carpentier-Edwards valve were 582 and 240 dynes/cm^2, respectively.

Figures 8 and 11 indicate that the flow field downstream of the Ionescu-Shiley is more symmetrical compared to that of the Carpentier-Edwards valve. In addition, the velocity defect type profiles observed in Figure 11 highlight the asymmetric opening characteristics of the leaflets of the Carpentier-Edwards valve. The asymmetric opening was also observed in the photography experiments. The symmetrical profiles observed with the Ionescu-Shiley prosthesis (Figure 8) reflect to a large extent the symmetric opening characteristics of that valve. Figures 9 and 12 indicate that even 70 mm downstream of the tissue valves the velocity profiles have not reached a steady state. They are still evolving toward turbulent pipe flow profiles.

The bulk turbulent shears (1000-3000 dynes/cm^2) created immediately downstream of the two tissue valves could damage blood components such as red-cells and platelets (Hellums and Brown, 1977; Anderson, et al. 1978, Ramstack, et al., 1979). The damage caused by such shear stresses to the blood components could be either sub-lethal or lethal. It should be noted that even sub-lethal damage to red-cells and platelets could over a long enough time period lead to hemolysis and thromboembolic complications. In addition, the jet type flow which emerges from the two bioprostheses could damage the wall of the ascending aorta if the jet impinges on the wall. Depending on the anatomy of the valve patient's ascending aorta, the impinging jet could cause considerable damage to the wall in and around the location of impingement. Damage to the endothelial lining of the ascending aorta could in turn lead to thromboembolic complications and hemolysis.

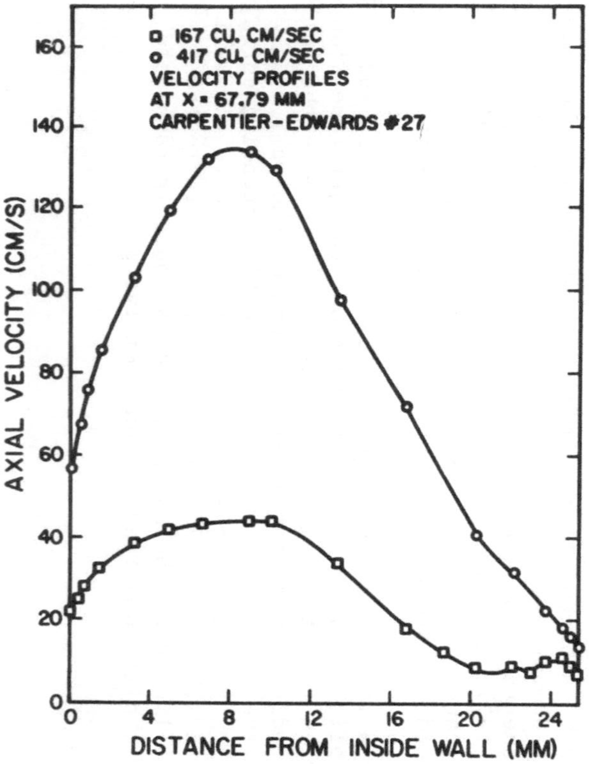

Fig. 12. Velocity profiles 67.79 mm down-stream form the Carpentier-Edwards valve.

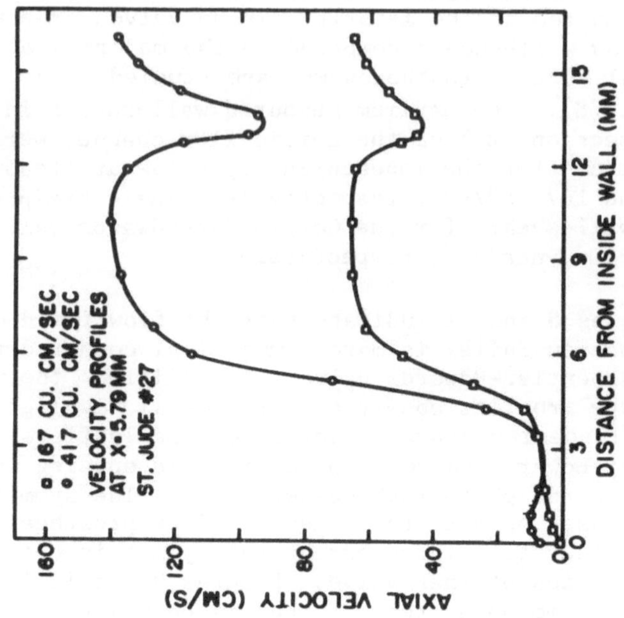

Fig. 13. Velocity profiles 5.79 mm down-stream from the St. Jude valve.

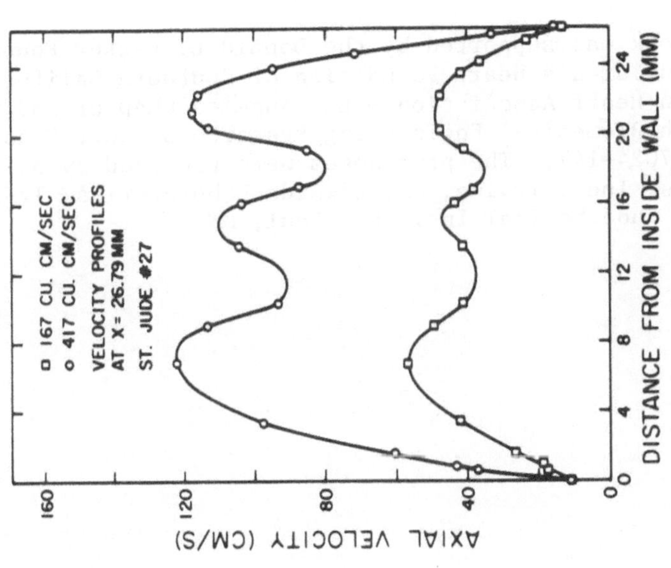

Fig. 14. Velocity profiles 26.79 mm downstream from the St. Jude valve.

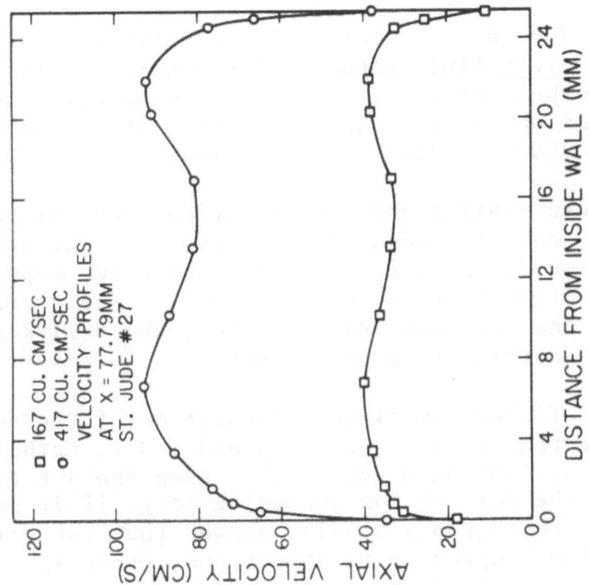

Fig. 15. Velocity profiles 77.79 mm downstream form the St. Jude valve.

CONCLUSIONS

The St. Jude aortic heart valve appears to have very favorable *in vitro* fluid mechanical characteristics when compared to other aortic valves we have studied. The mechanical and durability characteristics of the valve are questionable and have yet to be determined.

The Ionescu-Shiley and Carpentier-Edwards aortic bioprostheses do not have ideal fluid dynamic characteristics. The steady flow pressure drops across these two types of valves were larger than those observed with the corresponding sizes of the St. Jude, Bjork-Shiley, and Hall-Kaster low profile, mechanical aortic prostheses.

The flow fields immediately downstream of the two tissue valves were turbulent and jet like, and led to turbulent shears on the order of 1000-3000 dynes/cm^2. Also the jet type flow could damage the wall of the ascending aorta if it impinges on the wall. Photography studies showed that the valve leaflets of the two bioprostheses do not open properly. The opening characteristics of the two valves, especially that of the Carpentier-Edwards prostheses leave much to be desired.

ACKNOWLEDGEMENTS

This work was supported by the Donald E. Baxter Foundation, the Children's Heart Foundation of Southern California, the American Heart Association - Los Angeles Chapter,and the Georgia Tech Biomedical Engineering Research Support Grant (#5 S07 RR07024-14). The prostheses were provided by Shiley Laboratories, Inc., Irvine, CA, Edwards Laboratories, Irvine, CA, and St. Jude Medical Inc., St. Paul, MN.

REFERENCES

Anderson, G. H., Hellums, J. D., Moake, J., and Alfrey, C.P. (1978) Platelet response to shear stress: changes in serotonin release, and ADP induced aggregation, Thrombosis Res. 13, 1039-1047.

Broom, N. D. (1978) Fatigue induced damage in glutaraldehyde-preserved heart valve tissue, J. Thorac. Cardiovasc. Surg. 76, 202-211.

Dellsperger, K. C., and Wieting, D. W. (1979) Presented at the 14th Annual AAMI Meeting, Las Vegas.

Ferrans, V. J., Spray, T. L., Billingham, M. E., and Roberts, W. C. (1978) Structural changes in gluteraldehyde-treated porcine hetrografts used as substitute cardiac vlaves, Am. J. Cardiol. 41, 1159-1184.

Gabbay, S., McQueen, D. M., Yellin, E. L., and Frater, R. W. M. (1979) In vitro hydrodynamic comparison of mitral valve bioprostheses. To be published in Supplement to Circulation, Cardiovascular Surgery.

Hellums,J. D., and Brown III, C. H. (1977) Blood cell damage by mechanical forces. Cardiovascular Flow Dynamics (Edited by N. H. C. Hwang and N. A. Norman) University Park Press, Baltimore, Maryland.

Ionescu, M. I., Tanden, A. P., Mary, D. A. S., Abid, A. (1977) Heart valve replacement with the Ionescu-Shiley pericardial xenografts.J. Thorac. Cardiovasc. Surg. 73, 31-42.

Ramstack, J. M., Zuckerman, L., and Mockros, L. F. (1979) Shear induced activation of platelets, J. Biomech. 12, 113-125.

Spray, T. L., and Roberts, W. C. (1977) Structural changes in porcine xenografts used as substitute cardiac valves, Am. J. Cardiol. 40, 319-330.

Stinson, E. B., Griepp, R. B., Oyer, P. E., and Shumway, N.E. (1977) Long-term experience with porcine aortic valve xenografts, J. Thorac. Cardiovasc. Surg. 73,54-63.

Yoganathan, A. P., (1978) Cardiovascular fluid mechanics, Ph.D. Thesis, California Institute of Technology.

Yoganathan, A. P., Corcoran, W. H., and Harrison, E. C. (1978) Wall shear stress measurements in the near vicinity of prosthetic aortic heart valves, J. Bioeng. 2, 369-379.

Yoganathan, A. P., Corcoran, W. H., and Harrison, E.C. (1979a) In vitro velocity measurements in the vicinity of aortic prostheses. J. Biomech. 12, 135-152.

Yoganathan, A. P., Reamer, H. H., Corcoran, W.H., and Harrison, E. C. (1979b) A laser-Doppler anemometer to study velocity fields in the vicinity of prosthetic heart valves, Med. Biol. Eng. & Comput. 17, 38-44.

Yoganathan, A. P., Corcoran, W. H., and Harrison, E. C.(1979c) Pressure drops across prosthetic aortic heart valves under steady and pulsatile flow - in vitro measurements, J. Biomech 12, 153-164.

THE ROLE OF THE TRAPPED SINUS VORTEX IN AORTIC VALVE CLOSURE

A.A. van Steenhoven and M.E.H. van Dongen

Eindhoven University of Technology

Eindhoven, the Netherlands

INTRODUCTION

The aortic valve is one of the four valves controlling the fluid motion through the heart. It is positioned at the outlet of the left ventricle, which pumps blood into the aorta. This valve is shown diagrammatically in figure 1. It has three leaflets (cusps) and behind each leaflet there is a cavity, the sinus of Valsalva. The leaflets are very thin (0.1-0.3 mm), non-muscular and very flexible in the axial direction.

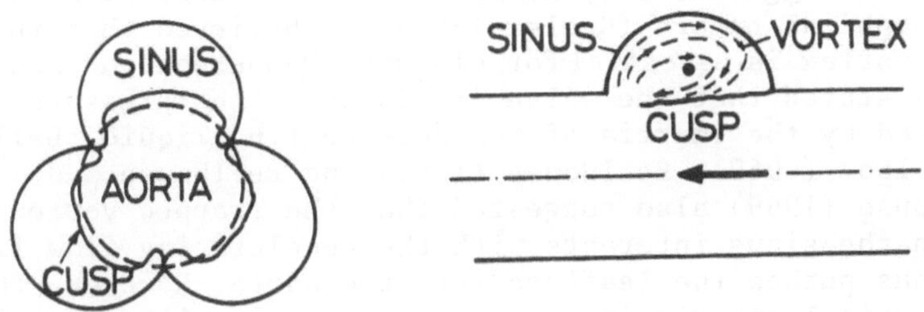

Figure 1.
Diagram of the aortic valve and the sinus of Valsalva.

The typical shape of the aortic valve and the role of the sinuses of Valsalva for a proper valve function have attracted scientific interest ever since Leonardo da Vinci made his first reports in 1513 (Keele, 1952). He examined the structure of the valve and made speculative drawings of flow patterns, which by now prove to be remarkably realistic. He was probably the first to mention the existence of a trapped vortex in the sinus during systole, although the author did not have any knowledge of the blood circulation. In 1912, Henderson and Johnson performed some simple experiments to study cardiac valve behaviour. They found that under pulsatile flow conditions a valve moves gradually towards closure during flow deceleration and finally closes without any regurgitation. More recently, Bellhouse and Talbot (1969) recognized the important function of the sinus cavity in closing of the aortic valve. From their model experiments, they concluded that indeed closing of the valve already starts during the deceleration phase of systolic aortic flow. At the end of systole the valve is almost closed. Early in diastole a small reversed flow of about 2% of the total forward flow then readily completes the closure. In the absence of sinuses the ratio of reversed and forward flow increases to about 25% (Bellhouse and Talbot, 1969). The significance of systolic gradual valve closure for the mechanical load on the valve leaflets has been discussed by Spaan et al. (1975).

The mechanism of the onset of valve closure during the deceleration of the main flow is not yet fully understood. The first suggestions by Leonardo da Vinci were of a rather philosophical nature (Keele, 1952). He believed that the sinus vortex is essential for closure. Henderson and Johnson (1912) stated that the valve is closed by the pressure produced by the inertia of the decelerating liquid. Bellhouse and Talbot (1969), Bellhouse (1969) and Bellhouse and Bellhouse (1969) also suggested that the trapped vortex within the sinus interacts with the decelerating flow field and thus pushes the leaflets into the aorta. However, their mathematical description of this interaction is not entirely satisfactory; their model results in pressure differences across the cusps which seem to be quite large in view of the small mass of the leaflets. Talukder et al. (1977) stated that aortic valve closure is primarily accomplished by the adverse pressure gradient, which simultaneously decelerates the flow and makes the valve to close gradually. They also suggest that the sinus vortices do not appear to play any essential role in valve closure. Swanson et al. (1978) studied the behaviour of a free moving potential vortex within a bounded stationary mainstream, without taking into account the geometry of the sinuses and the possible effects of the trapped vortices. They suggested that the unsteady vortex motion is one of the initiating mechanisms of valve closure during systolic aortic flow deceleration. Van Steenhoven and van Dongen (1979) investigated experimentally the closing behaviour of the aortic valve in model studies. They found that the flow phenomena and the cusp motions are strongly dependent on the Strouhal number and proposed simple theoretical models based on the phenomena observed. They suggest that valve closure is mainly induced by the adverse pressure gradient during flow deceleration and that the trapped vortex does not affect valve closure essentially. Similar results were reported by Lee and Talbot (1979), who focussed their attention mainly on the closing behaviour of the mitral valve.

In the present paper attention is focussed on the influence of the sinus vortex on valve closure. To that end, experiments were performed in a two-dimensional geometry. First some work has been done to characterize the steady flow field in the sinus. Next, the influence of the vortex strength on valve closure was studied. The cusp displacement during deceleration of the mainstream was recorded for two conditions in which the strength of the sinus vortex was apparently different.

THE EXPERIMENTAL SET-UP

The two-dimensional analogue of the aortic valve has been designed as a rectangular duct with a half-cylindrical cavity in the upper wall, both made of lucite, as shown in figure 2. A foil, 2 μm thick (Makrofol KG), is used as cusp material. The cusp length is 1.5 times the sinus radius R. The test fluid is water and the channel has a fixed height of 4.5 cm and a width of 12 cm. The Reynolds number for the analogue, defined as Uh_o/ν, where U denotes the maximum velocity of the mainstream, h_o the channel height and ν the kinematic viscosity, has a value of about 3000 which corresponds approximately to the physiological value. With the aid of a time-dependent fluid-dynamic resistance at the downstream end of the duct, the mainstream can be decelerated

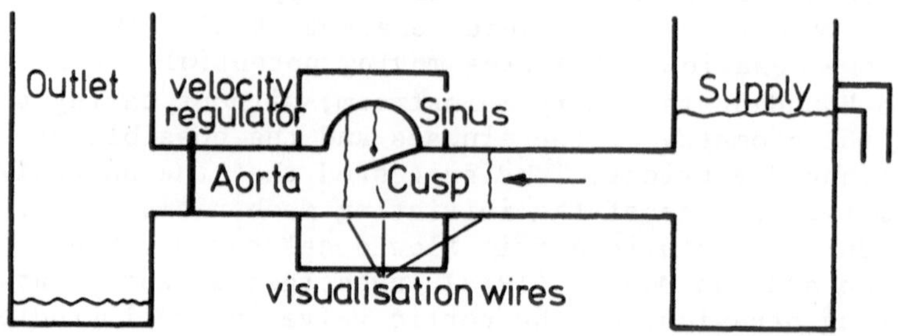

Figure 2.
Diagram of the two-dimensional set-up of the aortic valve.

at a constant rate from the value U at t=0 to zero at t=τ. For the present experiments the deceleration time for the analogue τ is taken such, that the Strouhal number R/Uτ corresponds to the physiological situation (τ=10s).

The velocity profiles and instantaneous velocities in the mainstream and sinus were measured using an anemometer system (TSI-1054B). The sensor is a cone-shaped hot-film model (1462) with a maximum diameter of 1.5 mm. The control system maintains a constant temperature difference of 40°C between sensor and its environment. Environmental temperature drift is automatically compensated by means of compensation probe 1310. The anemometer system is provided with a linearized output. To avoid problems with the directional sensitivity the sensor was placed perpendicular to the streamlines. The flow direction for the steady state could be determined by means of hydrogen-bubble flow visualization.

The cusp motion was recorded on ciné film. To quantify the cusp displacements, we introduced a closing parameter λ, defined as the ratio of the instantaneous opening beneath the cusp and the orifice of the channel. When this parameter equals one the valve is completely open, when it is zero the valve is closed. The parameter λ was measured at distinct times by means of a film motion analyser.

THE INFLUENCE OF THE VORTEX STRENGTH ON VALVE CLOSURE

In a steady state, the sinus fluid shows a vortex-like motion which can more or less be described as effectively inviscid with a constant vorticity except in the shear layers along the boundaries. This is in agreement with Batchelor's model for recirculating flow (Batchelor, 1956; Nallasamy and Krishna Prasad, 1977). The velocities in the steady sinus vortex are small compared to the mainstream velocity, even in the absence of a cusp. An example of such a velocity profile in the sinus is shown in figure 3. The maximum sinus velocity appears to be about 15% of the mainstream velocity; when the cusp is present this ratio is only 5%.

A much stronger sinus vortex can be obtained by accelerating the mainstream. The viscous forces along the wall and the leaflet then generate a strong vortex which is trapped in the sinus cacity. This flow field can be described numerically by representing the trapped fluid within the sinus cavity by a number of modified point vortices which are convected by the fluid and undergo a random walk representing viscous diffusion (Peskin and Wolfe, 1978).

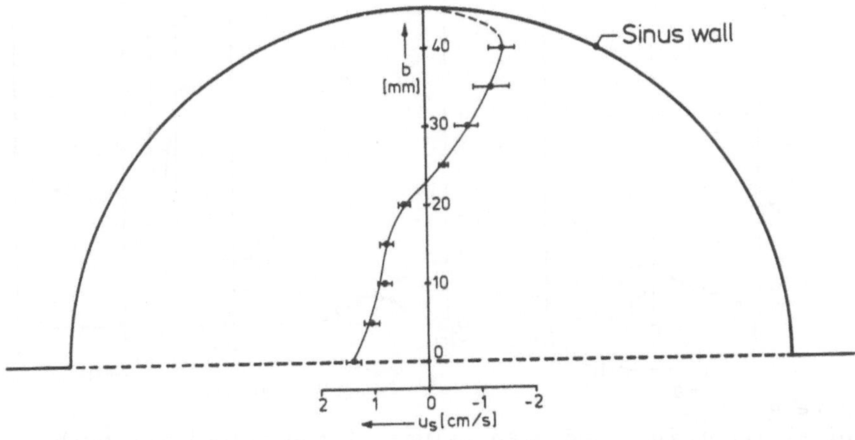

Figure 3.
Velocity distribution within the sinus in the absence of a cusp.
(U=7 cm/s, R=4.5 cm).

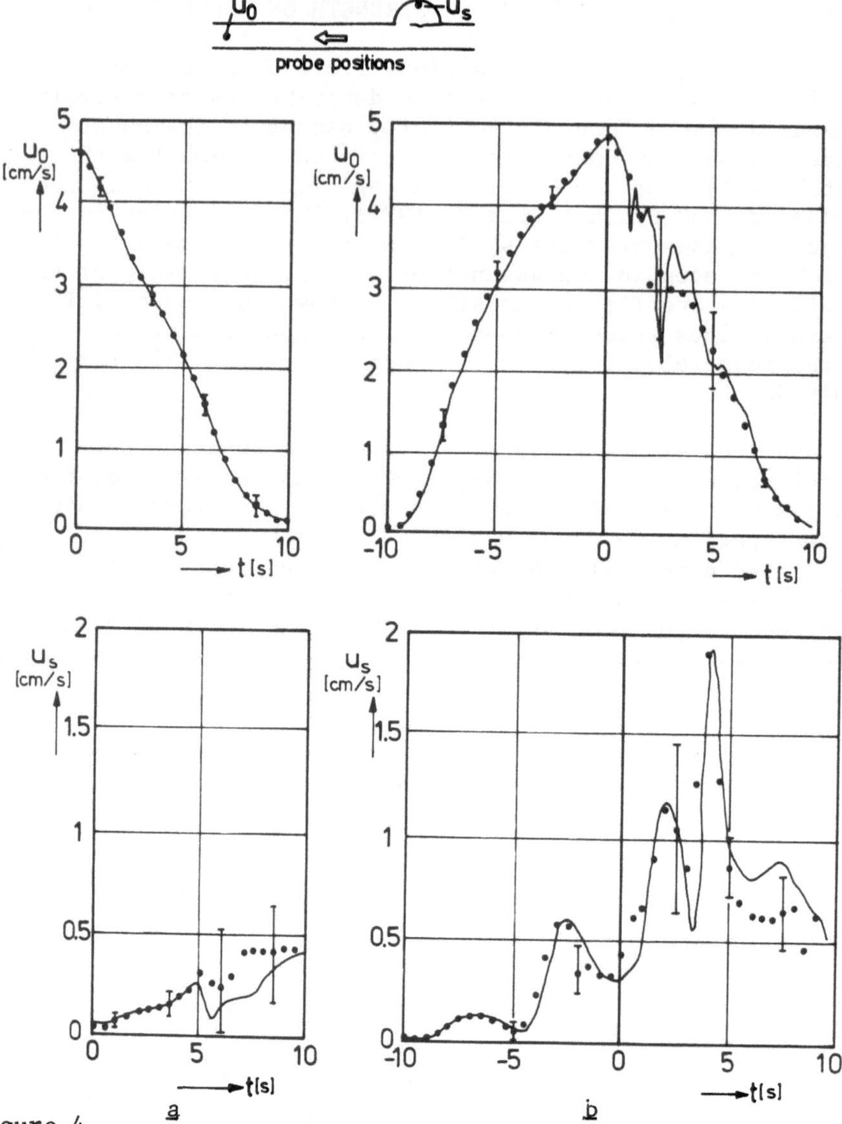

Figure 4.
Typical recordings and mean values of the velocities both of the mainstream and at the top of the sinus for:

a. deceleration of the mainstream starting from a steady state,

b. deceleration of the mainstream preceded by an acceleration. (R=4.5 cm).

To investigate the influence of the strength of the sinus vortex on the closing behaviour two different experiments have been done, namely:

i) A deceleration starting from the steady state.

ii) The same deceleration preceded by a practically constant acceleration of the mainstream fluid. The acceleration and deceleration times are equal.

Typical recordings of the instantaneous mainstream and sinus velocities are shown as the solid lines in figure 4. The points with their 95% confidence interval correspond to the mean of four repeated measurements. This graph shows that during the deceleration phase the strength of the sinus vortex, in case that the deceleration follows after an acceleration, exceeds the vortex strength if starting from a steady state.

The experimentally determined cusp displacements, in terms of λ as determined from four measurements ($\Delta\lambda=0.02$), are shown in figure 5 for both conditions mentioned. From this graph it is concluded that the strength of the vortex does not essentially affect the onset of valve closure.

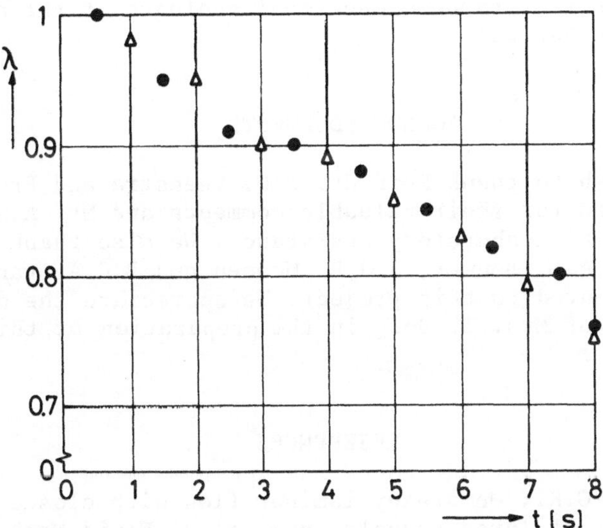

Figure 5.
Valve closure for the conditions: a (•) and b (Δ) as explained in figure 4.
($U=5$ cm/s, $\tau=10$s, $R=4.5$ cm).

CONCLUDING DISCUSSION

The closing behaviour as a result of flow deceleration has been studied for two different initial conditions. When the deceleration is preceded by a steady state, a relatively weak vortex is present in the sinus. However, when the fluid flow is accelerated before the onset of deceleration a much stronger vortex will be trapped in the sinus cavity. The closing rate is not significantly different for both situations. Therefore it can be concluded that the strength of the vortex is not an important parameter in the description of valvular closure. This conclusion is supported by the results of the experiments of van Steenhoven and van Dongen (1979), showing that the sinus depth can be remarkably decreased without a noticeable effect on valve closure. On the other hand, this conclusion is only justified for the two-dimensional case. According to Bellhouse and Talbot (1969), the fluid in the three-dimensional cavity enters the sinus along the plane of symmetry of each aortic leaflet and leaves the sinus near the commissures after being swept around the vortex. It has still to be established whether this three-dimensionality of the flow field is essential for the fluid dynamics of aortic valve closure. Therefore, similar experiments in a three-dimensional analogue of the aortic valve are under way.

ACKNOWLEDGEMENTS

We wish to thank Prof.dr. P.C. Veenstra and Prof.dr. R.S. Reneman for their valuable comments and Mr. A.A.M. Wasser for his laboratory assistance. We also thank the students T.A.M. Beumer, J.J.E. Moonen and B.C.A. van de Pas, who contributed to this project. We appreciate the dedicated assistance of Mrs. I. Borg in the preparation of this manuscript.

REFERENCES

Batchelor, G.K.: On steady laminar flow with closed streamlines at large Reynolds number. J. Fluid Mech. 1 (1956), pp. 177-190.

Bellhouse, B.J.: Velocity and pressure distributions in the aortic valve. J. Fluid Mech. 37 (1969), pp. 587-600.

Bellhouse, B.J. and Bellhouse, F.H.: Fluid mechanics of model normal and stenosed aortic valves. Circ. Res. 25 (1969), pp. 693-704.

Bellhouse, B.J. and Talbot, L.: The fluid mechanics of the aortic valve. J. Fluid Mech. 35 (1969), pp. 721-735.

Henderson, Y and Johnson, F.E.: Two modes of closure of the heart valves. Heart 4 (1912), pp. 69-82.

Keele, K.D.: Leonardo da Vinci on movement of the heart and blood. London, Harvey en Blythe Ltd., 1952.

Lee, C.S.F. and Talbot, L.: A fluid-mechanical study of the closure of heart valves. J. Fluid Mech. 91 (1979), pp.41-63.

Nallasamy, M. and Krishna Prasad, K: On cavity flow at high Reynolds numbers. J. Fluid Mech. 79 (1977), pp. 391-414.

Peskin, C.S. and Wolfe, A.W.: The aortic sinus vortex. Fed. Proc. 37 (1978), pp. 2784-2792.

Spaan, J.A.E., Steenhoven, A.A. van, Schaar, P.J. van der, Dongen, M.E.H. van, Smulders, P.T. and Leliveld, W.H.: Hydrodynamical factors causing large mechanical tension peaks in leaflets of artificial triple leaflet valves. Trans. Am. Soc. Artif. Int. Organs 21 (1975), pp. 396-403.

Steenhoven, A.A. van, and Dongen, M.E.H. van: Model studies of the closing behaviour of the aortic valve. J. Fluid Mech. 90 (1979), pp. 21-32.

Swanson, W.M., Ou, S.A. and Clark, R.E.: Vortex motion and induced pressures in a model of the aortic valve. J. Biomech. Engng. 100 (1978), pp. 216-222.

Talukder, N., Reul, H. and Müller, E.W.: Fluid mechanics of the natural aortic valve. In: Cardiovascular and pulmonary dynamics, ed. M.Y. Jaffrin. Paris, INSERM-Euromech. 92 (1977), pp. 335-350.

THE USE OF A CIRCUIT SIMULATION PROGRAM (SPICE 2) TO MODEL THE MICROCIRCULATION

D. C. Mikulecky

Department of Physiology
Medical College of Virginia
Virginia Commonwealth University
Richmond, VA 23298

It has been possible to model and simulate a large variety of dynamic physiological systems using network thermodynamics (1) - (10). In this work, a model of the glomerular Filtration apparatus of the kidney (1) - (3) is adapted to capillaries in the microcirculation outside the kidney. For concreteness, the examples given will focus on studies by Gore and his coworkers (12) - (14) of single capillaries in the intestinal circulation of the rat. It should be emphasized that this system was chosen because it is well characterized and can serve as a benchmark for the network model. Since the network model is so easily modified, the study of this specific system can be readily extended to other aspects of the microvasculature. A number of similar studies on the entire cardiovascular system have been made by Rideout (4). Once a physiological system is translated into a network, it can easily be simulated on a circuit simulation program. We have found the program SPICE 2 (Simulation Package with Integrated Circuit Emphasis) to be very useful for this purpose (11). The program is easy to use, readily available for most computers at nominal cost, and versatile enough to simulate a wide variety of physiological processes.

Since the most obvious differences between the general microcirculation and the glomerular capillary are that in the latter there is a lack of an appreciable pressure drop across the capillary and the hydraulic conductivity of its membrane is a number of orders of magnitude higher. Both factors tend to produce a significant reduction in the amount of fluid which crosses the capillary membrane in a non-glomerular capillary. For this reason, the variation in the concentration of solutes and red blood cells is much smaller and the resultant resistance and viscosity changes are capable of being ignored. For more information about these effects and how they can be introduced into the network model, the reader is referred to the glomerulus model (1) - (3). In this work, the flows will represent plasma and solute flow only, with the red blood cells omitted.

For the purposes of this example, the capillary will be idealized to some extent. In reality, the microvasculature involves complex structures which include branching and other anatomical detail which are beyond this first generation model and are left for future investigation. The structure modelled here is a straight, unbranched capillary interposed between arterial and venous vascular beds. This is in keeping with many other microvascular models and is the approach used by Gore and coworkers for modelling this same system (12) - (14).

The rat has a heart rate of 375-425 beats per minute and it is assumed that systolic and diastolic intervals are approximately equal and that an experimentally measured mean pressure reflects the true mean pressure. Mean flow and pressures are used throughout this stage of modelling and any possible capacitive effects due to elasticity of vessel walls, etc. are also ignored. Thus, the model is a true steady-state model. In some preliminary investigations of the glomerulus model, a sinusoidal arterial pressure was used to simulate pulsatile flow and results of this and its counterpart in the microcirculation will be forthcoming.

The pressure of the tissue spaces surrounding the capillary is the sum of a hydrostatic and an oncotic pressure and is held constant at -6 mm Hg for the basic model. It is also assumed that protein leak from capillary to tissue is negligible.

The model is a reticulation of the continuous capillary into 10 subcircuits each of which is identical. The accuracy of this type of discrete representation of a continuous structure has now been well established (1), (6).

The network model is shown in Fig. 1 and 2. The details of the subcircuits (Fig. 2) in each of the boxes in the main network (Fig. 1) are specified only once in the

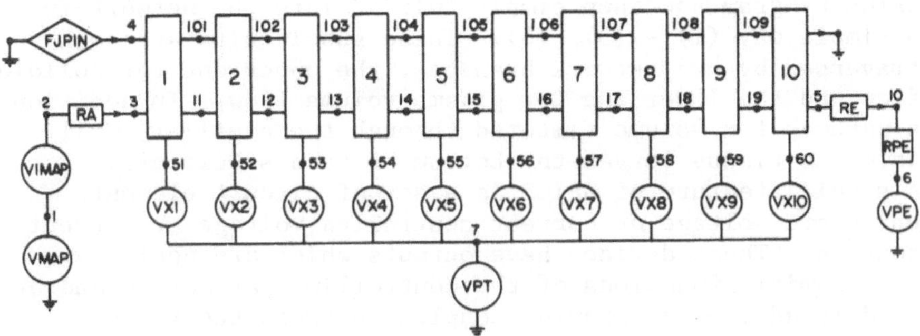

Fig. 1 The network model of the capillary. Details of the structures in boxes 1-10 are shown in Fig. 2. Solid lines indicate actual flow pathways, dotted lines are information flow pathways.

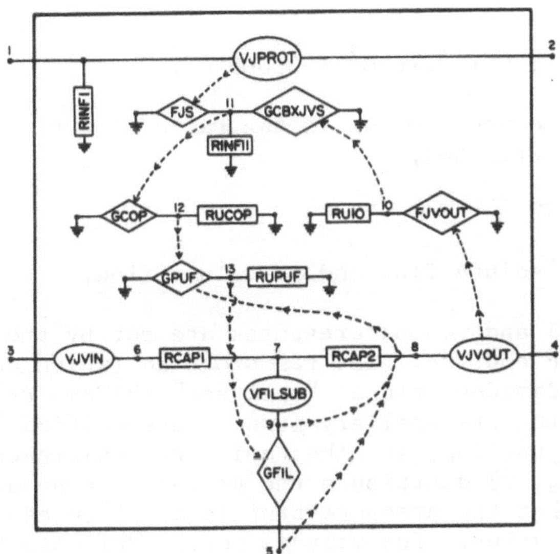

Fig. 2 Details of subcircuits in Fig. 1.

SPICE Program and then simply "wired" into the network in a simple way (1) - (3), (6). These subcircuits are each traversed by two network branches, the upper one for colloid flow and the lower one for plasma volume flow. In addition, a pathway for volume filtered through the capillary wall into the tissue leaves the bottom of each subcircuit. A A special feature of SPICE is a set of circuit elements which are voltage or current controlled voltage or current sources. These devices have outputs which are nonlinear (polynomial) functions of the controlling parameters and are readily adapted to provide coupling between the flows according to the equation

$$JV = LP \cdot PUF, \tag{1}$$

where LP is the filtration coefficient or hydraulic conductivity of the capillary wall and PUF is the net ultrafiltration pressure:

$$PUF = (PCAP - COP - PT). \tag{2}$$

PCAP is the average hydrostatic pressure in the capillary segment, PT is the tissue pressure (hydrostatis minus colloid osmotic), and COP is the colloid osmotic pressure and is in turn a cubic function of colloid concentration, C,

$$COP = 2.1\ C + 0.16\ C^2 + 0.009\ C^3. \tag{3}$$

The colloid concentration is the ratio of the flows in the two network branches,

$$C = JS/JV, \tag{4}$$

where JS is solute flow and JV volume flow.

Arterial and venous pressures are set by the voltage sources VMAP and VPE. The resistors RA (which can be further subdivided into an "upstream" RPA AND RA which represents the precapillary structure specifically), RE and RPE, in conjunction with the capillary resistance (RCAP1 and RCAP2 in Fig. 2) distribute the pressure drops over the system so that the pressure profile and flow can be set by any desired value. The voltage source VPT sets the tissue pressure. The current controlled current source FJPIN (in SPICE, the first letter of the name of any element designates its type; the remainder of its name distinguishes it from others of the same type. Hence, F designates a current controlled current source, G a voltage controlled source, etc.) calculates the flow of colloid from its initial concentration using equation (4). Table 1 gives the basic values of the various parameters which were used to define the three populations of capillaries studied. Some of the basic values will be systematically varied to investigate the model's behavior in detail. The values set by these devices are those of the independent variables in the system. All other voltages and currents are computed by the program. It is easy to study the way that variables in the system which are computed by the program depend on those set. This gives a whole series of results in one computer run. The simplest case is the variation of a pressure held constant in the basic model. The independent sources have some optional properties which allow their output to be programmed as a function of time. The programs available for these elements are a piecewise linear function, a pulse, a sinusoidal signal, and an exponential signal (11). For example, by replacing the independent voltage source setting tissue pressure (element VPT) at a constant value by a programmed time dependent value, the tissue pressure can be varied from an initial value to a series of new values as a piecewise linear function of time. At each point in time called for in the output instructions all of the other parameters requested will be printed in the output. By following the instructions and examples in the user's guide, it is equally straightforward to produce a pulse of given delay, rise and fall time, duration, height and period for repeat. A sinusoidal voltage output with a given magnitude, displacement from zero baseline, frequency and damping factor is another alternative. The only other alteration in the basic program needed is a set of a few control cards calling for the transient mode and the selected output (1) - (10).

One of the most interesting things that can be done with this model is the investigation of changes in RA, RE, and LP on the floating parameters. This is a somewhat more involved procedure than ramping a source, but still very straightforward. For example, to control a resistance by any parameter (or parameters) it is necessary to replace the resistance by a controlled source and to use a controlling polynominal of dimension one greater than the

Table 1: Parameter Values for the Basic Model

Common Values:

Parameter	Device Setting Value	Value
LP	GFIL	5x10-^{4}nl/mmHg.min (per segment)
C	FJPIN	5g/100 ml
PT	VPT	-6mmHg
MAP (mean arterial pressure)	VMAP	100 mmHg
PE (Venule Pressure)	VPE	9 mmHg
Efferent Resistance	RPE + RE	0.5/1 $\frac{\text{mmHg.min}}{\text{nl}}$

Values Defining Three Populations of Capillaries Filtering (f), Intermediate (I), and Absorbing (A)

Parameter	Device Setting Value	F	I	A
Afferent Resistance	RA	5/10*	10/20	13/26
Capillary Resistance	RCAP1‡ & RCAP2	0.5/1.0	0.25/0.25	0.1/0.2

* The two values represent those for an initial plasma flow of 2 nl/min and 1 nl/min respectively.

‡ The capillary resistance in each segment is divided into two parts so that the value of capillary pressure fed into the calculation of PUF can be the average value which is at the node between the two resistors (node 7 in Fig. 2). Each resistor is then 1/20th of the total capillary resistance.

number of controlling parameters. For example, with one controlling parameter a polynominal of 2 arguments is needed. The controlled source has its output set equal to the product of the voltage across it multiplied by some other parameter designated by a voltage in the system. This is equivalent to a parameterized conductance. One convenient parameter for this purpose is the output of a time dependent voltage source used as a clock to vary the parameter systematically between initial and final values. In cases where the resistance is dependent on a flow or other parameters in the system such as colloid concentration or hematocrit variations, these parameters are used in place of or in conjunction with a "clock". The appropriate polynominal dependence is then programmed into the controlled source acting as the controlled resistor/conductor.

Using these techniques, the various parameters can be varied singly or in combinations to explore their effect on model behavior. The result is the ability to perform a series of simulations in one computer run. The time of these runs is longer than a simple run without parameter variations (17), (6), but generally such procedures are are more efficient than using one computer run for each new parameter value. The results of some representative parameter variations are presented here as an illustration of what can be done with the model.

Figure 3 shows the amount of filtration occurring in each of the 10 segments of capillary type A2 at colloid concentrations from 4 g/100 ml to 8 g/100 ml. This capillary model has the same hydraulic conductivity in each segment and the amount filtered reflects the linear variation in hydrostatic pressure along its length (19.86-9.98 mm Hg for C = 4 to 20.12-10.02 for C = 8). The tissue pressure, PT, constant (-6 mm Hg) and the colloid osmotic pressure, COP, varies nonlinearly across the capillary, but not enough to influence the result (from a minimum of 20.38 at end to a maximum of 20.53 in the center, at C - 6).

One interesting finding of Gore and coworkers was that their experimental findings require that the hydraulic conductivity, LP, must vary along the length of the capillary (12). In order to investigate the effect of this variation on the model's behavior the capillary was modified as shown in Fig. 4. The upper portion of Fig. 4 represents part of one of the subcircuits (the third) and its connection to

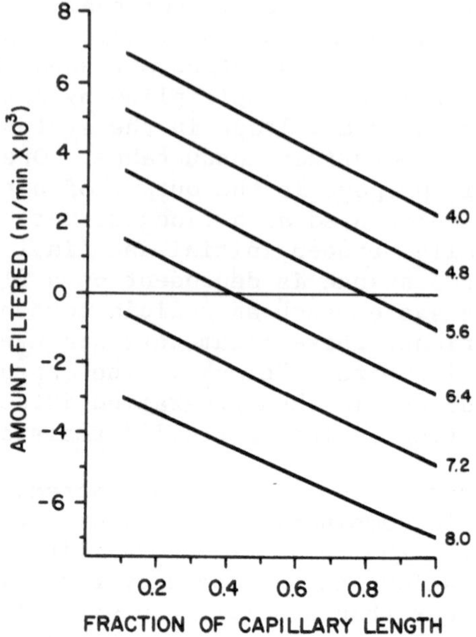

Fig. 3 Amount filtered in each of the ten segments at a series of colloid concentrations for capillary A2.

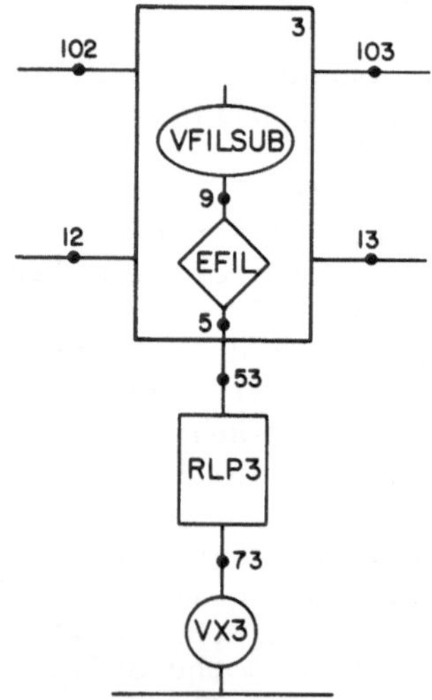

Fig. 4 Modification of a single segment of the model to incorporate spatially dependent hydraulic conductivity (RLP3).

VX3 as modified for allowing spatial variation in LP (RLP3). In the subcircuit, the small calculating circuit made up of elements GPUF and RUPUF and node 13 is eliminated and its function taken over by the voltage controlled voltage source EFIL. This element calculates a voltage which equals to PUF minus PT and its output is at node 53. The value of PT is the voltage at node 73 so that if the resistor RLP3 has its value at 20/LP, equation (1) is followed in each segment of the capillary with a different value for LP.

Three of the different sets of spatially varying hydraulic conductivities used for simulation appear in Table 2. Using the LP values in Table 2 for capillary L1 and all other model parameter values set at the values used for capillary I2, the profile of filtration/absorption is shown in Fig. 5. Increasing the gradient of LP, as in capillary L2, leads to the profiles in Fig. 6. Using an exponentially varying LP (12) leads to the profiles in Fig. 7. In each case, the hydrostatic pressure profile and initial flow rate did not differ significantly from the values in capillary I2 with constant LP. The main effect of the variation in LP along the length of the capillary is to supress filtration and enhance reabsorption, since the venous end is so much more permeable than the arterial end. This can be seen in the overall performance of the capillary as a shift in the filtration fraction towards absorption. Fig. 8 is a comparison of the filtration fraction as a function of colloid concentration for capillary F2 and its counterpart having LP values of L1, and capillary I2 and its counterpart having the L1 set of LP values. In both pairs of capillaries, the effect of the variation in LP is to move the curve toward greater absorption.

Another consideration of the effect of hydraulic conductivity is the manner in which the capillary behavior depends on this parameter when it is not spatially dependent.

For the largest values of LP, the values of the capillary pressure, PCAP, and initial flow were altered slightly due to the filtration and/or absorption acting as a parallel flow path. For example, in capillary A2 the hydrostatic pressure drop was 14.00-10.00 mm Hg at an LP of 5 X 10^{-5} nl/min mm Hg per segment while the values rose to 14.10-10.46 mm Hg at an LP value of 5 X 10^{-3} nl/min mm Hg per segment. For the same capillary, the initial plasma flow rates were 2.000 nl/min and 1.998 nl/min at the low and high values of

Table 2: Values of LP and RLP for each segment in three types of the Capillary Model.

Type		Segment 1	2	3	4	5	6	7	8	9	10
L1	$LPx10^4$	2.5	3.0	3.5	4.0	4.5	5.0	6.5	6.0	6.5	7.0
	RLP	4000	3300	2850	2500	2220	2000	1810	1670	1540	1430
L2	$LPx10^4$	1	2	3	4	5	6	7	8	9	10
	RLP	10000	5000	3330	2500	2000	1670	1430	1250	1110	1000
X	$LPx10^4$	4.40	4.88	5.38	5.92	6.58	7.25	8.00	8.85	9.80	10.81
	RLP	2270	2050	1860	1690	1520	1380	1250	1130	1020	925

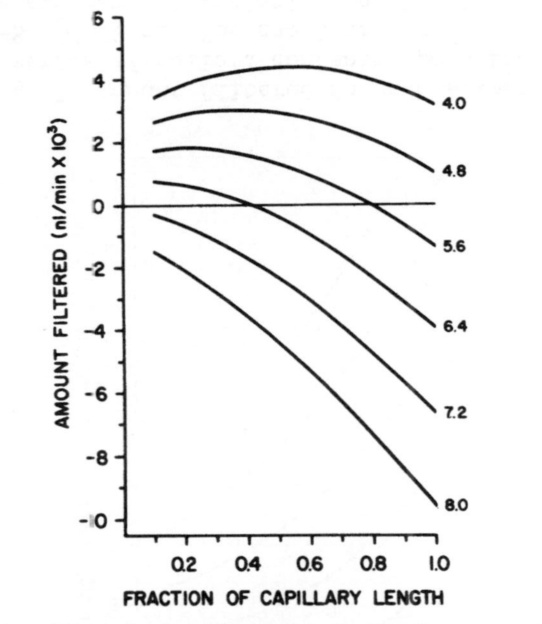

Fig. 5 Amount filtered in each segment at a series of colloid concentrations from 4-8 g/100 ml for a capillary with varying hydraulic conductivity along its length. Capillary type L1 in Table 2, with all other model parameters set at the values of capillary I2.

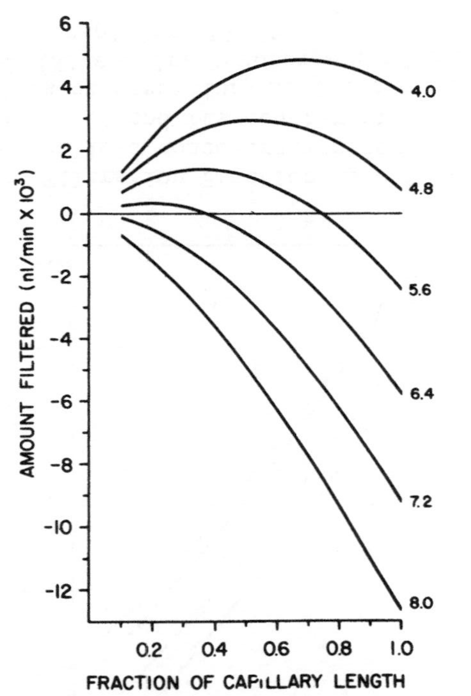

Fig. 6 Amount filtered in each segment at a series of colloid concentrations from 4-8 g/100 ml for capillary type L2 in Table 2, with all other parameters after capillary I2.

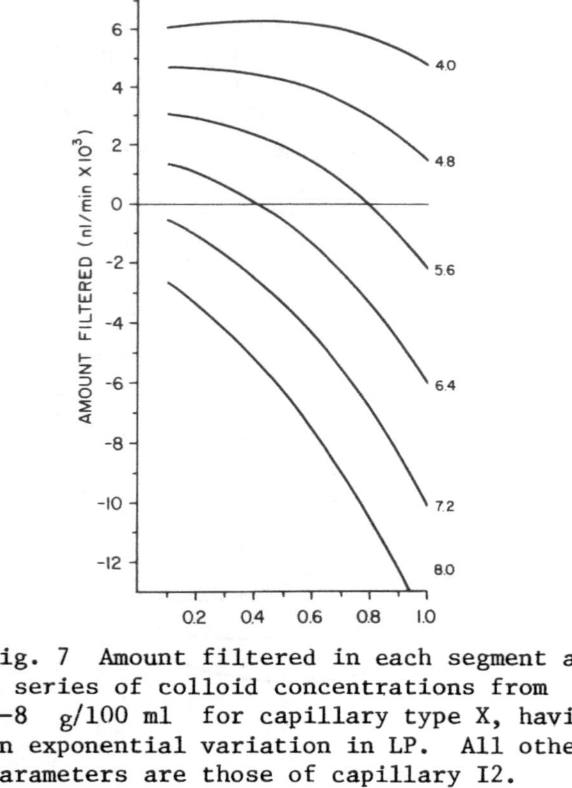

Fig. 7 Amount filtered in each segment at a series of colloid concentrations from 4-8 g/100 ml for capillary type X, having an exponential variation in LP. All other parameters are those of capillary I2.

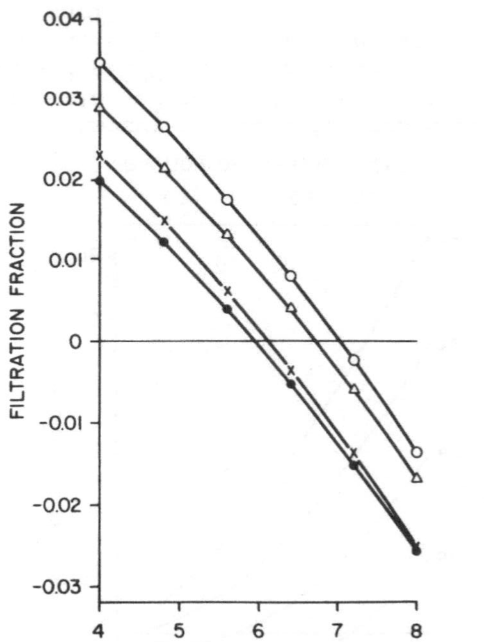

Fig. 8 Filtration Fraction as a function of colloid concentration for capillary F2 (0) and its counterpart with LP values of cappillary L1 (Δ) and capillary A2 (X) and its counterpart with LP values of capillary L1 (●).

Fig. 9 is the dependence of filtration fraction on colloid concentration for a series of capillaries of type A2 with the LP values given in the figure legend. In general, the effect is to amplify the effect of the capillary colloid osmotic pressure, COP, as a counterforce to the hydrostatic pressure in the capillary, PCAP.

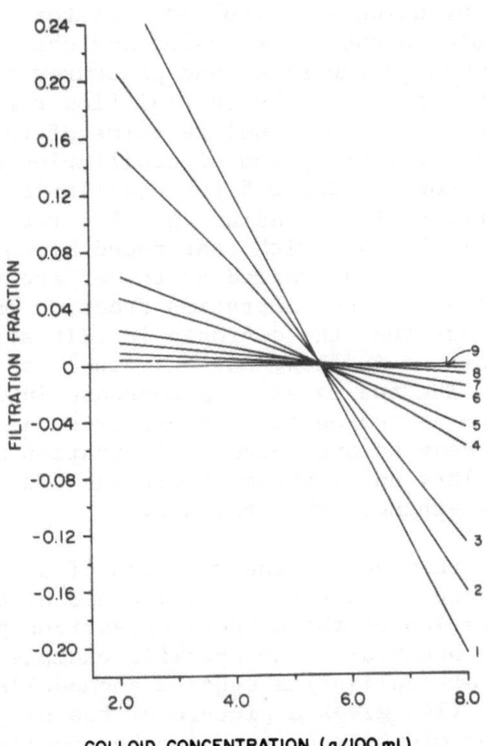

Fig. 9 Filtration fraction as a function of colloid concentration for capillary A2. LP values X 10^3 are 1-5, 2-3.5, 3-2.5, 4-1, 5-0.75, 6-0.375, 7-0.25, 8-0.125, 9-0.05 in units of nl min^{-1} mm Hg^{-1} per segment of capillary.

LP, respectively.

Table 3 is the result of changing the tissue pressure and colloid concentrations on the filtration fraction of the capillaries with a flow rate of 1 nl/min. When the flow rate is 2 nl/min, the filtration fraction for the same set of colloid concentrations and tissue pressures is shown in Table 4. The tissue pressure uniformly increases the tendency for reabsorption as it is made more positive and acts in harmony with the colloid concentration in the capillary. By using a controlled resistor in place of RA, it is possible to change its value and observe the effect of such changes on the flows and pressures in the capillary. Increasing RA decreases the initial flow rate and simultaneously increases the absolute value of the filtration fraction (with the exception of capillaries F2 and I1 at C= 6) as is shown in Table 5 for capillaries F1, I1, and A1. This effect of decreasing the flow rate on filtration fraction is in harmony with that found by Gore, et al (12) when the flow rate was varied at the efferent (venous) end of the capillary. The filtration fraction increase reflects the fact that the decrease in filtration is smaller than the decrease in flow, not an actual increase in filtration. The two cases of a decrease in filtration fraction with decreased flow (capillaries A1 and I1 at C = 6) reflect the near balance between filtration and absorption in the capillary under these conditions and that the decreased flow enhances that balance.

The network model of the function of a single capillary has proved to be very versatile and allows for a complete exploration of the effects of various parameters on the system's behavior. The specific example of the three populations of capillaries studied andmodelled by Gore, et al (12) - (14) gives a picture of how single capillaries in the microvasculature can be modelled using this technique. Using the controlled sources in SPICE, the transport by convection and diffusion of solutes other than the colloids can also be modelled and this method has already proven to be useful in other network models of coupled volume and solute flow (5), (9).

A very important advantage to using SPICE in conunction with network models is the ease in which programs can be written and modified with a minimum of mathematical

Table 3: Dependence of Filtration Fraction x 10^2 on Tissue Pressure and Initial Colloid Concentration with Initial Flow Rate = 1 nl./min.

Type of Capillary	Initial Colloid Conc. ($\frac{g}{100\ ml.}$)	Tissue Pressure (mmHg)				
		-6	-3	0	3	6
F1	2	9.86	8.47	7.07	5.66	4.25
	4	6.56	5.19	3.82	2.44	1.06
	6	2.40	1.07	-0.27	-1.62	-2.97
	8	-2.61	-3.90	-5.19	-6.50	-7.81
I1	2	7.70	6.27	4.83	3.40	1.95
	4	4.39	2.99	1.59	0.18	-1.23
	6	0.25	-1.11	-2.47	-3.84	-5.21
	8	-4.70	-6.01	-7.33	-8.65	-9.98
A1	2	6.35	4.89	3.44	1.98	0.52
	4	3.05	1.63	0.21	-1.22	-2.65
	6	-1.08	-2.45	-3.83	-5.21	-6.60
	8	-5.99	-7.31	8.64	-9.97	-11.13

Table 4: Dependence of Filtration Fraction $X10^2$ on Tissue Pressure and Initial Colloid Concentration with Initial Flow Rate = 2 nl./min.

Type of Capillary	Initial Colloid Conc. $(\frac{g}{100\ ml.})$	Tissue Pressure (mmHg)				
		-6	-3	0	3	6
F2	2	5.10	4.38	3.65	2.93	2.20
	4	3.44	2.73	2.01	1.29	0.57
	6	1.31	0.60	-0.11	-0.82	-1.53
	8	-1.36	2.05	-2.74	-3.44	-4.14
I2	2	0.15	-0.57	-1.28	-1.99	-2.71
	4	2.28	1.56	0.83	0.10	-0.62
	6	0.15	-0.56	-1.28	-1.99	-2.71
	8	-2.50	-3.19	-3.89	-4.59	-5.30
A2	2	3.22	2.48	1.74	1.01	0.27
	4	1.57	0.84	0.11	-0.62	-1.35
	6	-0.56	-1.27	-1.99	-2.71	-3.43
	8	-3.19	-3.89	-4.59	-5.29	-6.00

Table 5: Dependence of Filtration Fraction (F.F.) and Initial Flow (I.F., nl./min) X10^2 on Afferent Resistence (A) and Initial Colloid Concentration

Type of Capillary	Initial Colloid Conc. ($\frac{g}{100\ ml.}$)		Afferent Resistance (RA, mmHg.min./nl.) 10.0	12.5	15.0	20.0	30.0
FI	2	F.F.	9.864	9.989	10.11	10.36	10.86
		I.F.	1.014	0.9865	0.9607	0.9127	0.8300
	4	F.F.	6.562	6.596	6.631	6.699	6.835
		I.F.	1.010	0.9826	0.9568	0.9091	0.8267
	6	F.F.	2.40	2.32	2.25	2.11	1.82
		I.F.	1.005	0.9777	0.9520	0.9046	0.8225
	8	F.F.	-2.607	-2.798	-2.987	-3.363	-4.099
		I.F.	0.9985	0.9716	0.9462	0.8990	0.8176
A1	2	F.F.	6.347	6.688	7.029	7.707	9.053
		I.F.	1.002	0.9353	0.8768	0.7794	0.6376
	4	F.F.	3.047	3.154	3.260	3.470	3.882
		I.F.	1.001	0.9343	0.8759	0.7785	0.6369
	6	F.F.	-1.076	-1.244	-1.410	-1.737	-2.370
		I.F.	0.9997	0.9330	0.8747	0.7775	0.6361
	8	F.F.	-5.989	-6.457	-6.917	-7.817	-9.537
		I.F.	0.9981	0.9315	0.8733	0.7762	0.6351
I1	2	F.F.	7.698	7.956	8.213	8.726	9.747
		I.F.	1.006	0.9531	0.9058	0.8240	0.6980
	4	F.F.	4.394	4.471	4.548	4.700	4.999
		I.F.	1.003	0.9510	0.9038	0.822	0.6945
	6	F.F.	0.251	0.116	-0.018	-0.283	-0.801
		I.F.	1.001	0.9484	0.9013	0.8200	0.6946
	8	F.F.	-4.699	-5.067	-5.430	-6.143	-7.520
		I.F.	0.9973	0.9452	0.8984	0.8173	0.6924

manipulation. With a few hours of familiarization with the concepts of network modelling and the "rules" for using SPICE, the average experimental scientist can be creating and altering his or her own model of an experimental system. This is generally not the case if classical numerical methods are to be used.

In this demonstration, the assumption that the low filtration fraction and high capillary hydrostatic pressure gradient minimizes the effect of the colloid concentration changes due to filtration and absorption was validated. This fact makes the use of plasma flow as the main flow variable a reasonable simplification. On the other hand, for studies of blood gas transport, and in particular, oxygen transport, the inclusion of the red blood cells and the kinetics of oxygen uptake by the cells would be a necessary refinement for an accurate model. The techniques for this have been worked out in a number of studies already completed. The inclusion of the red blood cell flow has been achieved in the glomerulus model (1). The modelling of rate laws for non-linear chemical reactions and binding of ligands has also been worked out in detail (6), (10). This preliminary study of the volume and colloid flows paves the way for more elaborate network models of the microvasculature.

Further details of the program including a copy, will be supplied upon request.

Thanks to P. J. Baker, Peggy Godette and Frances Barker for the typing of the manuscript.

REFERENCES

1. Oken, D.E., S.R. Thomas, and D.C. Mikulecky. Submitted to Kidney Int. for Publication.

2. Thomas, S.R., D.C. Mikulecky, and D.E. Oken. Fed. Proc. 38: 901 (1979).

3. Mikulecky, D.C. and S.R. Thomas. Proc. 7th New England Bioeng. Conf., 459-462, Troy, N.Y. (1979).

4. Rideout, V.C. IEEE. Biomed. Eng. 19: 101-107 (1972).

5. Thomas, S.R. and D.C. Mikulecky. Am. J. Physiol. 235: F638-F648 (1978).

6. Wyatt, J.L., Jr., D.C. Mikulecky, and J.A. DeSimone. Chem. Eng. Sci. (in press).

7. Mikulecky, D.C. and S.R. Thomas. J. Franklin Inst. 308: 309-326 (1979).

8. Mikulecky, D.C., E.G. Huf and S.R. Thomas. Biophys, J. 25: 87-106 (1979).

9. Mikulecky, D.C. Biophys. J. 25: 323-340 (1979).

10. White, J.C. J. Biol. Chem. 254: 10889-10895 (1979)

11. Cohen, E. ERL Memo No. ERL-M592, Electronics Research Laboratory, University of California, Berkeley (1976).

12. Papenfuss, H.D., R.W.Gore, and J.F. Gross. Microvasc. Res. (in press)

13. Gore, R.W. W. Schoknecht, and H.G. Bohlen. Microcirculation Vol. 1: pp. 331-332 (J. Grayson and W. Zingg, eds.) Plenum, N.Y. 1976.

14. Gore, R.W. and G. Bohlen. Am. J. Physiol. 233: H685-H693 (1977).

15. Chua, L.O. Introduction to Non-linear Network Theory McGraw-Hill, N.Y. (1969).

16. Landis, E.M. and J.R. Pappenheimer. In: Handbook of Physiology, Circulation, sec 2, vol 2, pp. 961-1034 (W.F. Hamilton and P. Dow, eds.), American Physiological Society, Washington, D.C. (1963).

QUANTIFICATION OF THE MICROCIRCULATORY RESPONSE TO FREEZING

Kenneth R. Diller, C. Dudley Evans, and
John P. Parsons

Bio-Medical Engineering Program
Departments of Electrical & Mechanical Engr.
The University of Texas at Austin
Austin, Texas 78712, USA

Extensive clinical and experimental data verify that freezing and thawing processes can produce trauma and necrosis in living tissue. The primary effects associated with this phenomena have been summarized in several recent reviews[1-3]. The response of tissue to freezing injury is intimately related to modifications produced in the functional state of the localized circulatory system. Although subzero tissue temperatures are requisite to freezing insult, they may not be damaging *per se*; concommitant vascular collaspe may be the ultimate factor in tissue death. For example, skin, frozen to the extent that if left in place it will inevitably die, has been shown to survive if transplanted to a normal vascular bed[4,5]. Thus, under the proper conditions, damage caused by freezing can be reversed if adequate blood circulation is maintained.

In the present study, the leakage of macromolecules from vascular elements in the microcirculation of frozen/thawed tissue has been measured. The extravasation process is monitored by fluorescence microscopy using fluorescein isothiocyanate (FITC) tagged dextran as the marker molecule. The objective of the investigation is to quantify alterations to the microvascular bed in terms of the transport properties in the interstitial tissue at the site of injury.

EXPERIMENTAL APPARATUS

The experimental investigation was performed on a Zeiss Universal microscope equipped with a specially designed intravital stage capable of freezing and burning tissue in the field of view[6]. The stage has a transparent viewing platform on which the hamster gut mesentery can be displayed. Microcirculatory blood flow can be observed continuously during and subsequent to the injury event.

The tissue temperature is monitored by a 25 μm copper constanan thermocouple (Omega Engineering, INC., Stamford, Connecticut 06907) permanently mounted on the viewing platform. The thermocouple signal is used as the input to a digital controller on which a desired thermal protocol can be programmed[6]. Protocol parameters for freezing include initial temperature, cooling rate, minimum temperature and hold time, and warming rate. Regulation of the tissue temperature is effected by balancing the magnitudes of simultaneous refrigeration and heating fluxes in the stage. Refrigeration is produced by convection with a steady stream of chilled nitrogen gas flowing through the stage interior. Heating is produced by electrical energy dissipation in a transparent film resistor, spray deposited on the underside of the glass coverslip which is structurally the top surface of the platform on which the tissue rests. By varying the voltage applied across the resistor, the heater input can be controlled with a very short time constant. The entire apparatus forms a simple proportional control system. In its present configuration, linear cooling rates can be created from 1 to 120°K/min at temperatures down to -50°C. A sample temperature/time history recorded with an animal on the stage is shown in Figure 1. Further details of the stage design and operation are described in a recent publication[6].

EXPERIMENTAL PROCEDURE

The experimental model used in this study is the gut mesentery of the golden hamster. Thirteen hamsters (Charles River Breeding Labs, 251 Ballardvale Street, Wilmington, Massachusetts, 01887), six to ten weeks old, weighing between 70 and 100 gm, were anesthetized with 3% sodium pentobarbitol administered intraperitoneally at 30 mg/kg and supplemented as needed. A loop of the small intestine was exteriorized through a midline abdominal incision into a bath of buffered saline solution. The mesentery was kept immersed and bathed in the saline throughout the duration of the experiment to avoid contact with air. The intestine was draped gently over the viewing platform of the microscope stage, and a suitable area for viewing the microcirculation was located in the mesentery. A canula was placed in the external jugular vein for subsequent injection of tracer solution.

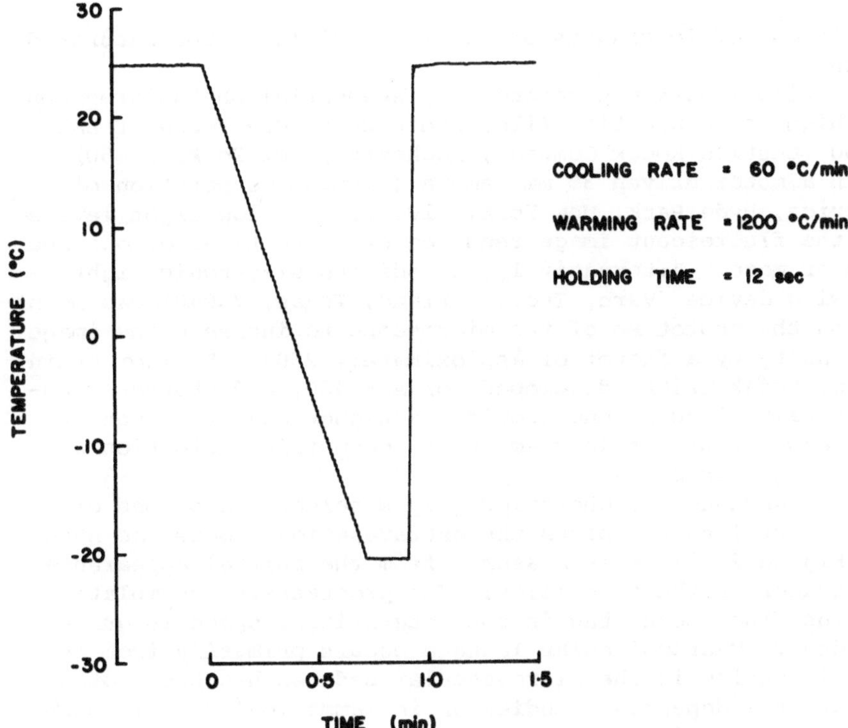

Figure 1. Tissue temperature-time history recorded during a digitally programmed freeze/thaw protocol.

The tissue was subjected to a predetermined freeze/thaw protocol using the digital temperature programmer/controller. Freezing regimens of limited severity were chosen for these tests so the blood flow would be restored after return to physiological temperatures. The entire freezing process was observed through the microscope using transillumination. Within one minute after completion of thawing, 0.5 ml of a 10% solution of FITC tagged 3000 MW dextran (Pharmacia Fine Chemicals, 800 Centennial Avenue, Piscataway, New Jersey, 08854) was injected through the jugular vein canula. The microscope was switched to epi illumination with a Zeiss FITC exciter/barrier filter pack (Carl Zeiss, Inc. New York, New York, 10018). A timer was started coinciding with the initial appearance of fluorescent dye in the field of view. The progressive extravasation of the marker was followed for a minimum of 20 minutes or until the field became saturated with dye.

Visual data was recorded by sequential photomicrography on high speed negative film, Kodak 2484, developed at ASA 3200 (Eastman Kodak Company, Rochester, New York, 14650) with a motor driven 35 mm camera (Olympus Corporation of America, Hyde Park, New York, 11040). The low light levels of the fluorescent image required exposure times of one second or more. Alternatively, a modified electronic night viewing device (Varo, Inc., Garland, Texas, 75040) was mounted on the phototube of the microscope to increase the image intensity by a factor of approximately 2500. A finer grain film, Kodak Tri-X, developed for ASA 400, and shorter exposure time, 1/30 second provided a higher quality image at the expense of a slight amount of geometric distortion at the image periphery.

A sequence of photomicrographs selected from one of the control runs depicts the extravasation process, as shown in Figure 2. Time is measured from the initial appearance of tracer in the arterioles. The progressive accumulation of the fluorescent tag in the interstitial space is quite evident. Macromolecular leakage occurs primarily from the small venules in the microvascular bed, as has been noted in other independent studies on inflammation[7,8]. Although the extravasation appears very graphic qualitatively, our objective was to use this phenomenon as a quantitative evaluation of freeze/thaw injury. To this end, a technique of densitometric analysis of the photomicrographs has been designed, based on converting the images to digital format for subsequent computer processing.

The hardware for digitizing photomicrographs is shown in Figure 3. The heart of this apparatus is a CVI 270A Video Digitizer (Colorado Video, Inc., Boulder, Colorado, 80302). The digitizer is linked via a high speed line to a host PDP 11/34 computer (Digital Equipment Corporation, Maynard, Massachusetts, 01754), so that the entire digitizing procedure is performed under direct computer control. 35 mm slides or film strips are placed in a bellows close-up attachment interfaced to a video camera and the magnification adjusted so that one film frame completely fills the video image. The photomicrograph is transilluminated by a D.C. powered fluorescent light source to provide a field of uniform brightness. The analog signal from the video camera is transferred directly to the digitizer. The

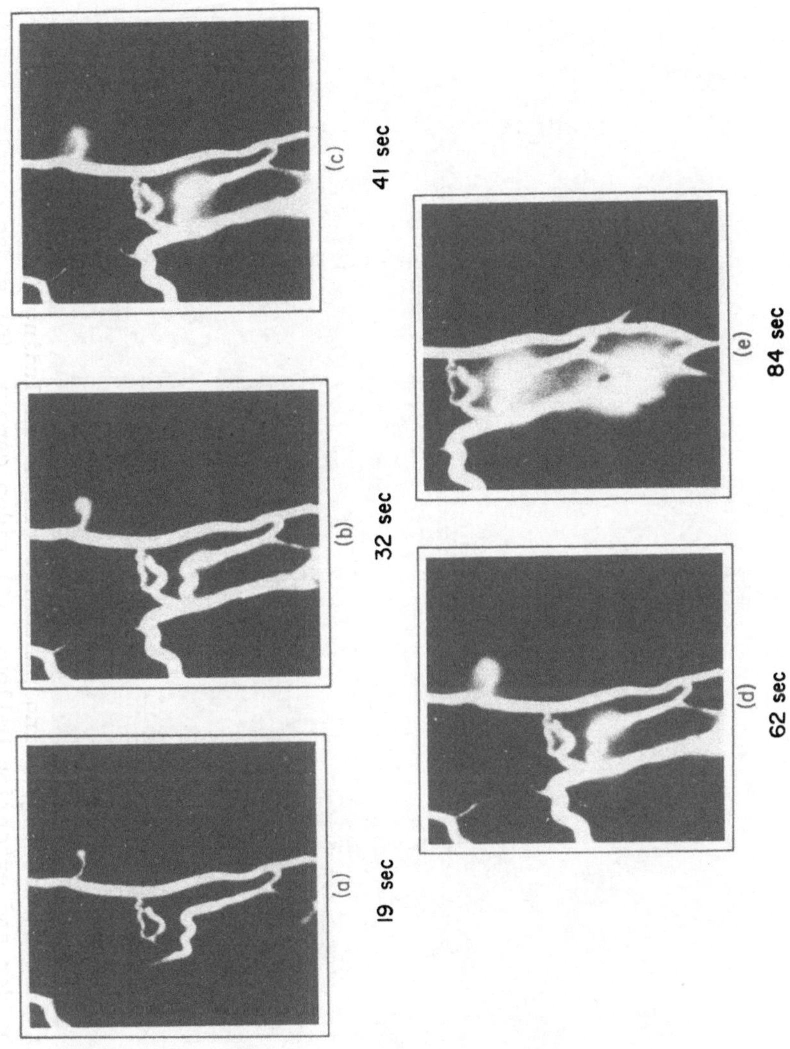

Figure 2. Progressive extravasation of 3000 MW FITC dextran in the hamster gut mesentery.

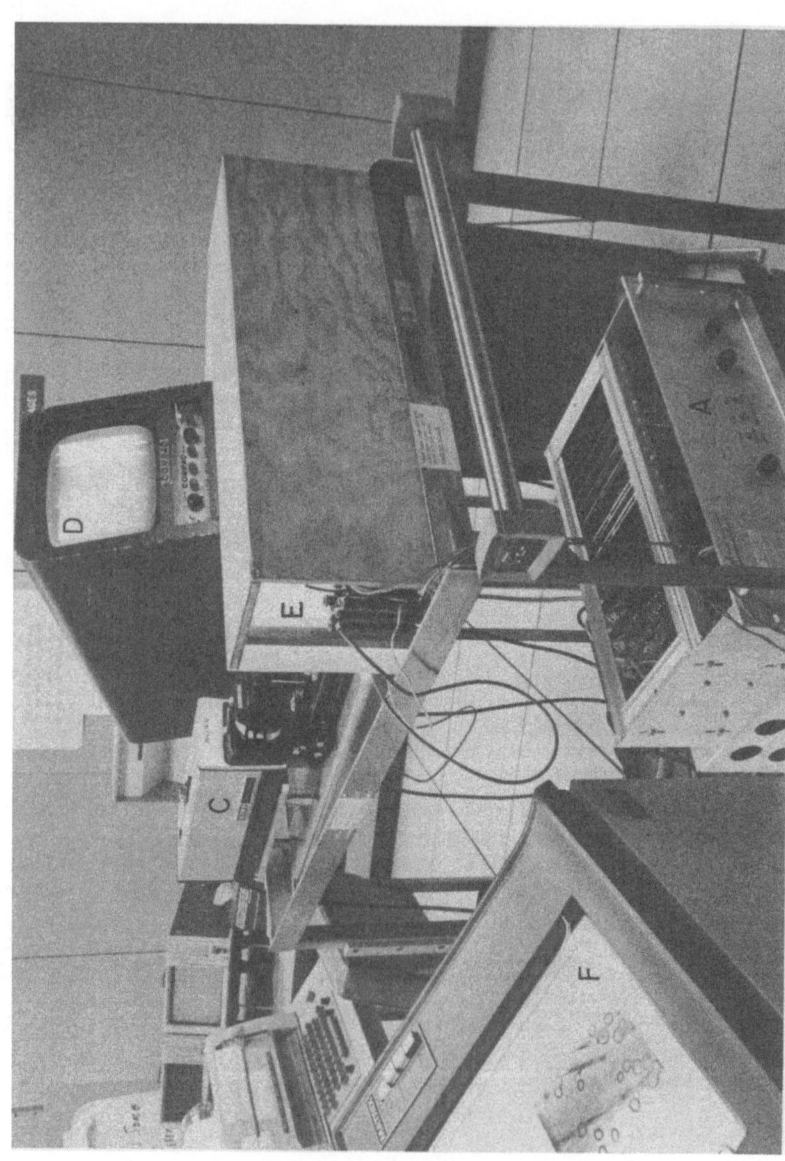

Figure 3. Apparatus for digitizing photomicrographs including (A) video digitizer, (B) film holder and extension bellows, (C) video camera, (D) video display monitor, (E) D.C. light source, (F) digital hard copy device.

digitizing procedure consists of eight rapid, repetitive scans of each picute element (pixel) followed by individual pixel averaging to obtain a minimum effective intensity resolution of 64 gray levels (6 bits of information). The scans are performed on a spatial coordinate matrix of 240 x 256 pixels. The matrix represents only 1/4 the capability of the system; it was chosen because it provided data adequate for analyzing the images and minimized the computer memory and processing time requirements.

Digital renditions of the photomicrographs were obtained on an electrostatic printer (Versatec, Inc., 2805 Bowers Avenue, Santa Clara, California, 95051) using a 16 gray level dot matrix pattern. Figure 4 presents a digital version of the pictures in Figure 2. It should be noted that the actual gray level information in the data base is considerably greater than is represented on plotter output in Figure 4. This fine degree of gray level resolution is an important factor in performing the digital densitometric analysis.

Determination of an optimum algorithm for processing the digital data posed a formidable problem. Early attempts at integrating the total extravascular fluorescent activity were unsuccessful due to difficulties in maintaining uniform control conditions[9]. The localized rate of macromolecular leakage is not only a function of the severity of trauma, but is also intimately dependent on the constitution of the microvasculature under observation, as characterized, for example, by the vessel density and the percentage and size of venules, as well as including significant inter-animal variations.

Consequently, a technique was developed by which the diffusion of extravasated macromolecules through the interstitium could be quantified based on measured transient concentration profiles. Information concerning the spatial concentration variations was obtained directly from the digital picture data base as the gray level plot along a record normal to a leakage site on a selected vessel. An example of the resulting plots is illustrated in Figure 5. The loci of four lines are identified on the photomicrograph in Figure 5a, from which the four corresponding gray level intensity plots show in Figure 5b were generated. The location of vessels in the image is easily verified by comparison of the intensity curves with the micrograph.

The interstitial transport of the tracer molecule can be characterized by a diffusivity coefficient D (cm^2/

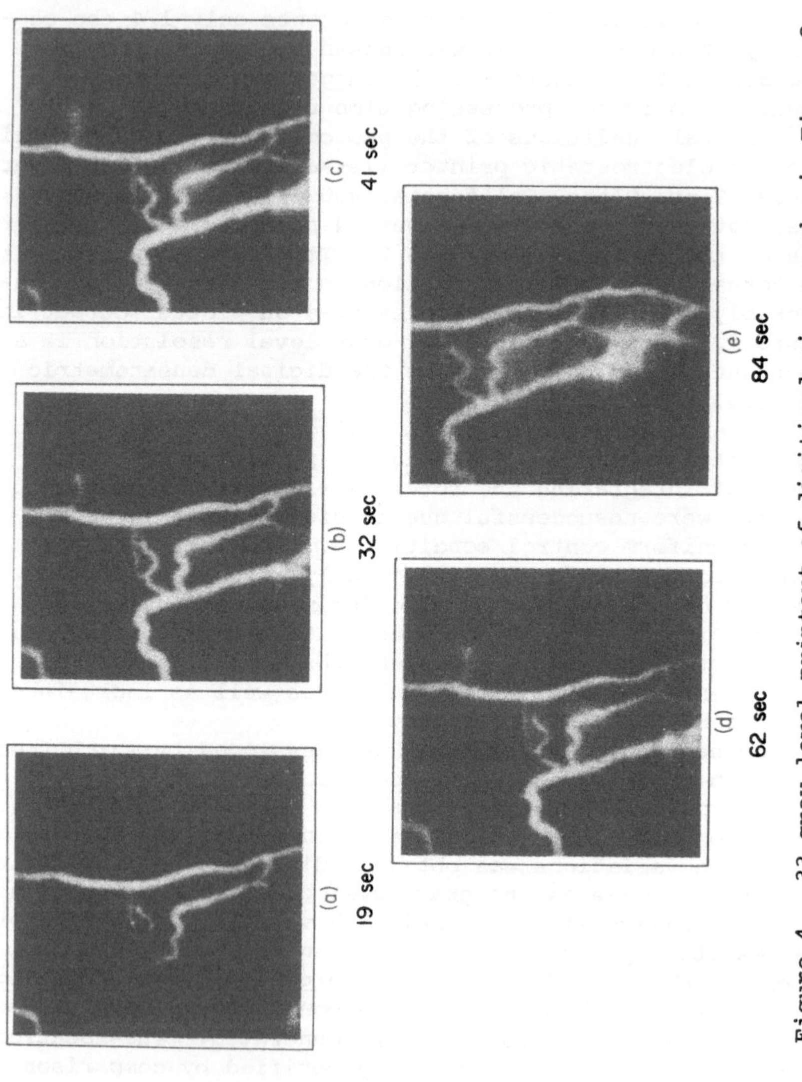

Figure 4. 33 gray level printout of digitized micrographs in Figure 2.

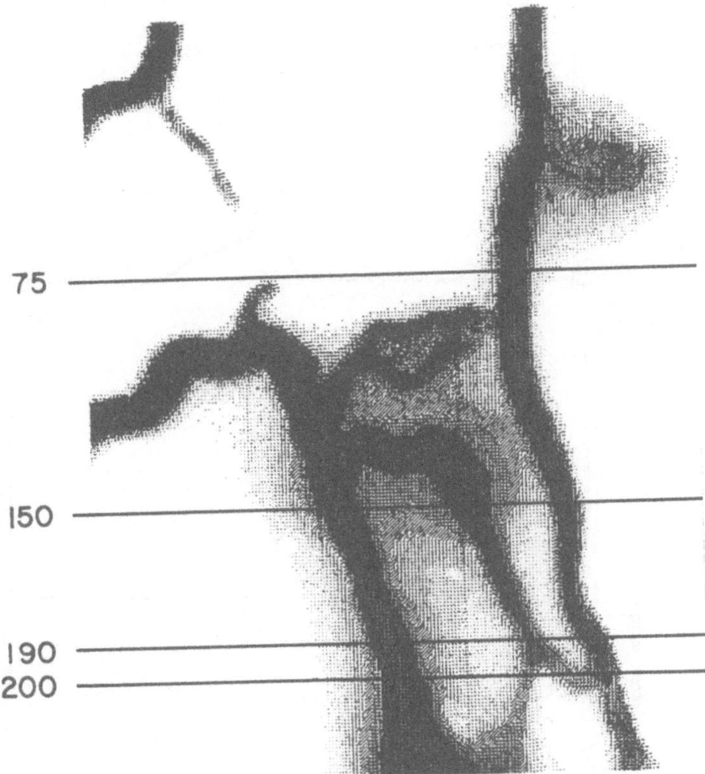

Figure 5a. 16 gray level negative digital printout of Figure 4d, indicating the locations of four raster scan records.

sec), a parameter dependent on the properties of both the tissue and the macromolecule. Apparent values for the diffusivity can be calculated from the data by fitting the experimental curves to an appropriate analytical model. Since the tracer concentration profile changes continuously with time as the diffusion process progresses, it is necessary to specify an initial condition and the temporal development of boundary conditions from the experimental data to be able to solve the model equations and determine the diffusivity parameter.

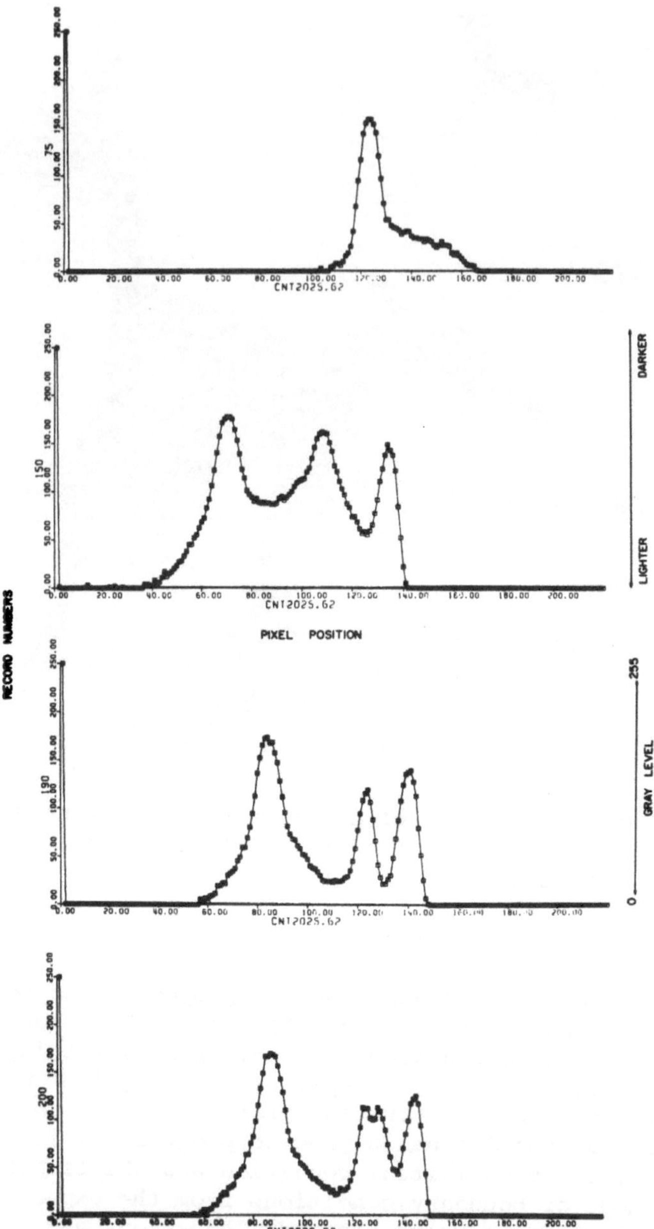

Figure 5b. Gray level plots along the four raster scan records identified in Figure 5a, from which concentration distribution data is obtained.

ANALYTICAL MODEL

An analytical model has been developed to describe the diffusion of tracer in the gut mesentery. A very simplified rectangular, one dimensional geometry has been assumed in which the flux is unidirectional, normal to the leaking vessel and parallel to the tissue surface, Figure 6. This geometry is justified based on the similar thicknesses of the tissue and vascular elements[8]. In prior studies on the hamster cheek pouch, a model of cylindrical geometry was used in accordance with the large tissue/vessel diameter ratio[10]. Data in the gut mesentery fit much better to a rectangular rather than cylindrical model. The tissue was assumed to be of uniform thickness and homogeneous in transport properties. The curvature of the vessel was neglected.

The transient concentration profile for the tracer diffusing through the tissue can be described by the Fourier equation:

$$\frac{\partial c}{\partial t} = \frac{1}{D}\frac{\partial^2 c}{\partial x^2} \qquad (1)$$

For an initial concentration of zero and a step increase in concentration of tracer at the vessel wall to c_o at time t=o, the solution to equation (1) is in the form of a complimentary error function[11]

$$c = c_o \operatorname{erfc} \frac{x}{2\sqrt{Dt}} \qquad (2)$$

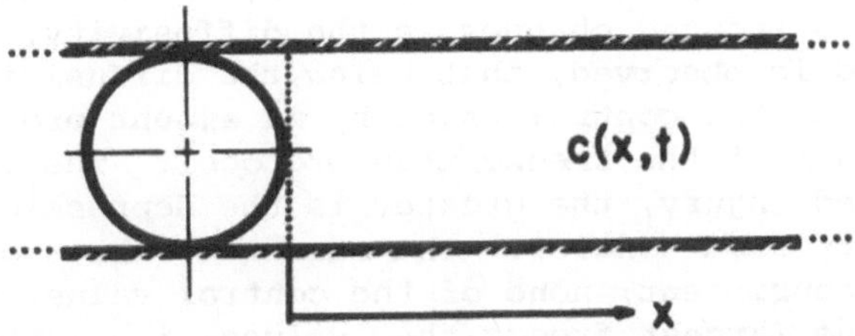

Figure 6. One dimensional model for interstitial diffusion of extravasated marker molecules. Tissue surfaces are assumed impervious to the marker. End on view of the vessel is shown; vessel radius of curvature is neglected. Marker concentration at x=0 is c_o for $t \geq 0$.

where x is measured from the vessel wall. The step change assumption is known to be inaccurate from recent measurements in our experimental model[12]. Nonetheless, it has been used in these initial studies for the convenience of obtaining a closed form mathematical solution.

Equation (2) is presented in graphical format in Figure 7. The set of curves is plotted for a spectrum of constant values of the product of D·t. Each curve can be considered to represent the concentration distribution of tracer at a time t in a medium with diffusivity D. These curves correspond to experimental data illustrated in Figure 5b. The parameters c, c_o, t, and x were all measured explicitly in the experimental procedure, and plotted concurrently with the model as in Figure 7. The best fit of the experimental data to the analytical model was used to determine a value of the diffusivity, D, for each measured concentration profile. In general, the fit of the data to the model was quite good.

RESULTS

The diffusivity parameter for 3000 MW FITC dextran was evaluated in 10 control animals. Measurements were made for each animal at various locations in the gut mesentery and for various times. In all, 38 values for D were determined to obtain for control $D=2.8\pm0.4\times10^{-6}cm^2/sec$. This value is very nearly equal to that for the free diffusion in water[13,14]. Freezing studies were conducted on 3 animals to perform a total of 16 measurements of D. The results are presented in Table 1, showing significant freeze/thaw induced changes in the diffusivity. A general trend is observed, that being the diffusivity decreases from the control value by an amount proportional to the rigor of the freeze/thaw protocol. The more severe the induced injury, the greater is the depression of the diffusivity from control. The measured values of D were all self-consistent; none of the control values were smaller than the largest freeze/thaw values, i.e., there was zero overlap between the controls and the trials. Very similar results have been obtained for a burn injury model[12].

The physiological basis for the freeze/thaw induced reduction of interstitial macromolecular diffusivity re-

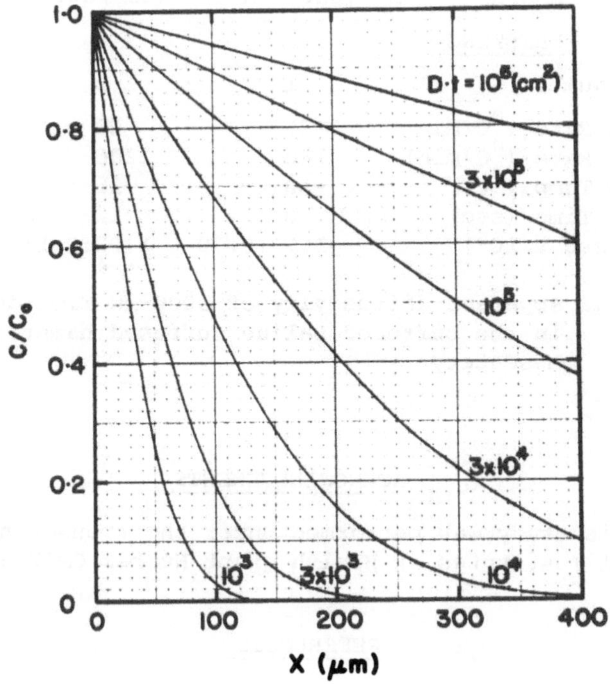

Figure 7. Plot of equation (2) showing effects of varying diffusivity, D, on the concentration profile at a given time.

mains to be identified. It may be conjectured that the freezing process causes denaturation of interstitial proteins resulting in an overall constriction. At the dimensions of the diffusive pathways are reduced, the resistance to macromolecular transport would be increased, causing the observed change in D. Clearly, further trauma experiments are necessary to define the diffusivity depression phenomenon. Basic studies on the molecular mechanisms of interstitial transport and of thermal injury in living tissue could provide information very helpful in explaining the reported data.

CONTROL	D=2.8±0.4 x 10^{-6} (cm^2/sec)		n=38
FREEZING			
Animal Number	1	2	3
Cooling Rate (^oC/min)	25	60	60
Warming Rate (^oC/min)	130	1200	20
Holding Temp (^oC)	-10	-15	-10
Holding Time (sec)	0	12	24
D (cm^2/sec x 10^6)	1.13±0.30	1.02±0.11	0.70±0.20
n	7	4	5

Table 1. Apparent diffusivity of 3000 MW FITC dextran in the buffered saline suffused hamster gut mesentery.

ACKNOWLEDGEMENTS

This study was performed under the sponsorship of National Institutes of Health grant Number GM22693-04.

REFERENCES

1. Meryman, H.T., ed. "Cryobiology", Academic Press London, 1966.
2. Smith, A.U., ed "Current Trends in Cryobiology", Plenum Press, New York, 1970.
3. Mazur, P. "Cryobiology: The Freezing of Biological Systems", Science 168, 1970, 939-949.
4. Kreyberg, L. and Hanssen, O.E. "Necrosis of Whole Mouse Skin In Situ and Survival of Transplanted Epithelium after Freezing to -78°C and -190°C., Scand. J. Clin. Invest., 2, 1950, 168-170.
5. Weatherly White, R.C.A., Sjostiom, B. and Paton, B.C. "Experimental Studies in Cold Injury. II. Pathogenesis of Frostbite", J. Surg. Res. 4, 1965, 22.
6. Evans, C.D. and Diller, K.R. "A Programmable, Controlled Temperature Stage for Burn and Freezing Studies in the Microcirculation", Microvas. Res., 1980, submitted.
7. Zweifach, B.W., Grant, L., and McCluskey, R.T. "The Inflammatory Process", Academic Press, New York, 1974.
8. Fox, J.R. and Wayland, H. "Interstitial Diffusion of

Macromolecules in the Rat Mesentery", Microvas. Res., 18, 1979, 255-276.

9. Green, D.M. and Diller, K.R. "Measurement of Burn Induced Leakage of Macromolecules in Living Tissue", J. Biomech. Engr. 100, 1978, 153-158.

10. Evans, C.D., Diller, K.R. and Green, D.M. "Interstitial Macromolecular Diffusivity in Burned Tissue", in 1979 Biomechanics Symposium, ed. by W.C. Van Buskirk, ASME, New York, 1979, 93-96.

11. Carslaw, H.S. and Jaeger, J.C. "Conduction of Heat in Solids, 2nd. Ed.", Oxford University Press, London, 1959.

12. Evans, C.D., Diller, K.R. and Parsons, J.P. "Burn Induced Alterations of Interstitial Diffusion in Mesentery Tissue", in 1979 Advances in Bioengineering, ed. by M.K.Wells, ASME, New York, 1979, 161-164.

13. Garlick, D.G. and Renkin, E.M. "Transport of Large Molecules from Plasma to Interstitial Fluid and Lymph in Dogs", Amer. J. Physiol., 219, 1970, 1595-1605.

14. Nakamura, Y. and Wayland, H. "Macromolecular Transport in the Cat Mesentery", Microvas. Res. 9, 1975, 1-21.

LINEAR SYSTEMS ANALYSIS APPLICATIONS IN THE STUDY OF ARTERIAL HEMODYNAMICS

S.Laxminarayan, R.Laxminarayan*, S.Chatterjee, O.Mills, J.Ronda, & E.D.Weitzman

Neurology, Montefiore Hosp. & Einstein College of Med.&*Chalmers Univ. Goteborg
111 E. 210 Street, Bronx, NY 10467, USA

INTRODUCTION

In recent years, the applications of linear systems analysis techniques have become increasingly valuable in the study and understanding of the arterial hemodynamics. The aim of this paper is to review these concepts as applied to the characterization of pressure-flow relationships in the ascending aorta.

The arterial tree is an intricate system consisting of a complex branched network of tubes. The heart can be considered as a generator of pressure and flow waves travelling with a finite velocity, through such a system. The large arteries, the aorta in particular, act as a so-called "wind-kessel", storing by its capacitance the stroke volume from the cardiac output, for delivery at a steady rate through the peripheral circulation. The wave shape of the pressure pulse depends upon the location of the measuring site. For example, the more peripheral the pressure wave is measured, the more delayed it is with respect to the aortic pressure. While travelling towards the periphery, the wave characteristics are such that the pressure wave becomes smoother, with the pulse amplitude increasing, whereas the flow amplitude decreases and its shape becomes broader and smoother. The pressure-flow relationship measured in the ascending aorta offers a very useful means of characterizing the entire arterial tree.

SYSTEM LINEARITY IN THE STUDY OF THE ARTERIAL SYSTEM

Several investigators have shown that the arterial system is sufficiently linear so as to make use of the linear systems analysis concepts. For example, Dick et al. (1968), measured the non-linearity in the arterial system of the dog and their results have indicated that the principle of superposition is valid for the arterial system. Nobel et al. (1967) demonstrated in their experiments on conscious dogs, that the pressure-flow relationship measured in the ascending aorta, was little affected by changes in heart rate, implying a linear system. Typical non-linear elements in the cardiovascular system are the heart valves. Abel (1971) has pointed out however that these do not contribute to significant errors in the system transfer function. Experimental studies have shown that the pressure-volume relationship for a segment of an artery behaves in a non-linear manner (Bergel, 1961). Comparison of the results of model studies of the entire arterial system by making use of linear systems theory with the real system has provided, however, satisfactory indications to justify the use of linear theory approach, in understanding the characteristics of the arterial system. Although, measurements by several investigators have yielded somewhat contradictory results as to the degree of non-linearity in the cardiovascular systems of the human and the dog, most studies have shown the non-linearity to be small.

MATHEMATICAL CONSIDERATIONS

Since the arterial system can be considered to be linear, certain important mathematical techniques can be utilized in order to characterize the system.

Expressed mathematically, the essential operation of a linear system (Gabel and Roberts, 1973) is given by the superposition convolution, which, for a causal system relates the output p(t) to a given input f(t), by means of an integral equation:

$$p(t) = \int_0^t f(t-\tau)\cdot h(\tau)\cdot d\tau = f(t)*h(t) \quad .....(1)$$

Here h(t) is the impulse response function of the system, and the asterisk denotes the convolution operation.

For the case of the arterial system, p(t) and f(t) can be visualized as the pressure drop over and flow through the system. The pressure drop is the difference between the aortic pressure and the venous pressure and flow is measured in the aorta. In most cases, the venous pressure is significantly small and may be disregarded. However, in instances where the venous pressure is high, the necessary subtraction should be included. The impulse response function of the arterial system is the response of the system in terms of pressure, resulting from flow that is a unit impulse function (infinitely short duration and infinitely large with unit area).

An equivalent expression for the impulse response function h(t), in the frequency domain can be derived by using the convolution theorem and would read:

$$P(j\omega) = F(j\omega)\cdot H(j\omega) \quad ....(2)$$

where $P(j\omega)$, $F(j\omega)$ and $H(j\omega)$ are respectively the Fourier transforms of p(t), f(t) and h(t); that is,

$$H(j\omega) = \frac{P(j\omega)}{F(j\omega)} \quad ....(3)$$

$H(j\omega)$ is the input impedance of the arterial system, with the modulus, $H(\omega)$, given by the ratio of modulii of pressure and flow harmonics and the phase, $\Phi(\omega)$, of the input impedance, given by the the difference in phases between those of pressure and flow. Due to the periodicity of pressure and flow that results from the regularly beating heart, the continuous Fourier spectrum reduces to a Fourier series. The heart period (T) determines the spacing (ω) between the harmonics. The frequency of the lowest harmonic is equal to the frequency of the heart beat.

Equations 1 and 3 imply that with a knowledge of measured pressure and flow waves, we should be able to evaluate the system transfer function both in time and in frequency domains. Historically, the frequency domain description, i.e. the input impedance of the arterial tree has received considerable attention in the past; whereas, the time domain approach, i.e. the impulse response function of the entire arterial system has been only recently studied (Laxminarayan et al., 1978a).

ARTERIAL IMPEDANCE

The concept of impedance has proved to be extremely useful in the past, as a description of the pressure-flow relationship. With a knowledge of the impedance spectrum, one can predict the variations in arterial pressure produced by any possible flow. Use of impedances has also allowed a systematic study of wave reflections which provide insight into the important influence that the branching of the arterial tree contributes towards determining the arterial load on the heart.

Concept of Longitudinal, Transverse and Characteristic Impedances

In order to understand the behavior of the input impedance of the entire arterial system, it is useful to briefly define the concepts of local impedances. The local impedances (which depend on the artery under consideration only) characterize the pressure-flow relationships in small uniform segments of artery. These local impedances are referred to as longitudinal, transverse and characteristic impedances. While the longitudinal impedance relates the oscillating pressure gradient over and flow through a segment of artery, the transverse impedance expresses the relationship between the transmural pressure and volume change of a section of artery. Based on Womersley's (1957) derivation for flow in a stiff tube of circular cross section of radius R, the expression for the longitudinal impedance for large arteries and at high frequencies, can be written as

$$H_1 = j\omega(\rho/\pi R^2) = j\omega L' \quad \ldots\ldots\ldots\ldots\ldots(4)$$

L' in this expression is referred to as the effective mass of blood.

The transverse impedance is defined in a similar manner and is written as

$$H_t = 1/j\omega \cdot C'(\omega) \quad \ldots\ldots\ldots\ldots\ldots(5)$$

where C' is the compliance per length and equals dS/dP, with $S = \pi R^2$. C' depends on the radius and wall thickness of the artery and the young's modulus of elasticity of the wall material.

The concept of characteristic impedance, H_c, is related to an artery which is infinitely long and has uniform properties. For large arteries, H_c is a real number and is given by

$$H_c = (\rho/\pi R^2 \cdot dP/dS)^{1/2} \quad \ldots\ldots\ldots\ldots\ldots(6)$$

Obviously, H_c depends on the size of the artery and the properties of the wall material and of blood. Much literature is available about the significance of these local impedances (see eg. McDonald, 1974).

Input Impedance of the Arterial System

The input impedances of the systemic arterial tree have been determined in man and dog by several investigators((McDonald, 1964; Patel et al., 1963; Mills et al., 1970; O'Rourke et al., 1967). The major consideration in deriving the input impedance is to be able to produce sinusoidal pressure and flow waves in the arterial system. This implies using a sine wave pump to generate these waves, which is not practically feasible. However, if we consider heart as the natural generator of pressure and flow waves, then the measured signals can be subjected to Fourier analysis and thereby the use of equation 3. The application of Fourier analysis however requires that a number of assumptions be satisfied, such as the Drichlet conditions and the periodicity of the measured waveforms. Furthermore, the system under study should be constant, during the time of measurement. In hemodynamics, it has been well established that the measured pressure and flow waves are periodic in nature, with the period given by the heart rate and the signals do not suffer from the presence of discontinuities. In order to ensure that the input impedance is not affected by any change in system parameters, only steady state portions in the measured data are considered for analysis.

Estimation of Input Impedance

The input impedance of the arterial system can be obtained by measuring aortic pressure (with appropriate subtraction of venous pressure, if required) and aortic flow. Using digital techniques, the input impedance is calculated by applying Fourier analysis to the measured pressure and flow time histories. The ratio of the moduli of Fourier spectra of pressure and flow yields the modulus of input impedance and the difference in phases between those of pressure and flow gives the phase information. In estimating the input impedance by means of Fourier series analysis, the frequency resolution in the modulus and phase spectra of the input impedance is governed by the heart rate. Obviously, this implies that only those harmonics that are multiples of the heart rate will contain valid information about the impedance.

There has been considerable interest in recent years, however, to determine the behavior of input impedance at very low frequencies. It has been shown that by slowing the heart by means of vagal stimulation or by surgical heart block, the lowest frequency could be brought down to between 0.5 and 1 Hz. A more elegant method which is often considered in engineering systems analysis, would be to excite the system by means of random noise and then measure the response of the system. Such an approach, when applied to the arterial system, would facilitate estimating the input impedance at a wide range of frequencies. Taylor (1966) used this technique and by applying random stimulation to the right atrium, obtained the input impedance at frequencies as low as 0.00125 Hz.

The method of obtaining the input impedance of the arterial system at ultra low frequencies, by inducing randomness in the pressure flow signals, has however an important limitation. The method requires long stretches of data to be gathered, for analysis, which implies experiments have to be carried out over long periods of time. During these long experiments, the arterial system may not remain constant and hence, may give rise to a number of adverse effects. In order to overcome this, it may be worth exploring the application of other nonhormonic type excitation functions such as for example, the linear frequency swept sine wave (Laxminarayan et al., 1978b). It has been shown that the use of linear frequency swept sine wave has the advantage that with relatively short swept sine wave excitation, one can obtain almost rectangular modulus spectrum with relatively good frequency selectivity.

Quite independent of any form of external stimulation, another method based on transient analysis techniques, has been applied recently (Laxminarayan et al., 1979c), in extracting information at "intermediate" frequencies, from the measured pressure and flow signals. The method makes use of the theorem that the Fourier transform of a transient is also given by the integral of the Fourier transform of the differential of the original transient. (Papoulis, 1962)

Apart from the frequency resolution criteria that we have so far discussed, there may be a number of possible sources of errors that need to be considered in the calculation of input impedance. These errors stem from factors such as presence of noise in the measured signals, violation of steady state assumption for the arterial system, phase errors etc. The information contained in the measured signals at higher frequencies may be so small that the estimation of the input impedance at these frequencies is hampered by the presence of noise. In order to obtain a meaningful estimate of the input impedance, therefore, it is often necessary to consider the impedance values averaged over 10-15 heart beats. Averaging is performed by estimating the means of the real and imaginary parts of the input impedance (Westerhof et al., 1973).

The limitations, requirements and the possible errors that are associated with the procedures for determining the input impedance have been thoroughly studied by several investigators (Attinger et al., 1966).

Characteristic Features of Input Impedance

The concept of input impedance has proved to be extremely valuable in the understanding of arterial hemodynamics. The features of the input impedance can be explained from the physiological and anatomical properties of the arterial bed. The input impedance has the following characteristics.

The modulus of input impedance at zero Hz (which is the ratio of mean pressure to mean flow) is the peripheral resistance of the system and is high for steady flow. At higher frequencies, the value is about 5 to 15 percent of the peripheral resistance and approaches the characteristic impedance of the aorta (Bergel and Milnor, 1965). The presence of small minima and maxima in the input impedance as observed by some investigators (van den Bos et al., 1976) is related to the wave reflections in the arterial system. At the frequency of the first minimum, one quarter of the wavelength is equal to the length of the system. From a knowledge of this first minimum, f_{min} (McDonald, 1974) and the phase velocity c_{ph}, the length of the system 1, can be estimated as:

$$1 = c_{ph}/4\ f_{min}$$

Models of the Arterial System

The use of arterial models has provided much insight in the past, into our understanding of the behavior of the arterial system.

(a) Windkessel Model. It has been shown that the input impedance of the arterial system of man and animal during control situation may be modelled by means of a windkessel (Westerhof, 1968). The model (Fig. 1) consists of a resistor (R_c , equal to the characteristic impedance of the ascending aorta) in series with an inertia (L) and in series with a parallel arrangement of another resistor (R_p, equal to the peripheral resistance of the arterial system) and a capacitor (C, the total arterial compliance). It has also been demonstrated (Westerhof et al., 1973) that increased or decreased peripheral resistance can be modelled by means of a windkessel. That the hydrodynamic equivalent of the windkessel model can be used as a load for the heart, has been studied by Westerhof et al. (1971).

The theoretical input impedance of the model (Fig.1) as a function of frequency is given by

$$H(j\omega) = R_c + j\omega L + 1/C(j\omega + 1/R_pC) \quad .....(7)$$

where $\omega = 2\pi f$ (f = frequency).
We can rewrite equation 7 in terms of real and imaginary parts as

$$H(j\omega) = (R_c+R_p/1+(\omega R_pC)^2)- j(\omega R_p^2C/1+(\omega R_pC)^2-\omega L) \quad .....(8)$$

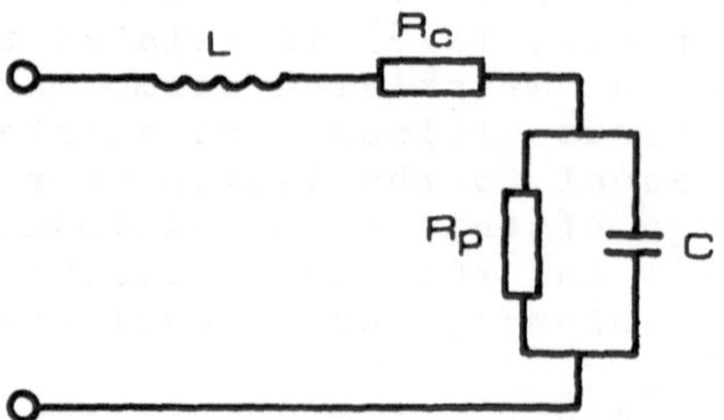

Fig. 1 Four-element windkessel model: R_p is the peripheral resistance, R_c is the characteristic impedance of aorta, L is an inertial term and C is the total arterial compliance.

From equation 8, we can obtain the frequency response vector plot (Laxminarayan et al., 1979b) for the model, by plotting the imaginary part of $H(j\omega)$ vs the real part of $H(j\omega)$.

From these equations, the d.c. absolute value of the input impedance is found to be

$$|H| = (R_p + R_c)$$

and the phase is zero degrees. For very high frequencies, it may be shown that

$$|H| = R_c$$

and the phase is back to zero again. Assuming the inertial component L, to be negligible in equation 8, (L = 0), the frequency response vector plot can be defined by the motion of a vector MD (Fig. 2) which follows a circular path with M as the center and the vector radius given by $(R_p/2)$. Using trigonometrical identities, we can compute the total arterial compliance C, from (Laxminarayan et al., 1979b).

$$C = (T/2\pi R_p)\cdot((1-\mathrm{Cos}\alpha_H)/(1+\mathrm{Cos}\alpha_H))^{1/2} \quad ....(9)$$

where T is the heart period in seconds.

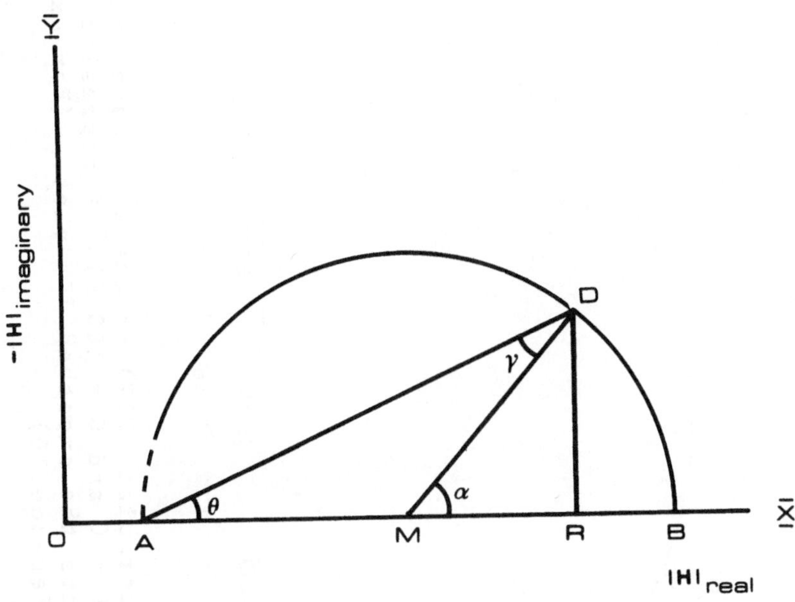

Fig. 2 Frequency response vector plot of the windkessel model.

$\mathrm{Cos}\,\alpha_H$ is calculated from the modulus and phase of input impedance at the fundamental frequency as

$$\mathrm{Cos}\alpha_H = \frac{\mathrm{Re}(H) - (R_p/2+R_c)}{\sqrt{((\mathrm{Re}(H)-(R_p/2+R_c))^2+(\mathrm{Im}(H))^2)}} \quad ....(10)$$

The frequency response vector plot calculated from "aortic pressure" and "aortic flow" as measured in a windkessel model subjected to random stimulation is shown in Figure 3.

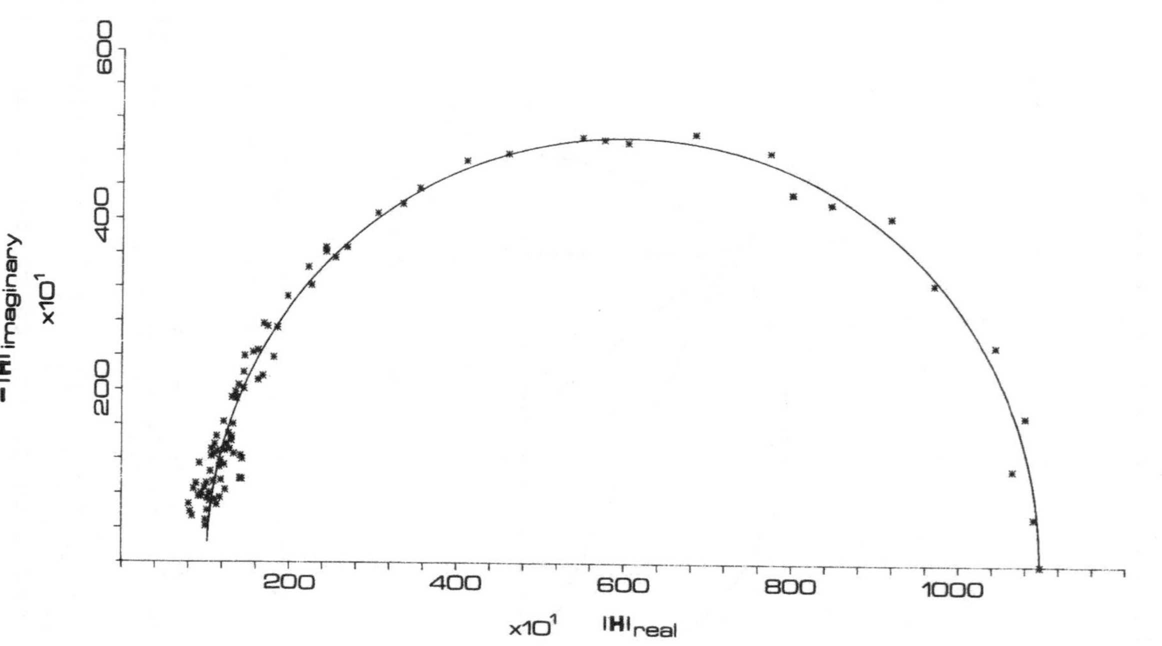

Fig. 3 Frequency response vector plot (FRV) of the windkessel model (R_p = 10kΩ, R_c = 1kΩ, L = 0 and C = 100μF). * represent measured and solid line represents the smoothed FRV plot. (For experimental details, see reference 10)

In the past, the total arterial compliance C has been often obtained by considering the peripheral resistance R_p and measuring the diastolic pressure decay that is assumed to be a negative exponential function with time constant equal to R_pC (Westerhof et al., 1973). The input impedance of the Windkessel model and the input impedance of man and animal during control situation have been shown to be similar (Fig. 4).

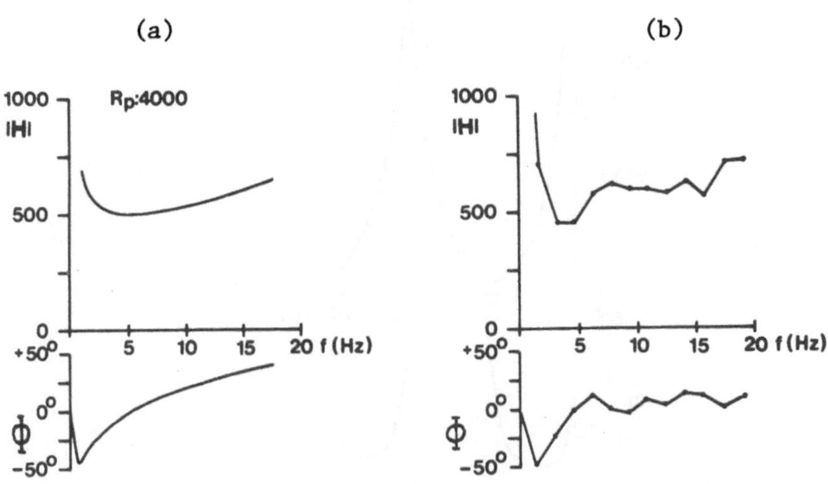

Fig. 4 (a) Input impedance of a windkessel model
(b) Input impedance of the arterial system of a control dog.

(b) Uniform Tube Model. The three element Windkessel is an appropriate model for the arterial system during control situation. In the presence of strong reflections during aortic occlusion, however, a more appropriate model would be the uniform tube with properties as those of the aorta and loaded with a peripheral resistance. Instead of a single tube, models based on two tubes either in series or in parallel, have been shown to be more similar in the input impedance characteristics, to the arterial tree (Wetterer and Kenner, 1968).

The input impedance of a uniform elastic tube (length l and phase velocity c_{ph}) loaded with a hydrodynamic resistor is shown in Fig. 5. It is seen that the presence of the reflection site at the end of the tube results in maxima and minima in the modulus and phase spectra of the input impedance. From a knowledge of the frequencies where the maxima and minima occur and the phase velocity, the length of the tube (which corres-

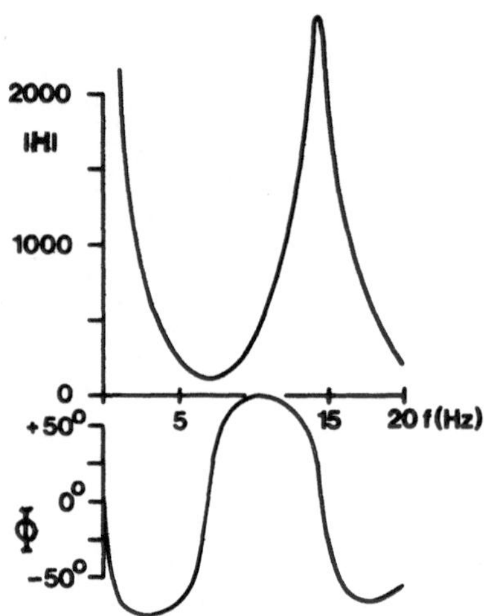

Fig. 5 Input impedance of a uniform tube model.

ponds to the distance between the heart and the location where the aorta is occluded) can be estimated (Sipkema and Westerhof, 1975.)

ARTERIAL IMPULSE RESPONSE FUNCTION

The impulse response function of the arterial system is the time domain description of the system (equation 1). It contains the same information as revealed in the input impedance, but emphasizes different aspects of the arterial tree. (Laxminarayan et al., 1978a) Considerable work has been done in estimating the impulse response function of the arterial system by making use of linear systems analysis concepts. In essence, the method consists of designing an "acceptable" solution for the convolution integral defined in equation 1.

We note from equation 1 that the experimental procedures for determining the impulse response function would involve producing a brief transient change in flow (unit impulse) and then measure the system response in terms of pressure. In practice, however, it is difficult to generate these transient changes, and therefore, we make use of the heart as the pump and resort to determining h(t) from the measured pressure and flow time histories. Several approaches have been investigated recently in search for a solution to the convolution integral. These are (a) by transforming the integral into a set of algebraic equations, (b) by applying matrix and numerical approximation methods, (c) by means of auto and cross correlation techniques, (d) by reducing the convolution integral into Volterra type of integral equations of the second kind and (e) by using transform techniques in frequency domain.

Methods (a) and (b) are extremely sensitive to the presence of noise in the data and it turns out that the initial value of the excitation time history, that is, the flow, is of crucial importance. The method of auto and cross correlation for the case of the arterial system fails, since the auto correlation of flow is not a delta function. Method (d) has been applied with some success, but involves extensive data smoothing (Laxminarayan, Laxminarayan and Jongbloed, 1979a). The last method, that is, the transform method is based on the inverse Fourier transformation of equation (3) and is detailed elsewhere (Laxminarayan, Sipkema and Westerhof, 1978a).

Estimation of Arterial Impulse Response Function Using the Transform Method

The input impedance and the impulse response function of the arterial tree are the Fourier transform pairs of each other. Therefore, we can obtain the arterial impulse response function by inverse Fourier transforming the input impedance which is given by equation 3. The limitation of Fourier transform process for obtaining the impulse response function from the input impedance is that the transformation involves an infinite integral, the evaluation of which requires a knowledge of $H(\omega)$ for all values of ω up to infinity. Truncation of $H(\omega)$ is valid provided the function that is to be transformed has reached steady zero values. However, the impedance values at higher frequencies approximate the characteristic impedance and never reaches zero values. If such a function is transformed, then effectively a window has been introduced, the effects of which are reflected as oscillations superimposed on the resulting impulse response function. In order to eliminate these truncation effects, the approach taken is to filter the higher harmonics so that the amplitudes at these higher harmonics taper off to negligible values. The filter that has been used in these studies is based on the Dolph-Chebychev function and the complete calculation methods to determine the impulse response function are given elsewhere (Laxminarayan, Sipkema and Westerhof, 1978a). The impulse response functions of a Windkessel model, with and without filtering, are illustrated in Figure 6.

Characteristic Features of the Impulse Response Function

The features of the arterial impulse response function are best illustrated by considering the arterial models.

The theoretical impulse response function of the 4-element windkessel model (Figure 1) is given by (for a periodic excitation of the model with a period T):

$$h(t) = R_c \cdot \delta(t) + L \cdot \delta'(t) + e^{-t/\tau}/(C(1-e^{-T/\tau})) \ldots.(11)$$

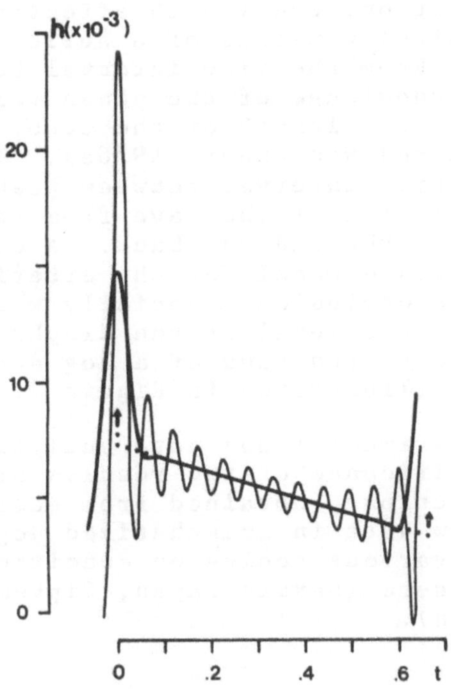

Fig. 6 Impulse response function of a 4-element windkessel model. Dotted line: theoretical impulse response function, oscillatory curve: impulse response function without filtering, smooth curve: impulse response function with filtering.

This implies that h(t) is composed of a delta function plus a doublet and an exponentially decaying function with time constant τ(See Figure 6). When the pressure-flow relations in the arterial system can be approximated with the windkessel model, the total arterial compliance can be obtained by estimating the height of the exponential decay in the impulse response function, extrapolated to time zero. The decay time of this exponential portion is the same as that of the diastolic part of aortic pressure tracing.

The impulse response function of a uniform tube closed at one end (which effectively forms a reflection site) consists of a series of equidistant peaks. From the time interval between these peaks and a knowledge of the phase velocity, we can calculate the length of the tube. (Laxminarayan, Sipkema and Westerhof, 1978a). This results because the time interval between peaks is equal to the travel time of the wave from the entrance of the tube to the end and back. A uniform tube is an appropriate model for the arterial system during aortic occlusion especially when the occlusion is at the level of the diaphragm. The impulse response function of a dog during aortic occlusion is illustrated in Figure 7.

The experimental and data analysis procedures and the significance of the results of the impulse response functions, obtained from measured presure and flow data in anesthetized dog, under control and various occlusion conditions, are discussed elsewhere (Laxminarayan, Sipkema and Westerhof, 1978a).

WAVE REFLECTIONS IN THE ARTERIAL SYSTEM

In cardiovascular dynamics, it is well known that the pressure and flow waves are not purely waves travelling outwards from the heart, but they include a number of reflections from bifurcations. In other words, all waves generated by the heart are partially reflected at many bifurcations. Although waves reflected from the different parts of the arterial system cannot be studied, we can determine the summated reflected wave. From the pressure and flow signals measured at a single location, we can determine the overall reflections

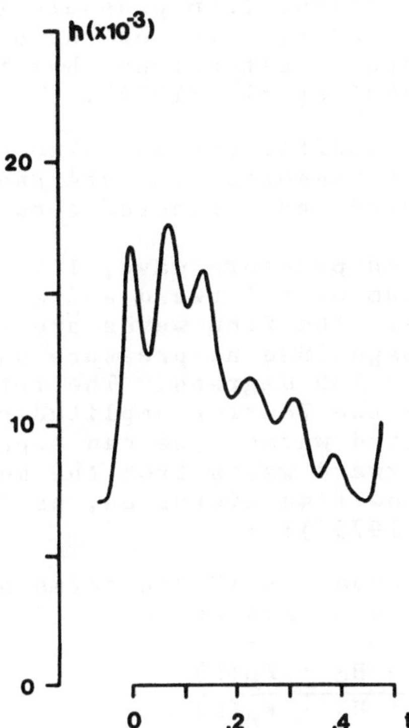

Fig. 7 Impulse response function of the arterial system of a dog during aortic occlusion (at the level of the diaphragm).

from points distal to this location. One way of characterizing the arterial system is by considering the reflection coefficient which can be calculated from

$$R(\omega) = \frac{H(\omega) - H_c}{H(\omega) + H_c} \qquad \text{........(12)}$$

where H_c is the characteristic impedance. Since we assume that the characteristic impedance is a real and frequency independent constant, it can be obtained from the modulus of input impedance at high frequencies (Bergel and Milnor, 1965). The reflection coefficient from pressure-flow measurements in the ascending aorta of the dog during various experimental situations, has been estimated by Westerhof et al. (1972).

Reflection coefficient can also be calculated by resolving the measured pressure and flow waves into their forward and reflected components.

The measured pressure wave, like the flow wave, is the sum of a forward and a reflected (backward) wave. The flow waves are reflected with the same magnitude as pressure waves, but the phase differs by 180 degrees. The reflection coefficient relates the Fourier amplitudes of the forward and reflected waves. We can separate the forward and backward waves from the measured pressure and flow time histories, as follows (Laxminarayan, 1979d):

We can rewrite equation 12, in terms of measured pressure and flow harmonics as:

$$R(f) = \frac{P_M(f) - H_c \cdot F_M(f)}{P_M(f) + H_c \cdot F_M(f)} \qquad \ldots\ldots\ldots\ldots(13)$$

where the subscript M denotes measured.

The harmonics of forward pressure are given by (Westerhof et al., 1972)

$$P_F(f) = \frac{P_M(f)}{1 + R(f)} \qquad \ldots\ldots\ldots\ldots(14)$$

Substitution of equation 13 in equation 14 yields

$$P_F(f) = 1/2[P_M(f) + H_c \cdot F_M(f)] \qquad \ldots\ldots\ldots\ldots(15)$$

Since $P_M(f) = P_F(f) + P_B(f)$, the backward flow reads

$$P_B(f) = 1/2[P_M(f) - H_c \cdot F_M(f)] \qquad \ldots\ldots\ldots\ldots(16)$$

Inverse Fourier transformation of 15 and 16 gives the time functions of backward and forward pressure:

$$p_F(t) = 1/2[p_M(t) + H_c \cdot f_M(t)] \quad ....(17)$$

$$p_B(t) = 1/2[p_M(t) - H_c \cdot f_M(t)] \quad ....(18)$$

Similar results can be obtained for forward and backward flow. The reflection coefficient is obtained by Fourier analyzing the forward and backward waves and computing the ratio of their Fourier amplitudes and phase differences. Forward and reflected waves have been calculated by Van den Bos et al. (1976) for the entire arterial tree during control and various interventions.

ACKNOWLEDGEMENTS

The authors wish to express their sincere gratitude to Miss Anne Marie Hardy for the meticulous typing of the different versions of the manuscript and to Miss Jeanne Cowan for her valuable assistance in the art work.

REFERENCES

1. Abel, F.L. (1971) Fourier analysis of left ventricular performance, Circ. Res. 28, 119-135.
2. Attinger, E.O., Sugarwara, H., Navarro, A., Ricetto, A., Martin, R. (1966) Pressure-flow relations in the dog arteries, Cir. Res., 19, 230-245.
3. Bergel, D.H. (1961) The dynamic elastic properties of the arterial wall, J. Physiol. 156, 458-469.
4. Bergal, D.H. and Milner, W.R. (1965), Pulmonary vascular impedance in the dog, Cir. Res. 16, 410-415.
5. Dick, D.E., Kendrick, J.E., Matson, G.L., and Rideout, V.C. (1968) Measurement of nonlinearity in the arterial system of the dog by a new method, Cir. Res. 22, 101-112.
6. Gabel, R.A. and Roberts, R.A. (1973) Signals and Linear Systems, John Wiley and Sons, New York, London, Sydney and Toronto.
7. Laxminarayan, S., Sipkema, P., and Westerhof, N. (1978a) Characterization of the arterial system in the time domain, IEEE Trans. Biomed. Eng., BME - 25, 2, 177-184.
8. Laxminarayan, S. and Laxminarayan, R. (1978b) Use of swept wine wave in physiological systems analysis, IEEE Trans. Biomed. Engng., BME - 24, 1, 103-105.
9. Laxminarayan, S. and Laxminarayan, R. (1979c) Application of a transient analysis method in the estimation of the input impedance of the arterial system, to be published.
10. Laxminarayan, S., Laxminarayan, R. Langewouters, G.J., and Vos A.v.D. (1979b) Computing total arterial compliance of the arterial system from its input impedance, Med. & Biol. Eng. & Comp., 17, 623-628.
11. Laxminarayan, S., Laxminarayan, R. and Jongbloed, A.A. (1979a) Linear systems analysis applications by deconvolution techniques in cardiovascular systems analysis, DECUS Europe symposium, Copenhagen, Denmark.

12. Laxminarayan, S. (1979d) The calculation of forward and backward waves in the arterial system, Med. & Biol. Eng. & Comp., 17, 130.
13. McDonald, D.A. (1964) Frequency dependence of vascular impedance, Pulsatile blood blow, Ed. Attinger, E.P., McGraw-Hill, New York.
14. McDonald, D.A. (1974) Blood flow in arteries, Arnold, London.
15. Mills, C.J., Gabe, I.T., Gault, J.H., Mason, D.T., Ross (Jun) J., Braunwald, E. and Shillingford, J.P. (1970) Pressure-flow relationships and vascular impedance in man, Cardiovas. Res. 4, 405-417.
16. Nobel, M.J.M., Gabe, I.T., Trenchard, D., and Guz, A. (1967) Blood pressure and flow in the ascending aorta of conscious dogs, Cardiovas. Res., 1, 9-20.
17. O'Rourke, M.F. and Taylor, M.G. (1967) Input impedance of the systemic circulation, Circ. Res. 20, 365-380.
18. Papoulis, A. (1962) The Fourier integral and its applications, McGraw-Hill, New York.
19. Patel, D.J., De Freitas, F.M., and Fry, D.L. (1963) Hydraulic input impedance to aorta and pulmonary artery in dogs, J. App. Phys. 18, 134-140.
20. Sipkema, P. and Westerhof, N. (1975) Effective length of the arterial system, Annals Biomed. Eng., 3, 296-307.
21. Taylor, M.G. (1966) Use of random excitation and spectral analysis in the study of frequency-dependent parameters of the cardiovascular system, Cir.Res. 18, 585-595.
22. Van den Bos, G.C. Westerhof, N., Elzinga, G., and Sipkema, P. (1976) Reflection in the systemic arterial system: effects of aortic and carotid occlusion, Cardiovasc. Res., 10, 565-573.
23. Westerhof, N. (1968) Analogue studies of human systemic arterial hemodynamics, Ph.D., thesis, University of Pennsylvania, USA.
24. Westerhof, N., Elzinga, G. and Sipkema, P. (1971) An artificial arterial system for pumping hearts, J. of App. Physiol. 31, 5, 776-781.

25. Westerhof, N., Elzinga, G., and Van den Bos, G.C. (1973) Influence of central and peripheral changes on the hydraulic input impedance of the systemic arterial tree, Med. & Biol. Eng. and Comp. 11, 710-722.
26. Westerhof, N., Sipkema, P., Van den Bos, G.C., and Elzinga, G. (1972) Forward and Backward waves in the arterial system, Cardiovasc. Res. 6, 648-656.
27. Wetterer, E. and Kenner, Th. (1968) Grundlagen der Dynamik des arterianpulsen, Springer - Verlag, Berlin-Heidelberg - N.Y.
28. Womersley (1957) An elastic tube theory of pulse transmission and oscillatory flow in mammalian arteries, WADC Tech. Rep. TR56-614.

LUMPED PARAMETER MODELING OF THE INCREASED HYDRAULIC RESISTANCE IN HYPERTENSION

P.M. Hutchins, V.L. Roddick and J.W. Dusseau

Departments of Biomedical Engineering
Physiology and Pharmacology
Bowman Gray School of Medicine
Winston-Salem, NC 27103

INTRODUCTION

Established hypertension is characterized by an elevated hydraulic resistance to blood flow through the organs of the body. The lack of understanding concerning the pathogenesis of essential hypertension may be due largely to the simplistic techniques utilized in analyzing vascular resistance. The inability to separate resistance into its constituent parts is of particular importance in this regard.

Previous analyses of the resistance vessel changes in hypertension have focused on the steady-state pressure and flow relationships in individual organs and the total body, [1,2,3] on the mechanical properties of the arterial wall, [4,5] or on the morphometric parameters of vascular tissue, [6,7]. The results of these experiments have provided valuable information concerning total resistance changes or mechanical and morphometric properties of maximally dilated vessels. However, little is known concerning the *in vivo* dimensions of the small arterioles and how they are altered in hypertension. The organization of the vascular tree is a fairly complex parallel-series arrangement of vessels of decreasing diameter and length and increasing number. How the internal diameter, number and length of each segment of the arterial tree contributes to the increased hydraulic resistance in hypertension has yet to be determined. Clearly, a method is needed which will allow the separation of total resistance into each of its constituent parameters and the localization of resistance changes in terms of site along the vascular tree.

METHODS

The hypertensive model used in our studies is the well-documented Spontaneously Hypertensive Rat (SHR), a highly in-bred strain of the Japanese Wistar rat [8]. The animals were bred and housed in our animal care facility with a 12:12 light:dark cycle and food and water *ad libidum*. The hypertension develops between the 3rd and 12th week of age, after which blood pressure plateaus at a stable value. We investigated animals before (3 weeks), during (6 weeks) and after (12 and 30 weeks) the development of hypertension.

In order to quantitate the vascular dimensions important in the determination of hydraulic resistance, we observed a microcirculatory bed with small arterioles of 100 μm and less internal diameter. The microcirculatory preparation used was a closed modification [9] of the cremaster preparation as described by Baez and Orkin [10]. The rats were anesthetized with urethane (450 mg/kg) and chloralose (120 mg/kg) injected intraperitoneally. Carotid arterial catheters were inserted for the measurement of systemic arterial pressure and heart rate.

As indicated in the Introduction the miscrovascular parameters deemed most important in the determination of hydraulic resistance were the length, internal diameter and number of parallel vessels in each arteriolar segment. Since the major cremasteric artery (approximately 100 μm internal diameter) undergoes, on the average, 3 successive branchings before emptying into capillaries, four series segments of arterioles were observed. The cremasteric artery was classified 1A or first order arteriole; the next branches were 2A; 3A branched off 2A; and 4A off 3A (Figure 1). The number of arterioles in each parallel segment was counted. Likewise, the average internal diameter of the arterioles in each segment was determined. The internal diameter was used since it is more directly related to hydraulic resistance than is the external diameter. We have previously shown that the length of arterioles is unchanged in hypertension[11], and thus they were not measured in this study but were assumed constant.

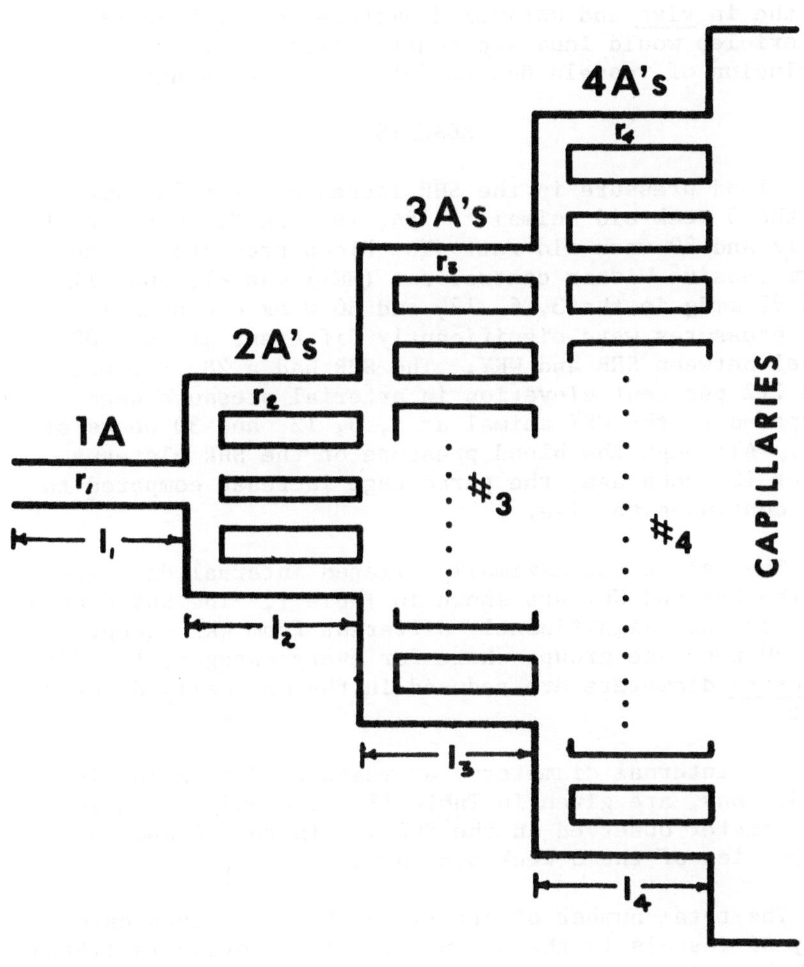

FIGURE 1

Model of Arteriolar Network

In order to assess the contribution of structural and functional factors, the microvasculature was quantitated *in vivo* and by latex injection after maximal vasodilatation. The difference between normotensive and hypertensive vasculatures when maximally dilated would be indicative of structural alterations. A comparison of the *in vivo* and maximum diameters and numbers of arterioles would indicate constriction and total occlusion of vessels due to functional influences.

RESULTS

Blood pressure in the SHR increased from 104 mmHg in the 3 week old animal to 154, 193 and 206 mmHg in the 6, 12 and 30 week old rat. The blood pressure of the normotensive Wistar control rat (WKY) was 81, 106, 119, and 97 mmHg in the 3, 6, 12, and 30 week old animal. The pressures were significantly different at the .005 level between SHR and WKY. The SHR had a 28, 45, 62, and 112 per cent elevation in arterial pressure when compared to the WKY animal at 3, 6, 12, and 30 weeks of age. Although the blood pressure of the SHR plateaus after 12 weeks age, the percentage increase compared to WKY continues to rise.

The values for maximally dilated internal diameters of the SHR and WKY are shown in Table I. The SHR diameters are not significantly different from WKY except in the 30 week age group, where for every category the SHR *internal* diameters are reduced in the *maximally dilated* state.

The internal diameters, as measured during *in vivo* conditions, are given in Table II. The only reduction in diameter observed in the SHR was in the 1A and 2A categories of the 3 week old animal.

The total number of arterioles found in each category of vessels in the maximally dilated state is listed in Table III. A reduction in the total number of parallel fourth order arterioles is seen in the SHR during the development of hypertension at 6 weeks of age. However, at 30 weeks of age this change has been reversed.

The number of arterioles open to blood flow (*in*

TABLE I

INTERNAL DIAMETER OF MAXIMALLY DILATED, LATEX INJECTED ARTERIOLES (MICROMETERS)

	ORDER OF ARTERIOLES							
	1A		2A		3A		4A	
	WKY	SHR	WKY	SHR	WKY	SHR	WKY	SHR
3 week	89.6	81.7	45.0	42.8	21.7	20.2	13.5	12.6
6 week	135.1	153.4	75.2	73.2	32.2	31.0	18.3	18.1
12 week	151.0	134.2	79.3	79.8	31.7	30.9	16.9	16.7
30 week	171.0	137.8*	97.6	81.3*	41.8	33.1*	21.2	17.8*

*Statistically significant at the $p < .025$ level when compared to the same vessel category of WKY at the same age.

TABLE II

INTERNAL DIAMETERS OF *IN VIVO* ARTERIOLES (MICROMETERS)

	ORDER OF ARTERIOLE							
	1A		2A		3A		4A	
	WKY	SHR	WKY	SHR	WKY	SHR	WKY	SHR
3 week	74.8	62.6*	33.8	28.7*	15.7	14.3	10.0	9.8
6 week	107.4	104.6	53.5	54.4	22.1	21.0	12.0	11.2
12 week	117.8	124.2	63.1	61.8	22.9	21.7	13.2	12.1
30 week	122.2	109.2	67.9	63.5	24.5	25.3	13.2	13.9

*Statistically significant at the $p < .05$ level when compared to the same vessel category of WKY at the same age.

TABLE III

TOTAL NUMBER OF ARTERIOLES IN MAXIMALLY DILATED, LATEX INJECTED PREPARATION
(Number of arterioles per mm^2)

	ORDER OF ARTERIOLE							
	1A		2A		3A		4A	
	WKY	SHR	WKY	SHR	WKY	SHR	WKY	SHR
3 week	-	-	0.10	0.11	0.65	0.72	1.75	2.03
6 week	-	-	0.04	0.04	0.34	0.30	1.47	0.97*
12 week	-	-	0.03	0.04	0.23	0.27	1.03	1.15
30 week	-	-	0.03	0.03	0.19	0.23	0.79	1.01*

*Statistically significant at the $p < .05$ level when compared to the same vessel category of WKY at the same age.

vivo state) is shown in Table IV. The most significant finding is a reduction in "density" of open 4A arterioles at 6 and 30 weeks. This is in spite of an increased total number of arterioles (latex) at 30 weeks of age (see Table III).

DISCUSSION

The resistance of a parallel grouping of arterioles is given by the Poiseuille Equation -

$$R = \frac{8\eta\ell}{\#\pi r^4}$$

where: η is the viscosity of the fluid, ℓ is the length of the arterioles in that segment, # is the total number of vessels in parallel, and r is the radius of the arterioles.

By assuming that all arterioles in a particular category (i.e., 1A, 2A, 3A, or 4A) have the same length and radius - the average for all arterioles in that category - and assuming that the entrance and exit pressures for each segment are equal for all arterioles in a category (see Figure 1), then a resistance value may be calculated for each segment. Since the segments are in series (Figure 1) the sum of the four segment resistances should approximate the total resistance of the arteriolar network. Microvascular pressure measurements have shown that the majority of vascular resistance resides in the arterioles.

In fact, most hydraulic resistance is due to the fourth order arterioles. With this in mind, a quantity directly related to the structural resistance of the fourth order arterioles was calculated from the diameter and number measurements derived from the latex injection studies (Table V). The equation used was:

$$R = \frac{10^8}{\#_{\text{Latex}} \cdot D^4_{\text{Latex}}}$$

This calculation assumes that the length of fourth order arterioles is not different between SHR and WKY, a finding demonstrated by us previously[11]. It is apparent

TABLE IV

NUMBER OF ARTERIOLES OPEN TO BLOOD FLOW - *IN VIVO* PREPARATION
(Number of Arterioles per mm^2)

	ORDER OF ARTERIOLE							
	1A		2A		3A		4A	
	WKY	SHR	WKY	SHR	WKY	SHR	WKY	SHR
3 week	-	-	0.07	0.09*	0.27	0.34	0.65	0.67
6 week	-	-	0.03	0.04*	0.15	0.12	0.43	0.26*
12 week	-	-	0.03	0.03	0.16	0.16	0.45	0.43
30 week	-	-	0.02	0.02	0.15	0.12	0.52	0.39*

*Statistically significant at the $p < .05$ level when compared to the same vessel category of WKY at the same age.

TABLE V

STRUCTURAL RESISTANCE (LATEX) CHANGES OF FOURTH ORDER ARTERIOLES DURING THE DEVELOPMENT OF HYPERTENSION (Arbitrary Units)

	3 weeks	6 weeks	12 weeks	30 weeks
WKY	7.2	2.9	5.9	2.8
SHR	7.2	6.4*	5.8	4.6*

*Indicates significant differences ($p < .05$) when compared to same age WKY.

TABLE VI

FUNCTIONAL RESISTANCE (*IN VIVO*) CHANGES OF FOURTH ORDER ARTERIOLES DURING THE DEVELOPMENT OF HYPERTENSION (Arbitrary Units)

	3 weeks	6 weeks	12 weeks	30 weeks
WKY	7.3	6.1	3.4	2.6
SHR	7.6	11.0*	5.1	2.7

*Indicates significant differences ($p < .05$) when compared to same age WKY.

from Table V that the "structural resistance" of SHR is elevated above WKY at 6 and 30 weeks of age. Analysis of Tables I and II indicates that the increase in structural resistance was number-based at 6 weeks and diameter-based at 30 weeks.

A similar quantity related to actual resistance to blood flow was calculated for the fourth order arterioles from the *in vivo* measured data (Tables II and IV). Again the "resistance" to blood flow was elevated at 6 weeks due primarily to the structural reduction in number of parallel vessels. When one compares the percentage of the total number of fourth order arterioles that are open to blood flow during the development of hypertension, a definite pattern is apparent. Whereas the WKY exhibits an increasing percentage open with age (from 30 to 60%), the SHR maintains approximately 35% of the total number open at all ages.

In order to assess the deviation from the maximally dilated structural resistance at different ages during hypertension development, a quantitation of "tone" due to vasoconstriction and "tone" due to vasoocclusion was derived. It was desired that each of these vary from 0 to 1: tone being 0 when the vessels were maximally dilated or maximally open, and tone equal to 1 when they were maximally constricted or occluded. Tone due to vasoconstriction was called "Lumenal Tone" and was calculated from:

$$\text{Lumenal Tone} = \frac{D^4_{MAX} - D^4_{\underline{IN\ VIVO}}}{D^4_{MAX}}$$

The latex diameter was assumed to be the maximum allowable structural diameter at the existing pressure. The diameters were raised to the fourth power since this is how they appear in the Poiseuille Equation. The most apparent difference (Table VII) is a reduction in Lumenal Tone at 30 weeks in the SHR. This is most likely due to the reduction in structural diameter of the 30 week SHR (see Table I), a finding consistent with morphometric studies of fixed tissue from older SHR[6,7].

The tone due to complete occlusion of some fourth

TABLE VII

LUMENAL TONE OF FOURTH ORDER ARTERIOLES DURING THE DEVELOPMENT OF HYPERTENSION

$$\frac{D^4_{MAX} - D^4_{\underline{IN}\ \underline{VIVO}}}{D^4_{MAX}}$$

	3 weeks	6 weeks	12 weeks	30 weeks
WKY	.62	.82	.63	.85
SHR	.63	.85	.72	.63

TABLE VIII

NUMERICAL TONE OF FOURTH ORDER ARTERIOLES DURING THE DEVELOPMENT OF HYPERTENSION

$$\frac{\#_{MAX} - \#_{\underline{IN}\ \underline{VIVO}}}{\#_{MAX}}$$

	3 weeks	6 weeks	12 weeks	30 weeks
WKY	.63	.71	.56	.41
SHR	.67	.74	.63	.61

order arterioles was calculated and referred to as Numerical Tone:

$$\text{Numerical Tone} = \frac{\#_{MAX} - \#_{\underline{IN}\ \underline{VIVO}}}{\#_{MAX}}$$

Again, the number of arterioles is entered with a power of one since that is how it appears in the Poiseuille Equation. Numerical tone is 0 when all vessels are open and 1 when all available arterioles are closed in the normal, *in vivo* state. The values for Numerical Tone are presented in Table VIII. The most outstanding difference is an elevation in Numerical Tone in the 30 week SHR. This was previously alluded to in the discussion of the greater percentage of open vessels in the 30 week WKY (60% for WKY, 39% for SHR).

It is particularly interesting to note that there is no increase in either Lumenal or Numerical Tone in the 6 week SHR. This further indicates that the increased resistance to blood flow at 6 weeks is structurally based and due to a reduction in number of fourth order arterioles.

SUMMARY

Hypertension has been shown to involve a structural decrease in number of smallest arterioles early in the disease and a functional reduction in number of these same arterioles after the hypertension had become established. A structural reduction in radius was observed only in the late phases of the disease.

ACKNOWLEDGMENTS

The authors gratefully acknowledge the excellent assistance of Ms. Stephanie Burgoyne in the preparation of this manuscript. This work was supported in part by PHS grants HL 13936,HL 05392 and grants from the N.C. Heart Association.

REFERENCES

1. Ferrone, R.A., G.M. Walsh, M. Tsuchiya, and E.D. Frohlich: Comparison of Hemodynamics in Conscious Spontaneous and Renal Hypertensive Rats. Amer. J. Physiol. 236(3):H403-H408, 1979.
2. Brock, T.A. and J.N. Diana: Effect of DOCA-NaCl Hypertension on Pre- and Postcapillary Resistance in Isolated Hindlimbs of Dogs. Amer. J. Physiol. 236(4):H586-H591, 1979.
3. Romanovska, L., I. Pierovsky, and J. Stribrna: Blood Flow and Vascular Resistance in Lower Limbs in Hypertensives at Rest and at Reactive Hyperemia. Cor Vasa 19(1):61-65, 1977.
4. Cox, R.H.: Comparison of Arterial Wall Mechanics in Normotensive and Spontaneously Hypertensive Rats. Amer. J. Physiol. 237(2):H159-H167, 1979.
5. Berry, C.L. and S.E. Greenwald: Effects of Hypertension on the Static Mechanical Properties and Chemical Composition of the Rat Aorta. Cardiovas. Res. 10(4):437-451, 1976.
6. Mulvany, M.J., P.K. Hansen, and C. Aalkjaer: Direct Evidence that the Greater Contractility of Resistance Vessels in Spontaneously Hypertensive Rats is Associated with a Narrowed Lumen, a Thickened Media, and an Increased Number of Smooth Muscle Cell Layers. Circ. Res. 43:854-864, 1978.
7. Furuyama, M.: Histometrical Investigations of Arteries in Reference to Arterial Hypertension. Tohoku J. Exp. Med. 76:388-414, 1962.
8. Okamoto, K. and K. Aoki: Development of a Strain of Spontaneously Hypertensive Rats. Jap. Circ. J. 27:282-293, 1963.
9. Hutchins, P.M., A.W. Greene, and T.D. Rains: Effect of Isoproterenol on the Blood Vessels of the Spontaneously Hypertensive Rat. Microvas. Res. 9: 101-106, 1975.
10. Baez, S. and L.R. Orkin: Microcirculatory Reactions to Chemical Denervation in the Anesthetized Rat. Bibl. Anat. 9:61-65, 1967.
11. Hutchins, P.M. and A.E. Darnell: Observations of a Decreased Number of Small Arterioles in Spontaneously Hypertensive Rats. Circ. Res. 34-35 (Supp. I):161-165, 1974.

INTERNAL VISCOSITY (RIGIDITY) OF THE RED CELL AND BLOOD VISCOSITY EQUATION; COUNTERACTION OF ERRORS DUE TO FLOW INSTABILITY OF PLASMA

L. Dintenfass

Haemorheology & Biorheology Dept., Medical Research KMI, Sydney Hospital, and Department of Medicine, University of Sydney, Sydney 2006, Australia

SUMMARY

The viscosity of blood depends on the apparent internal viscosity of the red cell, concentration of red cells and plasma viscosity as primary factors, and on aggregation of red cells. The blood viscosity equation based on these primary factors has been suggested: $\eta_r = (1 - CkT)$ exp -2.5, in which T is the Taylor's factor, and k the immobilized plasma on and between red cells. Problems arose in regard to determination of plasma viscosity which was much higher in rotational than in capillary viscometers; this is suggested to be due to flow instability and Taylor's vortices in plasma. Term Tk, obtained from the blood viscosity equation, can be used to characterize rigidity of red cell in various diseases. Blood viscosity equation can be modified and reviewed from the viewpoint of various possible axial ratios of red cells and aggregates of red cells, as also from the viewpoint of low-shear-rate aggregation of red cells. Such a modified equation contains two additional coefficients, α and β, and a subscript D indicating shear-rate dependence:

$$\eta_r = (1 - \alpha_D C k_D T_D)^{-2.5\beta_D}$$

INTRODUCTION

The viscosity of blood is complex since it depends not only on the concentration of red cells (haematocrit), plasma viscosity, aggregation of red cells, and of course on the shear rate, but also on the fluidity or rigidity of the red cell, and on the ratio of the internal viscosity of the red cell to the viscosity of plasma.

Although this is obvious and recognized to day, it was not so twenty years ago. Dintenfass suggested in 1962 (1) that as the viscosity of packed red blood cells is appr. 20 cP, and thus usable for transfusion, blood has to be treated as a suspension of fluid drops. Were the red cells rigid, it would not have been possible to attain volume concentrations as high as 0.98, and the consistency of the packed cells would resemble consistency of concrete. It did appear that blood behaves as an ideal emulsion in which the interfacial layers (red cell membrane) is of very little if any consequence.

Thus Dintenfass proposed that the theoretical work on emulsion by Taylor (2) and Oldroyd (3) should be applied to blood. This was specially justified as he noted that the relative viscosity of blood depends on the viscosity of plasma while in suspensions of rigid particles the relative viscosity is independent of the viscosity of the continuous phase. Utilizing Roscoe's equation for viscosity of suspension of spheres (4), Dintenfass suggested in 1968 (5) an equation for blood viscosity of the following type:

$$\eta_r = (1 - CkT)^{-2.5} \qquad \text{I.}$$

in which η_r is the relative viscosity of blood (that is, viscosity of blood measured at homogenous shear rates of about 150 to 250 reciprocal seconds under which conditions red cells are disaggregated or monodisperse) divided by the viscosity of plasma; C is the volume concentration of red cells; k is the packing coefficient of red cells which includes effects of plasma 'atmosphere' and of plasma which is entrapped between the red cells; T is the Taylor's factor as employed in the theory of ideal emulsions by Taylor and Oldroyd, and which relates to the ratio, p, of the internal viscosity of the fluid drop (fluid cell) and plasma viscosity in the following manner: $T = (p + 0.4)/(p + 1)$.

Taylor's studies took into account the internal circulation in the suspended fluid drops. This circulation is due to the transmission of the tangential and normal stresses across the interface between the continuous and disperse phases during shear flow. Fluid circulation inside the drops

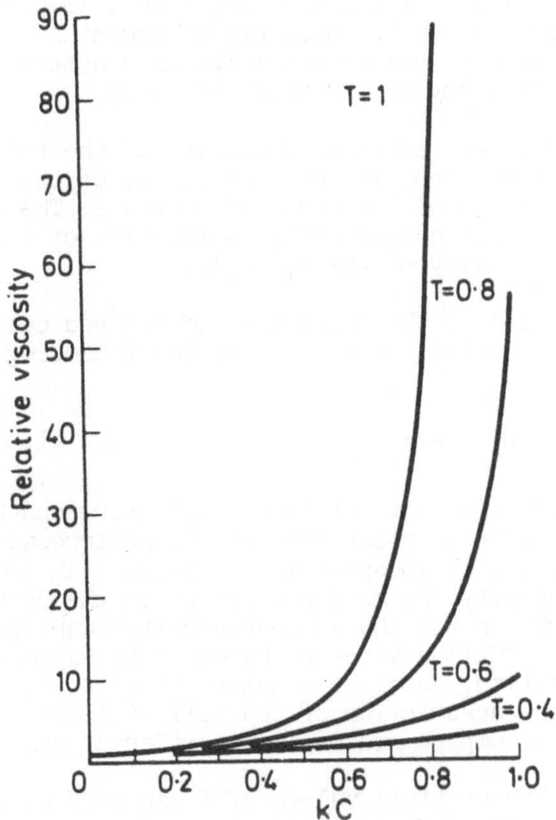

Fig. 1. Theoretical curves for the relative viscosity of blood at different concentrations of red cells and different values of Taylor's factor T.

reduces the distortion of the flow pattern outside the drops and reduces hence the volume fraction of the disperse phase by a factor T.

The interfacial film (in our case, the red cell membrane) will influence the transfer of tangential and normal stresses. If this interfacial film is rigid (or elastic, in mathematical sense), then the drop behaves as a rigid particle. This suggested that red cell membrane is quite fluid (6,7) and might even combine shear-thinning and shear-thickening characteristics. Dintenfass suggested a liquid crystalline layer restrained by two networks of proteins (6, 7) forming a highly dynamic and multiphase system capable of surface thixotropy and surface dilatancy.

The apparent internal viscosity of the red cell would be a sum of the true internal viscosity of the cell and of the contribution of the red cell membrane. The latter, most likely, will not add more than a fraction of 1 cP to the apparent viscosity of the red cell.

In practice, the 'rigidity' of the red cell was obtained from the equation I. by the following transformation:

$$CkT = (\eta_r^{0.4} - 1)/\eta_r^{0.4} \quad .... \quad II.$$

The relative viscosity of blood was determined by defining blood viscosity at shear rate of 180 reciprocal seconds in the rotational rhombospheroid viscometer (8, 9) while plasma viscosity was measured either in the British Standard Micro-U-tube or in the Harkness-Coulter Capillary Viscometer. Term 'CkT' divided by haematocrit expressed as a volume fraction, supplied a numerical value for kT which was taken as description of red cell rigidity (10, 11) and was used to compare red cells in different diseases.

While theoretical values of T can very only between 0.4 and 1, the term 'kT' has been observed to vary from below 0.4 to 1.35. Thus it is prudent to review some problems in determination of the main parameters of the blood viscosity equation.

REVIEW OF PARAMETERS OF THE BLOOD VISCOSITY EQUATION

A review of parameters and factors used and/or implied in the blood viscosity equation containing a term for the internal viscosity of the red cell will include haematocrit, plasma viscosity (and flow instability of plasma), exponents used in the equation in relation to the shape of suspended red cells (or aggregates), and effect of rate of shear. It must be stressed that, at this stage, all these considerations are mainly of theoretical importance and do not preclude use of equations I and II in clinical haemorheology.

Problems with Haematocrit

The actual volume concentration of red cells is not identical to the haematocrit (measured as volume fraction at high g values) and neither of these two is identical with the hydrodynamic or free volume of the red cells. While the volume concentration of red cells remains constant in any blood sample, the hydrodynamic or free volume of the red cells depends on the deformability of red cells and the mode of packing, on the plasma layers and plasma entrapped around the red cells and within the red cell aggregates should such aggregates be present.

The actual volume concentration of red cells is much smaller than the hydrodynamic or free volume of the red cells. Haematocrit is larger than the true volume concentration of red cells but smaller then the hydrodynamic volume of red cells. In the case of very fluid and deformable red cells, the difference between the actual volume concentration and haematocrit will be very small, of the order of 1 per cent only. In the case of rigid red cells, as for instance in the case of cells treated with formaldehyde, haematocrit value might be up to 35 per cent higher than the true volume concentration C. It is expected that the hydrodynamic volume will be close to haematocrit at high shear rates or high g-values. Free sedimentation in Westergren tubes showed volume fraction of red cells after two to four days of sedimentation to be only about ten per cent higher than the haematocrit in normal and many abnormal blood samples. It must be assumed that deformation of red cells during flow would be intermediate between that observed in sedimentation and that in centrifugation; at the same time, plasma trapping would be less during flow than, during free sedimentation.

Consequently, the only practical way to deal with multitude samples of blood is to assume that Hk (that is, haematocrit multiplied by a coefficient k) corresponds to the hydrodynamic volume of the disperse cells. Thus, the term used in the blood viscosity equation would be 'HkT'. As experimental definition of k is difficult, instead of the Taylor's factor T, we use term 'Tk', while haematocrit is measured in the usual way. Experimental studies showed there is a direct relationship between T and k ; values of k can vary from 1 to 1.65, while values of T can vary from 0.4 to 1.0 in a parallel and functionally related (Table 1)

Table 1. Values of Taylor's factor T, ratios of the internal viscosity and plasma viscosity p, and experimental values for k (coefficient related to plasma trapping) and Tk

T	p	k	Tk ;	T	p	k	Tk
0.400	0	1.014	0.405;	0.920	6.50	1.078	0.992
0.500	0.20	1.016	0.508;	0.940	9.00	1.099	1.033
0.600	0.50	1.020	0.612;	0.960	14.00	1.138	1.092
0.700	1.00	1.025	0.710;	0.980	29.00	1.227	1.202
0.800	2.00	1.035	0.828;	0.990	59.00	1.338	1.325
0.900	5.00	1.064	0.958;	1.000	∞	1.670	1.67

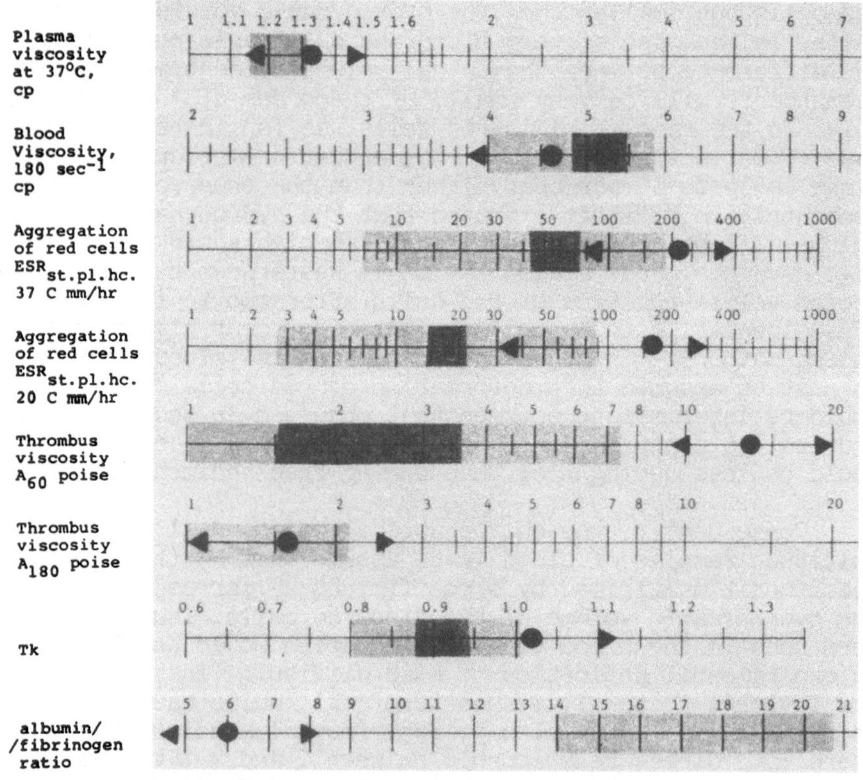

Fig. 2. Application of red cell rigidity Tk in the rheological characterization of patients with diabetes (12,13).

manner. Figure 2 shows application of Tk in characterization of a series of diabetes mellitus patients; other factors used include blood viscosity, plasma viscosity, aggregation of red cells, apparent viscosity of artificial in vitro thrombi, and albumin/fibrinogen ratio. Means and limits of one standard deviation are shown as dots and arrows, while the shaded areas correspond to normal values (two normal ranges are superimposed: the left one corresponds to the long distance runners, the right one to 'normal' normals).

Problems with Plasma Viscosity

It would appear that nothing could be simpler than to measure plasma viscosity; especially, that the greater part of plasma samples appears to be nearly Newtonian. However we encounter two sets of problems.

The first problem concerns differences between plasma viscosities measured in the capillary viscometers and in the rotational viscometers. In our experience, the values obtained in the Harkness viscometer were much lower than in the rhombo-spheroid viscometer. The difference was lowest when shear rate of about 20 rec. sec. was used in the rotational visco-meter, but much higher at shear rate around and above 100 rec. seconds. While first obvious response would be to attribute this discrepancy to the different ranges of shear rates in the two types of viscometer, the reason for it appears to rely on another mechanism: flow instability.

Flow instability under coaxial flow in a Couette visco-meter has been described first by Taylor (14), while the first experiments on blood were carried out by Phillips (15). The latter found appreciable flow instability and formation of Taylor's vortices at rotational speeds as low as 20 r.p.m. However he was using Couette of very large gaps; also he noted that that the critical Taylor number was greatly increased at increasing haematocrits. That is, formation of vortices was counteracted by the presence of red cells. At haematocrit of 60 per cent, Phillips obtained Taylor's number of 25000.

In Newtonian fluids instability is basically a result of complex interactions of gravitational and inertial forces. In the case of non-Newtonian fluids, the instability could be

Table 2. Flow instability expressed as a ratio of plasma viscosities measured in the rotational rhombospheroid viscometer and in the Harkness capillary viscometer.

Subjects	Flow instability at 100 r.p.m.			; at 250 r.p.m.		
	n	$\bar{x}$ ±	SD ;	n	$\bar{x}$ ±	SD
NORMALS	87	1.58 ±	0.31;	90	1.81 ±	0.27
MELANOMA	33	1.56	0.27;	33	1.82	0.33
ISCHAEMIA	51	1.49	0.32;	51	1.71	0.26

either that of the type analogous to that existing in a Newtonian fluid with superimposed non-Newtonian properties, or it might arise directly from the non-Newtonian characteristics of the fluid. A nonlinear rheological behaviour can make Couette flow unstable to small disturbances of the Taylor vortex type.

For Couette geometry, with the inner cylinder rotating, the critical Taylor number is 3390:

$$T = (2R_1 d^2 \Delta^3 \Omega^2)/\eta^2 \qquad \text{III}$$

in which R_1 is the radius of the inner cylinder, d is the density of plasma, Δ is the gap between the inner and outer cylinders, Ω is the rotational rate of the inner cylinder (= $2\pi/P$; in which P is the period of rotation in seconds), and η is the viscosity of plasma.

As we have no correct mathematical expression for the cone-in-cone or cone-plate geometries, we can only estimate the values of T, or rather we can assume that the value for T will be smaller in these geometries than in the Couette geometry. There is no question that such instability exists at 100 r.p.m. It was thus convenient to show as the measure of instability the ratios of viscosities obtained in the rotational and capillary viscometers; it was this ratio which directed our attention at the instability. At the same time we must assume after Phillips that such instability does not exist in blood at the usual level of haematocrits. (Table 2)

The second problem is of different nature. It is well known that viscosity of suspensions increase with increasing concentrations of the disperse phase (i.a., viscosity of blood increases with increasing haematocrit). Nevertheless, evidence mounted over the years that viscosity of blood in patients suffering from Waldenström's macroglobulinaemia can be less than the viscosity of plasma in the same patients (16, 17). One would be inclined to dismiss such results as absurd and inconsistent with theory if not for two facts: one, such data were well documented; second: a similar observation has been reported for colloidal suspensions and explained nearly two decades ago (9, 18, 19).

In the case of colloid suspensions, adsorption of the low molecular weight fraction of polymer onto the suspended powder led to changes in the molecular weight distribution of the polymer (in the continuous phase) resulting not only in an increase of viscosity of the liquid phase but also, in many cases, in a change of its rheology from Newtonian to dilatant (shear-thickening) type.

It has been thus suggested (17) that in the case of macroglobulinaemia there might exist an adsorption of the lower molecular weight proteins onto the red cells, with a formation of "plasma atmosphere": a concentration of low viscosity portion of plasma around the red cells. Thus, the viscosity of plasma after centrifuging represents only the free and unbound plasma. This "free" and unbound plasma may show higher viscosity than the total suspension without contradicting the basic law of physics.

Problem of Exponent of Viscosity Equation

The original equation of Taylor and Oldroyd, and of Roscoe, dealt with spherical particles and thus a question arose whether the exponent '2.5' (originally from the Einstein viscosity equation) was applicable to blood cells. Experimental work by Dintenfass (5) appeared to indicate that variations in shape between normal cells, crenated cells, and swollen cells did not affect the results of equation I. Indeed, according to Davies (20), the dynamic shape factor for prolate spheroids of axial ratio 2 is only 1.05, and the corresponding exponent would be 2.5 x 1.05 (= 2.625). Spindle--like cells, observed at very high shear rates by Sutera et

Table 3. Relative viscosities of blood calculated from equation $\eta_r = (1 - CT)^{-2.5\beta}$ at different volume concentrations (C) and different Taylor factors (T).

C	T	relative viscosity at exponents 2.5xβ				
		2.5	2.8	3.0	3.5	4.5
0.2	0.4	1.232	1.263	1.284	1.339	1.455
0.4	0.4	1.546	1.629	1.687	1.841	2.191
0.6	0.4	1.986	2.156	2.278	2.613	3.438
0.2	0.7	1.458	1.525	1.572	1.695	1.971
0.4	0.7	2.273	2.509	2.679	3.157	4.385
0.6	0.7	3.903	4.596	5.125	10.556	20.698
0.2	0.9	1.642	1.743	1.813	2.003	2.442
0.4	0.9	3.052	3.489	3.815	4.768	7.451
0.6	0.9	6.968	8.796	10.273	15.148	32.93
0.2	0.95	1.694	1.804	1.882	2.091	2.581
0.4	0.95	3.304	3.813	4.196	5.329	8.595
0.6	0.95	8.25	10.62	12.57	19.18	44.60

et al (21) would be requiring a higher exponent if not for the fact that such cells align in the field of flow. Davies (20) suggested that for prolate spheroids of axial ratio 10, the shape factor β would be 1.59. A series of shape factors was employed in construction of Table 3. A comparison of experimental blood viscosity values with theoretical values suggest that at medium shear rates (100-200 rec.sec.) the original exponent (=2.5) should be retained; but at higher shear rates perhaps a higher exponent might be applicable. However, in the latter case in order to retain blood viscosity values not higher than those observed at medium shear rates one would have to postulate internal shear-thinning or internal thixotropy of the red cell. Thus, the exponent might be shear rate dependent, and this could be indicated by a subscript D.

It must be underlined that from the viewpoint of clinical haemorheology the exponent 2.5 appears to be satisfactory for evaluation of the contribution of the internal viscosity of the red cell. (Fig. 3).

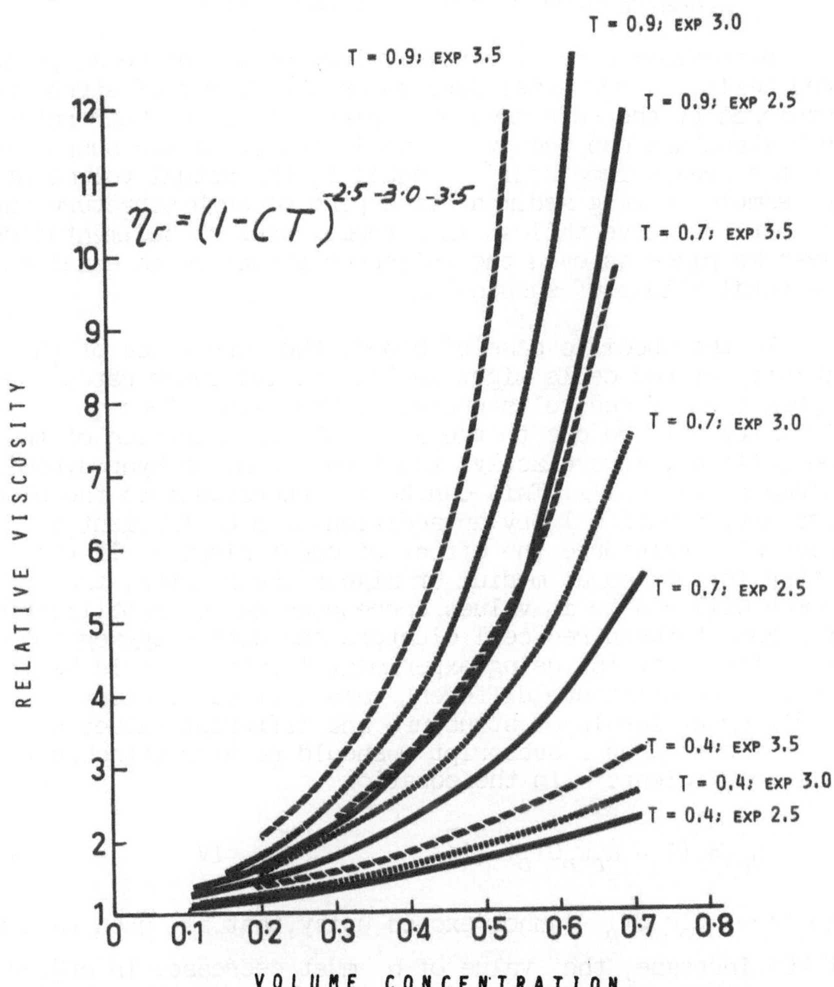

Fig. 3. Relative viscosity of suspensions of fluid drops as a function of volume concentration (volume fraction) of the fluid drops (in our case, blood cells). Theoretical curves are drawn for different values of Taylor's factor T, and different exponents of Dintenfass's equation for monodisperse suspensions. Although exponent -2.5 is used in the routine clinical-haemorheologic studies for comparison of various diseases or different patients, other exponents may have a theoretical justification depending on the shape of the red cells, their alignment, etc.

Aggregation of Red Cells and Viscosity

Hydrodynamic or effective volume of the disperse phase (red cells, in our case) depends on the degree of structure developed by the suspended particles. At lower shear rates such structure can occupy the whole volume of the suspension and the free volume would be equal to the actual volume of the sample. During sedimentation part of such structure can be lost, but nevertheless in extreme cases no sedimentation may take place as even the collapsed structure is equal to the total volume of suspension.

In the specific case of blood, the importance of the fluidity of red cells might be less at low shear rates when aggregation of red cells prevails. The main effect on viscosity will be due to the apparent concentration of the red cells or, more exactly, the free volume or hydrodynamic volume of red cells. This can be incorporated into the blood viscosity equation I by an addition of a coefficient α which will reinforce the effect of coefficient k. If the latter is defined at medium or higher shear rates, the former will add to it values corresponding to immobilization of plasma between red cell clusters and within aggregates of red cells. That is, using experimental data it would be possible to attribute different values of coefficient α to different levels of structure and different values of shear rates: thus a subscript D should be also attached to this coefficient. In the equation

$$\eta_r = (1 - \alpha_D k_D CT_D)^{-2.5} \qquad \text{IV}$$

the term $\alpha_D k_D CT_D$ cannot exceed unity, that is, when values of kCT increase, the value of α must decrease. In other words, the degree of aggregation is limited by the space available, and the potential for aggregation may be much larger than the actual aggregation depending on the constraint of physical space.

This is well known not only in the colloidal suspensions but also in blood: a high haematocrit (appr. 60 per cent) blood does not necessarily settle.

Equation IV implies that the Taylor's factor T might be shear rate dependent; that is, the apparent internal cell

Table 4. Equations required for analysis of blood viscosity.

---

1. relative viscosity of blood:

$$\eta_r = \eta_{blood} / \eta_{plasma}$$

Note: viscosity will depend on the rate of shear which might affect blood and plasma in different manner;

2. evaluation of the effective volume of red cells from relative viscosity:

$$CkT = (\eta_r^{0.4} - 1)/\eta_r^{0.4}$$

Note: term CkT will depend on the shear rate, but at higher shear rates will not include a component due to aggregation of red cells; packing coefficient k is lower the lower is fluidity and higher deformability of red cells, and both might be related by an empirical hyperbolic equation:

$$(k - 1.01)(1/T - 0.999) = 0.0034$$

3. evaluation of the effective volume of red cells which includes all effects due to aggregation of red cells may be carried out using the following equation:

$$\alpha CkT = (\eta_r^{0.4} - 1)/\eta_r^{0.4}$$

Note: the value of this term should be obtained at low shear rates; the value of α can be obtained from equation 3 divided by equation 2, when all the component values of equation 2 are extrapolated to the same shear rates at which equation 3 is defined; values of T and k are both increasing as the shear rate is decreasing;

4. the ratio of the internal viscosity of the red cell to the viscosity of plasma, p, can be obtained from the Taylor equation:

$$p = (0.4 - T)/(T - 1)$$

$$p = \eta_i/\eta_{plasma} = (\eta' + \eta'')/\eta_{plasma}$$

in which η' gives the true viscosity of the cell interior, and η'' supplies a contribution of the cell membrane;

---

viscosity might change with the changing shear rate. The methods of calculations of all coefficients have been described in the book 'Blood Microrheology, Viscosity Factors in Blood Flow, Ischaemia and Thrombosis' (page 75-76) of Dintenfass, as also in a paper (22). Basically the steps required include determination of relative viscosity of blood over a range of shear rates, calculation of term αkCT, determination of term T at medium shear rates at which the contribution of aggregation of red cells is nearly or totally absent, calculation of plasma trapping k at these medium and high shear rates; extrapolation of these values to low shear rates; and, finally, calculation of coefficient α. Table 4.

This approach opens a number of avenues, as the exponent of blood viscosity equation may be also changed.

CONCLUSIONS

Blood viscosity equation containing a term for the internal viscosity of the red cell presents an avenue for microrheological analysis of various blood systems over a range of shear rates and concentrations.

It is not denied that the number of avenues possible might also represent a drawback. However, from the viewpoint of clinical haemorheology, equation I appears to serve the needs. The complex equation of the type

$$\eta_r = (1 - \alpha_D C k_D T_D)^{-2.5\beta_D} \qquad \text{V}$$

might be of greater use to experimental rheologists.

While considering blood viscosity equation we have to realize that even simple terms such as plasma viscosity or concentration of red cells are not so simple. A presence of flow instability of plasma in rotational viscometers suggests that capillary viscometers should be preferred, unless some progress in instrumentation will counteract it different way. Concentration is modified by plasma trapping and by hydrodynamic effects. Aggregation of cells migh be constrained. However, a realization of these difficulties makes the challenge greater.

REFERENCES

1. Dintenfass, L.: Considerations of the internal viscosity of red cells and its effects on the viscosity of whole blood. Angiology 1962, 13: 333
2. Taylor, G.I.: The viscosity of a fluid containing small drops of another fluid. Proc. Roy. Soc. 1932, 138A: 41
3. Oldroyd, J.G.: The effect of interfacial stabilizing films on the elastic and viscous properties of emulsions. Proc. Roy. Soc. 1955, 232A: 567
4. Roscoe, R.: The viscosity of suspensions of rigid spheres. Brit. J. appl. Phys. 1952, 3: 267
5. Dintenfass, L.: Internal viscosity of the red cell and a blood viscosity equation. Nature (London) 1968, 219: 956
6. Dintenfass, L.: Molecular and rheological considerations of the red cell membrane in view of the internal fluidity of the red cell. Acta Haemat. 1964, 32: 299
7. Dintenfass, L.: The internal viscosity of the red cell and the structure of the red cell membrane. Considerations of the liquid crystalline structure of the red cell interior and membrane from rheological data. Mol. Cryst. 1969, 8: 101
8. Dintenfass, L.: A coaxial rhombo-spheroid viscometer: a further development of the cone-in-cone viscometer. Biorheology 1969, 6: 33
9. Dintenfass, L.: Blood Microrheology, Viscosity Factors in Blood Flow, Ischaemia and Thrombosis. Butterworths, London 1971
10. Dintenfass, L.: Internal viscosity of the red cell: problems associated with definition of plasma viscosity and effective volume of red cells in the blood viscosity equation. Biorheology 1975, 12: 253
11. Dintenfass, L.: Theoretical aspects and clinical applications of the blood viscosity equation containing a term for the internal viscosity of the red cell. Blood Cells 1977, 3: 367
12. Dintenfass, L.: Rheology of Blood in Diagnostic and Preventive Medicine. Butterworths, London 1976
13. Dintenfass, L.: Haemorheology of diabetes mellitus. Adv. Microcirc. 1979, 8: 14

14. Taylor, G. I.: Stability of a viscous liquid contained between two rotating cylinders. Phil. Trans. Roy. Soc. London, 1923, A223: 289

15. Phillips, W.M.: Viscometric techniques and the rheology of blood. Blood Cells, 1977, 3: 101

16. Somer, T.: The viscosity of blood, plasma and serum in dys and paraproteinemias. Acta med. scand. 1966, Suppl. 456: 1

17. Dintenfass, L., and Somer, T.: A hypothesis of plasma 'atmosphere' around the red cells in patients with Waldenström's macroglobulinaemia and multiple myeloma. Microvasc. Res. 1976, 11: 325

18. Dintenfass, L.: Thixotropy and dilatancy in complex multiphase emulsions and suspensions. In: Proc. 4th Int Congr. Rheology, John Wiley and Sons, New York 1964, p. 621

19. Dintenfass, L.: Micro-rheological classification and analysis of complex multiphase suspensions. In: Proc. Int. Symp. Second Order Effects in Elasticity, Plasticity and Fluid Dynamics, Haifa, p. 764; Pergamon Press, Oxford 1964

20. Davies, C.N.: Shape of small particles. Nature, London 1964, 201: 905

21. Sutera, S., Mehrjardi, M., and Mohandas, N.: Deformation of erythrocytes under shear. Blood Cells, 1975, 1: 369

22. Dintenfass, L.: Rheology of blood, Viscosity factors in blood flow, ischaemia and thrombosis. In: Proc. CHEMECA '70, p. 25. Butterworths, Sydney 1970

1980 UPDATES

IN OPTICAL AND THERMAL ANEMOMETRY

R. L. Humphrey

Disa Electronics
779 Susquehanna Avenue
Franklin Lakes, NJ 07417

ABSTRACT: LDA WITH VARIABLE FREQUENCY SHIFT

This paper will discuss various methods of optical and electronic frequency shift as used with a laser doppler anemometer system. Specifically, an opto-electronic system that allows a stable frequency shift variable from 10 KHz to 50 MHz will be covered. The system combines a fixed shift from a Bragg cell and an electronic shift; resulting from the mixing of the detector signal with a variable local oscillator. The advantages of using this system with tracker or counter processing will be explained.

INTRODUCTION

Frequency shifting of the laser beam is usually required for highly turbulent flows, reversing or oscillating flows, and vibrating surfaces. The various methods available are:

Acousto optic (Bragg cell)
Rotating disc (diffraction grating)
Electro optic (Kerr cell/Pockels cell)

In comparing these methods we find that the Bragg cell has several advantages over the others; however, to optimize these advantages the Bragg cell must be used in a specific manner. (See Figure 1.) (1)

Specifications	Moving Grating	Acousto-Optic Cell	Electro-Optic Cell
Efficiency	20%	80%	55%
Frequency shifting range	0-16^6 Hz	10^7 Hz	10^3-10^8 Hz
Range of velocity measurement	Low	High	All
Power consumption of the optical setting	Low	High	High
Costly	No	Yes	Yes
Highly accurate alignment needed	No	Yes	Yes
Bigger laser power than an ordinary LDV is required	Yes	Yes	Yes
Moving parts	Yes	No	No
Sensitive to zero velocity	Yes	Yes	Yes
Thermal effect	No	Yes	Yes
Bigger probe volume	No	No	No
Possibility of optical filtering	No	No	No
Affect by particles depolarization	No	No	No

Figure 1. Comparison Chart of Different Methods of Frequency Shifting

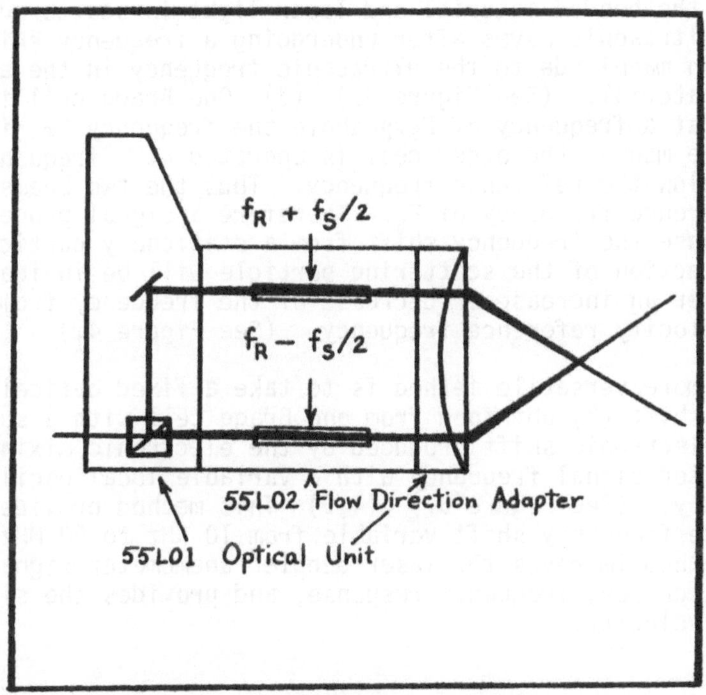

Figure 2. Schematic of the 55L02 flow direction adapter.

BRAGG CELL FREQUENCY SHIFT

The first and perhaps simplest method is to use two Bragg cells in the differential mode of an optical arrangement. (See Figure 2.)

The Bragg cells are a series of bonded piezoelectric transducers. These transducers via an electronic power supply produce a series of ultrasonic waves which propagate across the bonded material and laser light is scattered from these ultrasonic waves after undergoing a frequency shift equal in magnitude to the ultrasonic frequency in the acousto optic material. (See Figure 3.) (3) One Bragg cell is driven at a frequency of $F_{S/2}$ above the frequency F_r; in the same manner the other cell is operated at a frequency $F_{S/2}$ below the reference frequency. Thus the two beams have a difference frequency of F_S. Therefore a signal processor will sense the frequency shift from a stationary particle. The direction of the scattering particle will be indicated by either an increase or decrease of the frequency from the zero velocity reference frequency. (See Figure 4.)

A more versatile method is to take a fixed optical frequency shift (2) obtained from one Bragg cell with a subsequent electronic shift produced by the electronic mixing of a detector signal frequency with a variable local oscillator frequency. (See Figure 5.) (4,5) This method provides an accurate frequency shift variable from 10 KHz to 50 MHz. This method improves the laser doppler anemometer signal-to-noise accuracy, frequency response, and provides the sign of the velocity.

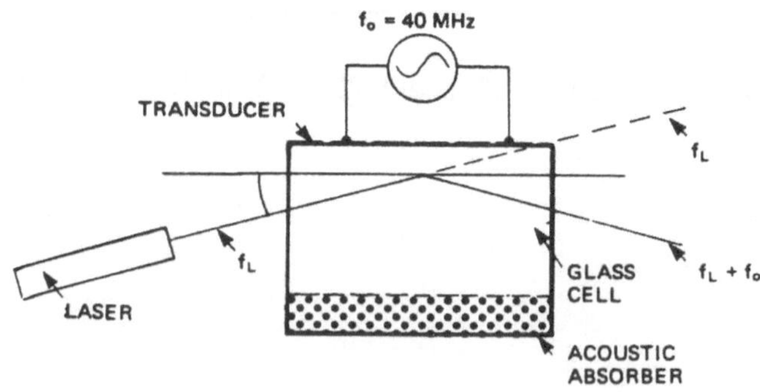

Figure 3. Acousto-optical frequency shifting of laser light.

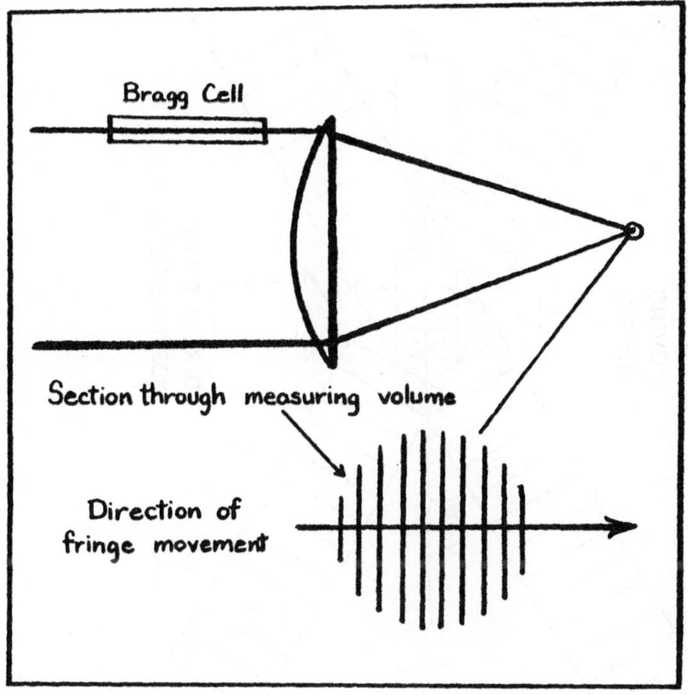

Figure 4a. Principle of frequency shifting.

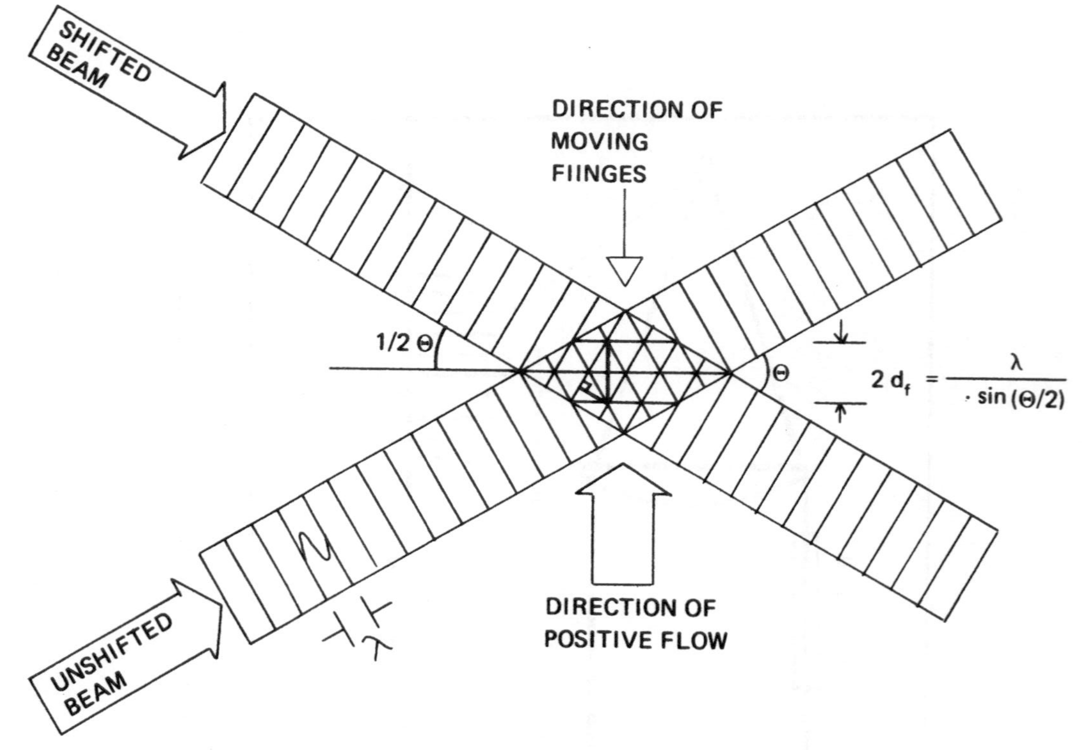

Figure 4b. Frequency shifting may be described as shifting of the fringes with the velocity $v_s = f_s \cdot df$

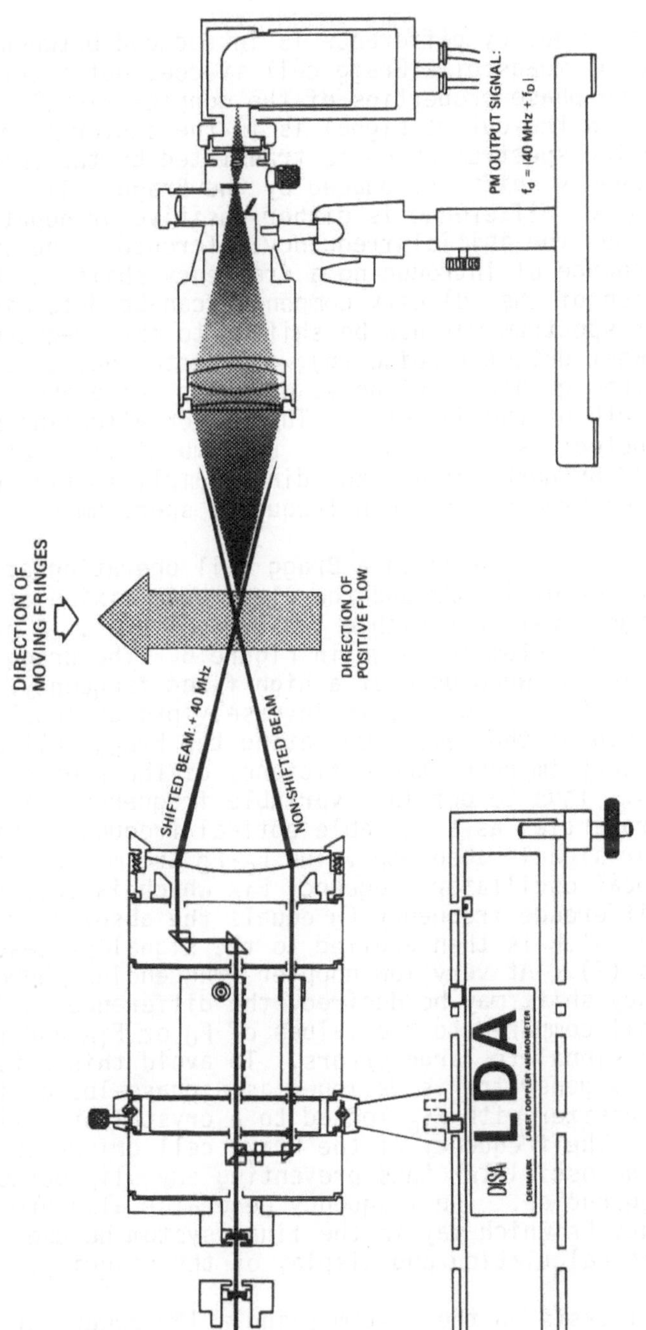

Figure 5. Orientation of X29 Bragg Cell and + 40 MHz Shifted Beam

When a frequency difference is introduced between two laser beams by means of a Bragg cell it does not alter the amplitude and phase properties of the doppler signal. The only change in the output signal is in the center frequency of the doppler spectrum which is translated by the amount of the frequency shift introduced by the Bragg cell. This shift frequency difference is either positive or negative depending upon the initial frequency reference. One immediate consequence of introducing a frequency shift is that the direction of the velocity component can be determined. The doppler spectrum may now be shifted to the frequency band of lowest detector noise (6). Detector noise results primarily from quantum shot noise which is white and within the band width of the detector. The factor affecting semiconductor detectors is $I/_F$ noise. Photomultiplier detectors are affected by dark current and dinode amplification noise. Each detector type has its own frequency spectrum.

This system is based on a Bragg cell operating at a fixed frequency of 40 MHz and the electronic mixing of the detector signal with a variable local oscillator. A block diagram of the system is shown in Figure 6. The Bragg cell is most efficient when used at a high fixed frequency. The efficiency of the Bragg cell is inversely proportional to the band width of the cell. Operating the Bragg cell at a fixed frequency improves the efficiency of the piezoelectric transducers. Thus to obtain a variable frequency shift with the same properties as a variable optical frequency shift, the detector signals absolute value F_o-F_D are mixed with the variable local oscillator frequency F_{lo} which is less than Fo. The difference frequency F_d equals the absolute value F_D+(F_o-F_{lo}) which is then applied to the signal processing electronics (7). At very low doppler frequencies where a low frequency shift may be desired, the difference F_o-F_{lo} will be small compared to the values of F_o or F_{lo} and thus subject the signal to large errors. To avoid this situation the frequency generator is designed as a phase-locked frequency synthesizer with F_{lo} locked to a crystal stabilized oscillator. The frequency of the Bragg cell driver is also locked to the oscillator thus preventing any slip between the two frequencies. The frequency generator also provides the frequency Fs which may in the final system be used for an automatic calculation and display of the velocity.

Several tests on the system using a low power narrow

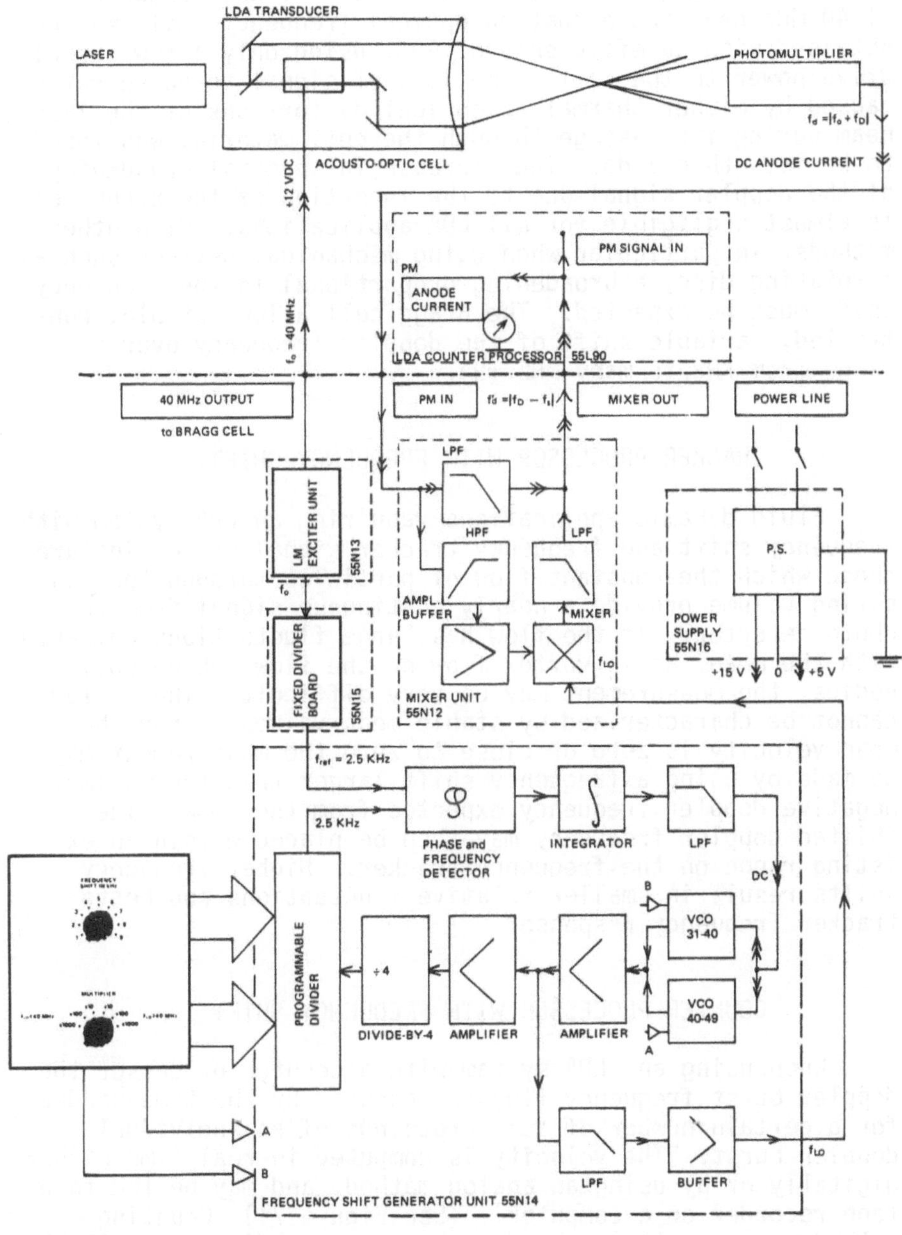

Figure 6. Block Diagram of LDA System with Variable Frequency Shift

band width Bragg cell (8) operating at a fixed frequency of 40 MHz have shown that an optical frequency shift may be obtained with an efficiency of 90%, using only 1.8 watts of drive power to the cell. The loss of signal-to-noise ratio caused by either thermal or optical distortions of the laser beam during its passage through the cell material was found to be less than 3 db. The increase in spectral broadening of the doppler signal due to the insertion of the Bragg cell is almost negligible for all LDA applications. With other methods, in particular when using mechanical devices such as a rotating disc, a broadening proportional to the frequency shift must be expected. The Bragg cell allows stable, controlled, variable shift of the doppler frequency over a range from KHz to many MHz (9).

TRACKER PROCESSOR WITH FREQUENCY SHIFT

Fluid dynamic applications requiring an LDA system with frequency shift and frequency tracker signal processing are those which the constant flow of particles through the measuring volume provide a nearly continuous signal from the photo detector. If the flow has large fluctuations compared with the mean, as in vortex flow or the flow behind solid bodies, the measurement may be very difficult. These flows cannot be characterized by stable mean values. When the mean velocity is zero or close to zero the measurement may be made by using a frequency shift larger than the maximum negative doppler frequency expected from the flow. The shifted doppler frequency may then be placed within an existing range on the frequency tracker. Higher frequency shifts result in smaller relative fluctuations and better tracker frequency response.

COUNTER PROCESSOR WITH FREQUENCY SHIFT

When using an LDA system with a counter processor the doppler burst frequency (10) is measured by the time needed for a certain number of zero crossings of an individual doppler burst. The velocity is computed in real time either digitally or by using an analog method, and may be fed to a tape recorder or a computer. (See Figure 7.) Counting methods are usually applied where spare seeding is available and when there is one or less particle in the measuring

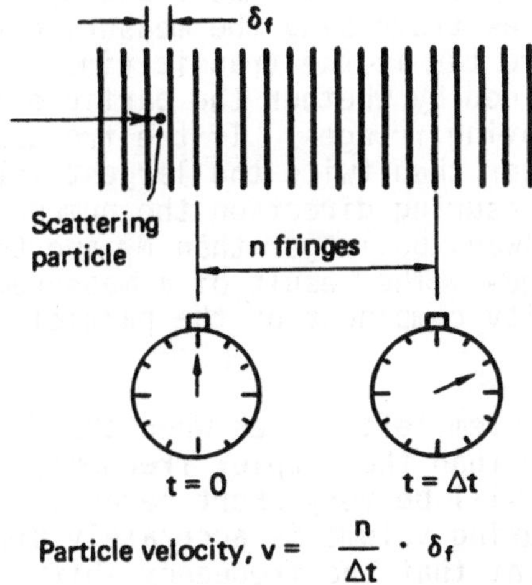

Figure 7. LDA COUNTER - Principle

volume at any one time. The counting method is not limited by the frequency range or slew rate as is the tracker method (11). The only range limitations to the doppler signal counter are those imposed by the doppler signal or by prefiltering to limit the noise band width. Within its frequency range the counter may accept all incoming doppler bursts above a certain level which satisfied previously set validation criteria. However, the counting method does suffer weak points such as the presence of dead zones (12). Dead zones are velocity directions which give an insufficient number of zero crossings for the proper operation of the counter. In order to obtain good accuracy in the determination of the time of flight through the measuring volume it is desirable to count a number of available fringes (40); however, the greater the ratio, the greater the dead zones.

The introduction of an optical frequency shift with a counter-based LDA system has the effect of totally removing the dead zones and of giving a relative cross-section of sigma equal to one for all particle trajectories. The effect of the frequency shift may be described in the fringe picture as a motion of the fringes along the measuring direction of the LDA with a velocity of $V_O=F_SD_f$ where F_S is the amount of frequency shift introduced and D_f is the fringe spacing. Particles traversing the measuring volume will emit periods where tau is the transit time. Plus or minus signs are determined by whether the particle is moving with or against the moving fringes. If the frequency shift velocity is greater than twice the largest velocity component along the measuring direction the number of fringe crossings will always be larger than N_f and thus allow a count of N_e fringes. The result of a measurement will be the actual velocity component of the particle and its direction.

The only problem left arises when the frequency shift F_S is much larger than the doppler frequency since the time of a measurement will be very short because only a small part of the measuring volume is accurately timed. It is therefore important that the frequency shift be adjusted to give the necessary number of counts so as not to reduce the accuracy of the measurement. Thus, we find a variable frequency shift indispensable in the use of an LDA system.

SUMMARY

In summary, this type of frequency shift does not increase the frequency broadening already present in the doppler signal and reduces the signal-to-noise of the detected signal less than 3 db.

The use of a frequency tracker offers the possibility of placing the doppler spectrum in one of several ranges and overcomes the difficulties connected with the lock range of a tracker. When using a counter the frequency shift allows the measurement of the sign of the velocity component and also removes the problem of dead zones. In most counting systems the accuracy of the measurement would be adversely affected by a large frequency shift. Therefore the shift should be of the same order of magnitude as the doppler frequency.

ABSTRACT: MULTI-CHANNEL CTA SIGNAL AND DATA PROCESSING

A new Constant Temperature Anemometer (CTA) and signal analysis system has been designed which incorporates some new concepts and features specifically designed to facilitate multi-channel measurements. These features include a miniaturization and simplification of the analog modules, a fast digitization and digital pre-processing equipment and and extensive facilities for computer interfacing and control.

The problems with the use of most fluid mechanics instrumentation, designed for single-channel operation, in multi-channel experiments are, in addition to the shear size and weight of the electronic equipment, the rather complex set-up and wiring of several channels compounded by grounding, shielding, and ground-loop problems for many, often remotely located sensors, plus the practical problems in connection with calibration and check-out of many channels. Furthermore, the amount of data from a multi-channel experiment often becomes overwhelming and requires large storage facilities in the form of analog or digital tape and the use of much computer time for data reduction. Finally, in high frequency applications, the cost of A/D conversion of sufficient speed and resolution for the digital processing of signals in most applications within fluid dynamics is still rather high and large buffer memories may be required for signal acquisition.

The improvements incorporated in this system are based on three main principles:

1) Miniaturization and simplification of the analog instruments while maintaining full stability and specifications. The analog system includes a CTA module, which consists of a single print and a 1" wide front panel. Thus a full cabinet for 19" racks may contain up to 16 CTA modules. Signal conditioner and other analog modules are of a similar small scale. The CTA bridge is a plug-in module which fits into the CTA module. The specifications allow replacement and interchangeability of bridges and probes which greatly facilitates the set up of a multi-channel system. (See Figure 8.)

2) A new, fast digital data pre-processing and analysis method based on a 1-bit signal quantization and fast, simplified digital circuitry capable of processing data sampled at a 1 MHz rate. The system includes modules designed for real-time computation of mean, mean-square and rms values

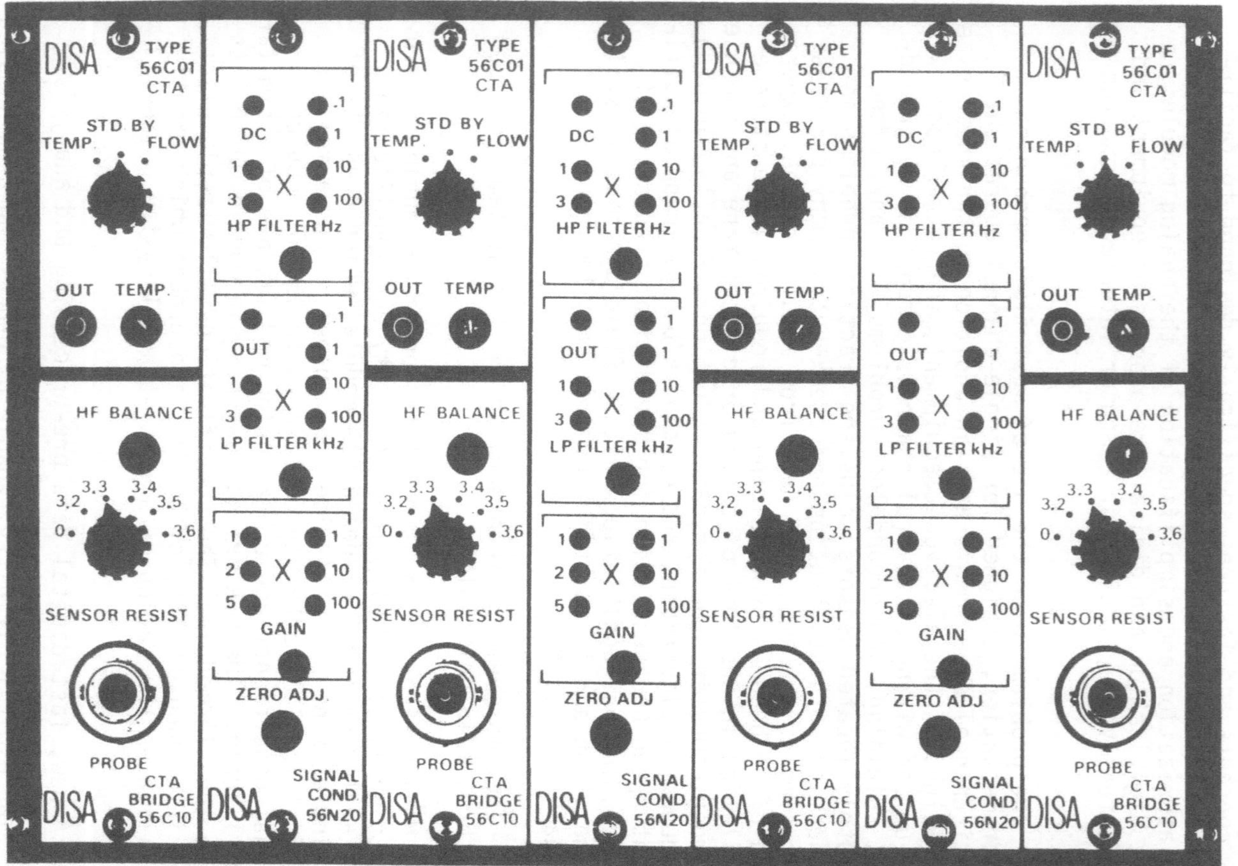

Figure 8. Multi-channel CTA Signal and Data Processing System

as well as units which allow the direct formation of shear (correlation) stress.

3) Extensive facilities for interfacing to external computer for remote set-up, control and read-out.

Further features and details will be presented in a forthcoming paper at the APS winter meeting.

SYMBOLS

F_s Difference frequency between two laser beams

F_r Reference frequency

F_o Optical frequency

F_D Detector frequency

F_d Difference frequency between optics and detector

F_{lo} Local oscillator

F_s Total frequency shift

V_o Velocity of shifted fringes

D_f Distance between fringes

N_f Total measuring volume fringes

N_e Particle crossing fringes

REFERENCES

1. Durst, R., Zare, M., "Removal of Pedestals and Directional Ambiguity of Optical Anemometer Signals", Purdue University International Short Course on Laser Velocimetry (1974)

2. Drain, L. E., Moss, B. C., "Opto-electronics" pp. 429-439 (1972)

3. Gordon, E. I., "Proc. IEEE", pp. 649-652 (1970)

4. Stevenson, W. H., "Applied Optics", pp. 649-652 (1970)

5. Dennison, E. B., Stevenson, W. H., "Rev. Sci. Instruments", pp. 1475-1478 (1970)

6. Mazumder, M. K., "Applied Phys. Letter", pp. 462-464 (1972)

7. Cummins, H. Z., Swimney, H. L., "Progress in Optics" VIII North Holland (1970)

8. M40 Acoustic-optic Modulator, Zenith Radio Corp., Chicago, IL, pp. 1-6 (1971)

9. Oldengarm, J., Van Krieken, A. H., Raterink, H. J., "Opt. Laser Technol", pp. 249-252 (1973)

10. Asher, J. A., "G. E. Tech. Information Series", Report No. 72 CRD 295 (1972)

11. Deighton, M. O., Sayle, E. A., "Disa Information 12", pp. 5-10 (1971)

12. Steenstrup, F. V., "Disa Information 18", pp. 21-25 (1975)

AN EXPERIMENTAL INVESTIGATION OF PULSATILE LAMINAR FLOW SEPARATION IN EXPONENTIALLY DIVERGING TUBES

Frederick J. Walburn and Daniel J. Schneck*

Departments of Medicine and Surgery
Henry Ford Hospital, Detroit, MI 48202 and
*Department of Engineering Science and Mechanics
Virginia Polytechnic Institute and
State University, Blacksburg, VA 24061

INTRODUCTION

Since the turn of the century, when Ludwig Prandtl formulated the theoretical concept of the boundary layer, the phenomenon of boundary layer or flow separation has been associated mostly with trouble, i.e., energy losses, reverse flow, vortex formation, wakes, stalling, increased drag, decreased lift, wall scouring effects, etc. One of the more serious problems with which flow separation has been associated in recent years is cardiovascular disease. There is an ever increasing abundance of evidence which suggests indirectly that the pathogenesis, localization and/or aggravation of certain forms of cardiovascular disease may be related to the incidence, persistence and consequences of internal, unsteady, laminar-flow separation. This very same evidence, because of its indirect speculative nature, has also brought to light an area of fluid mechanics that has hitherto been conspicuously neglected.

For over fifty years the criterion for the onset of flow separation was accepted to be that suggested by Prandtl in 1904 (1):

$$\partial u/\partial y = 0 \text{ at } y = 0 \qquad [1]$$

where y is distance measured normal to the bounding wall surface and u is the fluid velocity component directed parallel to the wall. In 1910 Blasius (2) successfully used this criterion in an experimental study of steady flow through an exponentially diverging tube. He reported that, for each axial location, there is a characteristic Reynolds number downstream of which the flow is completely separated and upstream of which the flow is laminar and attached. Sometime later, in 1934, Patterson (3, 4) conducted experiments using air which confirmed the results of Blasius. More recently, a number of investigators (5-12) have reported both the appearance of a "critical" Reynolds number for separation and an upstream displacement of the separation point with increasing Reynolds number. In 1956 it was demonstrated that Prandtl's criterion was valid only for cases of two-dimensional or axially-symmetric steady flow occurring over or within fixed boundaries (13-15). A more generalized criterion was suggested which was appropriate at least for the case of moving rigid walls:

$$\partial u/\partial y = 0 \text{ at } u = 0. \qquad [2]$$

For the flow situation to which it applies, equation [2] has been rather convincingly verified by experiment (16, 17). However, for the case where the unsteadiness in the flow is due to fluctuations not related to wall motion, Despard and Miller (18) have presented experimental evidence which suggests that separation in an oscillating, high Reynolds number (but laminar) flow is characterized by the existence of a wall station, downstream of which there is reverse flow throughout the entire cycle of oscillation. These investigators offered as a criterion:

$$\partial u/\partial y \leq 0 \text{ at } y = 0 \text{ for all } t. \qquad [3]$$

They found that the stationary separation point in a flow oscillating about a net mean value was located upstream of where it would be in a corresponding steady case having the same mean flow and that as the frequency of oscillation increased, the point of separation was displaced downstream, towards its steady state location. This observation clearly led to a paradox, for it implied an upstream separation point displacement with decreasing frequency of oscillation -- so that reducing the frequency of oscillation would tend to move the separation point further and further away from its steady state location. Recently, it has been reported that the presence of an oscillating component in the flow leads to the appearance of a secondary, steady streaming motion which acts to retard downstream flow near the wall of the tube and enhance such motion closer to midstream (19-24). These investigators suggest that the secondary streaming directly affects flow separation since as the streaming disappeared, the separation point moved gradually downstream with increasing oscillation frequency. The same paradox noted by Despard and Miller (18) was also observed in these studies.

That the separation point is displaced further and further upstream, away from its steady state location, as the frequency of flow oscillation is progressively decreased cannot possibly be true for all lower oscillation frequencies, since, in the limit of oscillation frequency going to zero, the point of flow separation must eventually appear back downstream at its corresponding steady-state location. In fact, it has been observed further that there exists a transition frequency (or frequencies) in unsteady flows (holding the mean Reynolds number constant) above which the separation point does behave as predicted by Despard and Miller (18) and Schneck, et al. (19-21), but below which it does not (25, 26). Hence, there is a need to define specifically the range of oscillation frequencies and Reynolds numbers for which various flow separation criteria are valid. It is the intent of this paper, therefore, to study the effect of Reynolds number, frequency and amplitude of flow oscillations upon the flow separation phenomenon for viscous flow in an exponentially diverging tube.

METHODS

The recirculating experimental arrangement used in this study is shown schematically in Fig. 1. The test fluid was pumped from the reservoir, such that flow to the damping tank could be regulated by the two glass stopcocks. Thus, the mean volumetric flow rate through the test section could be precisely adjusted from 0 to 3 liters/min within a tolerance of ± 2 ml/min. After passing through the damping tank, where undesirable large scale fluctuations were dissipated, the test fluid encountered a variable oscillator. This device was used to superimpose a pulsatile component on the mean flow. It consisted of a piston and cylinder arrangement which allowed for independent control of both the frequency and the amplitude of the oscillatory flow component. Hence, the mean flow (or Reynolds number) through the test section, as well as the amplitude ratio (Q_{max}/Q_{mean}) and

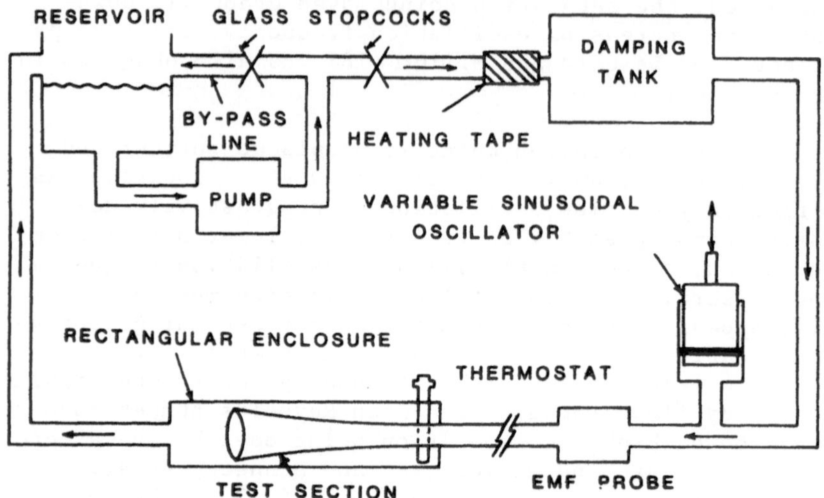

Figure 1. Schematic Diagram of the Experimental Arrangement

frequency (f) could be varied independently. The unsteady Reynolds number (α) is defined as $R_o\sqrt{2\pi f/\nu}$; where R_o is the inlet radius, ν is the kinematic viscosity, and f is the frequency of flow oscillation. An electromagnetic flow probe (Carolina Medical Electronics, King, North Carolina) was attached proximal to the entrance to the test section.

The test fluid was a saline solution (2% sodium chloride by weight) having a kinematic viscosity of 0.1022 cm^2/sec at a temperature of 26.67°C. The viscosity and refractive properties of this fluid were carefully controlled by maintaining a constant fluid temperature using the thermostat and heating tape components shown in Figure 1.

A single channel laser Doppler anemometer (DISA Electronics, Franklin Lakes, New Jersey) was used in this study. The mode of operation was the differential Doppler mode with forward scattering. The flow was seeded with silicon carbide particles having a mean diameter of 1.5 microns. The components of the laser Doppler anemometer were aligned and mounted on an optical bench which, in turn, was fastened to the carriage of a lathe. Thus, the measuring volume could be moved in minute increments along the x-y-z coordinate system defined in Figure 2. The longitudinal tube axis, x, was always kept perpendicular to the plane of the laser beams so that the laser Doppler anemometer measured only the axial component of fluid velocity at a "point" in the flow.

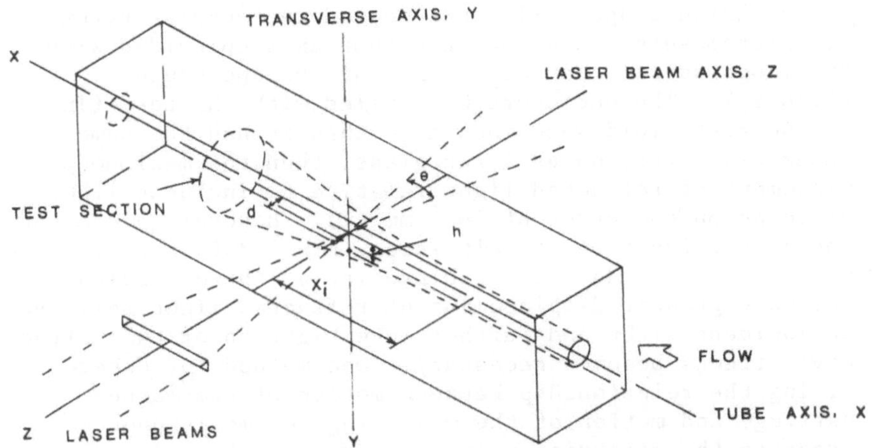

Figure 2. Schematic Representation of the Configuration of the Exponentially Diverging Circular Test Section Contained within a Rectangular Enclosure to Eliminate Adverse Effects of Laser Beam Refraction.

The test section used in this investigation is shown in Figure 3. It was constructed from Pyrex glass (refractive index = 1.5) by the glass blowing shop in the Chemistry Department (Virginia Polytechnic Institute and State University). The equations of the diverging wall of the test section are shown in Figure 3. Flow in the second downstream portion of the divergence was examined in this study.

The circular shape of the exponentially diverging Pyrex glass tube wall introduces some rather undesirable refractive effects as far as laser Doppler anemometry is concerned. The most undesirable of these is that the centerlines of the two converging laser beams could be prevented from intersecting exactly with each other. This must occur in order to have proper fringe formation. Moreover, due to refraction, an incremental movement of the lathe carriage will not produce the same movement of the measuring volume. In this study the method of Bentz and Evans (7) was used to minimize these effects. The diverging portion of the test section was mounted carefully inside a specially constructed rectangular Pyrex glass enclosure, such that the tube axis coincided with the longitudinal centroidal axis of the enclosure (Figure 2). The enclosure was filled with the test fluid. If the test fluid were chosen so that it had the same index of refraction as Pyrex glass, then the maximum displacement of refracted light relative to incident light would be on the order of 10^{-2} mm (7). However, since the index of refraction of salt water ($n = 1.336$) and Pyrex glass ($n = 1.5$) do not match precisely, there results a somewhat greater displacement of refracted light relative to incident light and further investigation of the refractive effects becomes necessary. One method for determining the relationship between motion of the lathe carriage and motion of the measuring volume is photographing the measuring volume before and after it has been moved by incrementing the lathe carriage. The actual displacement may be determined from the photographs and compared with the lathe carriage displacement (27). Another method used the output of the photomultiplier to determine when the measuring volume is completely in the glass wall or completely in the lumen of the tube (28).

The criterion used to determine the onset of laminar flow separation in steady flow was that proposed by Prandtl (equation [1]). Strictly speaking, Prandtl's criterion is a zero shear requirement (Appendix), which requires taking simultaneous velocity gradient measurements in more than one direction near the tube wall. Due to the limitations of our experimental facilities, however, and considering the special circumstances of the particular flow situation we examined (see Appendix), it was considered reasonable to reduce Prandtl's criterion (from equation [A.9]) to

$$\partial V_x / \partial z = 0 \text{ at } z = z_{dw} \qquad [4]$$

where V_x is the velocity component in the x-direction and z_{dw} defines the z coordinate of the distal wall of the diverging test section for any arbitrary value of x. (Note: z is considered to be positive moving towards the proximal wall from the distal wall at any x-station.)

The criterion used to determine the point of onset of unsteady laminar flow separation was that proposed by Despard and Miller (equation [3]). In terms of the coordinate system defined in Figure 2 and the same assumptions made in the Appendix, equation [3] becomes:

$$\partial V_x / \partial z \leq 0 \text{ at } z = z_{dw} \text{ for all t.} \qquad [5]$$

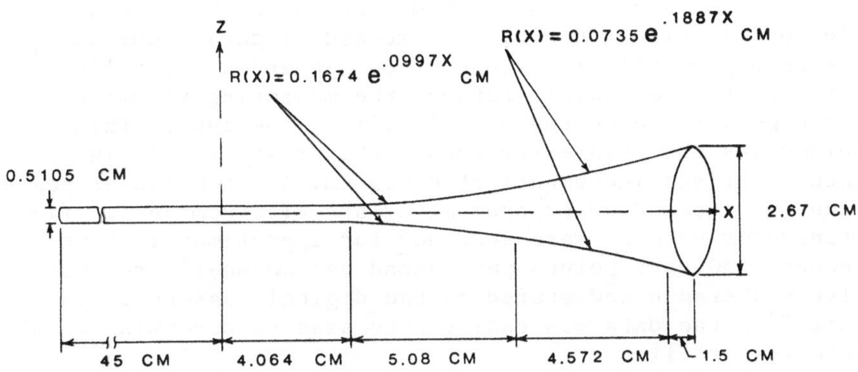

Figure 3 Exponentially Diverging Circular Test Section Used in the Flow Separation Studies

The symmetry of the separation patterns were examined with a flow visualization technique (28). A more detailed examination was made of the symmetry of the flow separation patterns as they occurred along the diametric x-z plane using the laser Doppler anemometer (28). From these observations it was concluded that there is a relatively slight asymmetry in the diverging flow, which separates first on the back wall of the tube (this behavior was also observed in oscillating flows) (28). Thus, it was decided that: 1) flow separation in the diverging tube may be reasonably characterized by examining fluid behavior in the diametric x-z plane, and 2) the flow need only be examined near the rear wall (28).

One of the methods used to record the laser Doppler anemometer and the electromagnetic flow transducer signals is shown schematically in Figure 4. Specifically, for each radial position, the outputs of the laser Doppler anemometer and electromagnetic flow transducer were processed by the multiplexor and the analog-to-digital (A/D) converter, and recorded on the digital tape cassette. Unfortunately, the data acquisition configuration was limited to A/D conversion of approximately 100 channels/sec. This is sufficient to digitize analog signals arising from steady and slowly oscillating flows, but the A/D conversion rate is too slow to represent properly the analog signals generated by moderate and high frequency flow oscillations. Hence, this method of data acquision was used primarily for steady flow situations as follows: for a particular axial station, the measuring volume was first positioned on the distal wall of the tube. This volume was then traversed toward the proximal wall in lathe carriage increments of 0.025 cm. At each radial position the laser Doppler anemometer and electromagnetic flow transducer outputs were recorded for approximately four seconds (50 data points per second per channel), respectively averaged and stored on the digital cassette. In this form the data was easily processed to determine axial velocity profiles.

To acquire data in an oscillating flow situation, a Hewlett-Packard two channel strip recorder was used. At each radial position, moving towards the proximal (front) wall, the laser Doppler anemometer and electromagnetic flow transducer outputs were fed continuously and simultaneously to the strip chart recorder. Data was recorded for a minimum of three complete flow cycles in order to detect possible artifacts in either the laser Doppler anemometer or electromagnetic flow transducer signals. The traversing process in the z direction was continued until axial velocities of 2 to 3 cm/sec were recorded, at which time the data acquisition was stopped. The measuring volume was then moved to another axial location, or the flow parameters were changed at the same axial location and additional data acquired.

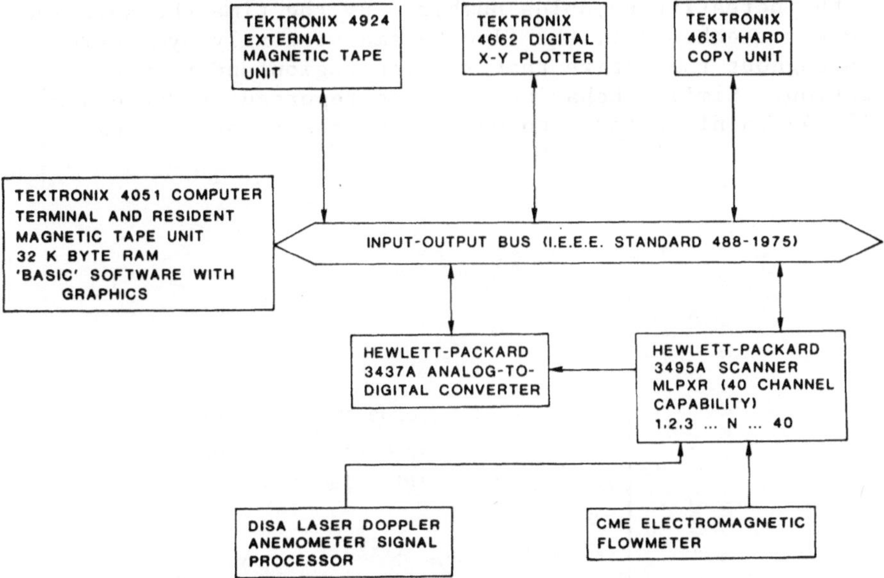

Figure 4 Tektronix 4051 Data Acquisition Configuration

RESULTS AND DISCUSSION

Steady Flow

It was found that in steady flow through a diverging circular channel that a threshold Reynolds number existed for which the flow separated initially at the furthest downstream end of the channel divergence. With increasing Reynolds number above threshold, the point of flow separation defined by Prandtl's criterion (equation [4]) moved upstream, such that for any given axial location x_i, there corresponded a characteristic Reynolds number at which the flow was completely separated downstream of x_i. Upstream, the flow remained laminar and attached. These results are in complete agreement with those presented analytically by Blasius (2) in 1910 and more recently, by Schneck (21).

Investigation of the symmetry of the separation pattern revealed a marked dependence on the Reynolds number (28). Specifically, the asymmetry which characterized flow separation in its early stages decreased nonlinearly with increasing Reynolds number. By the time the Reynolds number reached 325, the flow became virtually symmetric throughout the entire test section region under investigation. Similar behavior has been reported by Patterson (3, 4) in his efforts to verify the theory of Blasius.

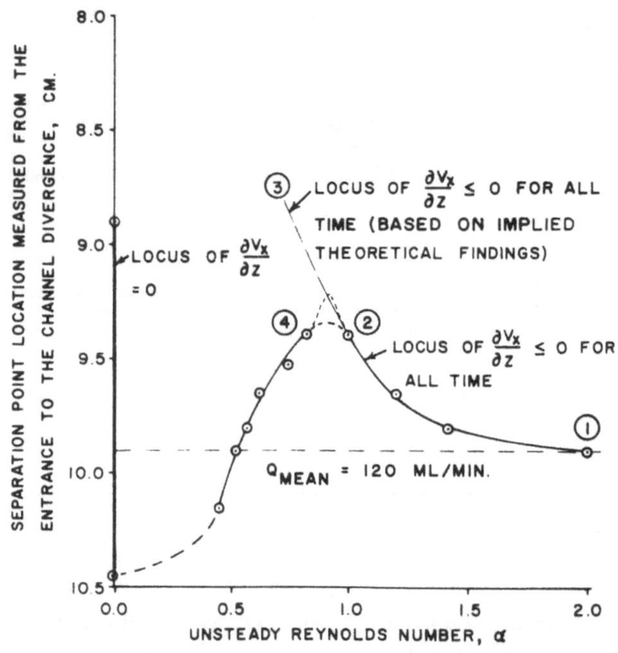

Figure 5 Trajectory of Despard and Miller's Separation Point as a Function of the Unsteady Reynolds Number, α.

The Separation Point Location vs α

The trajectory of the point of unsteady flow separation is shown in Figure 5 as a function of α, the unsteady Reynolds number. It can be seen clearly that for $0.52 \leq \alpha \leq 2.00$ the point of separation is located upstream of where it would be in a corresponding steady case having the same mean flow (shown by the dashed line parallel to the α axis). The data obtained for $\alpha = 0.46$ shows the separation point to be located downstream of its steady state location corresponding to Q = 120 ml/min. Also shown in this figure is the range of axial locations for which the flow will separate according to Prandtl's steady state criterion (i.e., $\alpha = 0$), the latter being applied at each instantaneous flow rate during a complete flow cycle. That is, if the flow were held steady at Q = 80 ml/min the separation point would be fixed at x = 10.41 cm, whereas for a steady flow of Q = 160 ml/min the separation point would be found at x = 8.89 cm.

When the flow oscillated very slowly, such that the unsteady Reynolds number, $\alpha << 1$, the point of flow separation as defined by Despard and Miller (equation [5]) was found near the point of steady flow separation as defined by Prandtl (equation [4]), the latter being applied for a steady volumetric flow rate equal to 80 ml/min (Q_{min}, Figure 5). This was also the furthest downstream position where Despard and Miller's separation criterion could still be defined, regardless of the value of α (if the mean Reynolds number was held constant). With increasing α, (Re_{mean} still constant), the effects of unsteady inertia and non-linear steady streaming began to appear (20). These effects caused Despard and Miller's point of separation to move upstream until by the time $\alpha = 0.52$, it was located precisely where it would be for a corresponding steady flow having a volumetric flow rate equal to 120 ml/min (Q_{mean} rather than Q_{min}, Figure 5). As the frequency of oscillation was increased still further, the backflow induced near the wall by steady streaming became larger. As a result, the point of pulsatile flow separation continued to move progressively upstream of its steady-state position. The maximum steady streaming effect, as well as the furthest upstream excursion of the separation point, both occurred for $0.82 < \alpha < 1.00$. For $\alpha > 1$, the streaming effect began to taper off, and with this, the point of separation started moving back towards its steady state location for $Q = Q_{mean}$. By the time $\alpha = 2$, the streaming effect had essentially disappeared and the point of pulsatile flow separation was virtually coincident with its corresponding mean flow steady state location. Moreover, it's location was now observed to be essentially independent of further increases in α .

The behavior of the point of flow separation in the range where it has been almost exclusively examined, i.e., $\alpha > 1.00$, gives rise to the paradox alluded to previously. That is, since Despard and Miller (18), Schneck (21), and others have observed oscillation frequencies for which $\alpha > 1$, their results would fall somewhere along curve 1 - 2 in Figure 5 (although perhaps on an extrapolation of the curve beyond $\alpha = 2.00$). In any case, the results of these investigators imply that the separation point is displaced continuously upstream with decreasing frequency of oscillation (or, α) which would tend to move this point further and further away from its steady state position as α is reduced. This behavior is shown in Figure 5 by the line extending curve 1-2 towards point 3. Although these authors recognized that the separation point must reappear at its steady state location in the limit of $\alpha \longrightarrow 0$, they were unable to explain how this occurred in terms of their results. The reason seems to be that the relatively high oscillation frequencies investigated by Despard and Miller (18) and Schenck (21) produced results that do not reflect the decrease in the streaming effect for $\alpha < 1$, or equivalently, the decreasing importance of unsteady inertial forces at these low values of the unsteady Reynolds number.

Transient Wake Region

Despard and Miller's criterion for the onset of pulsatile flow separation is characterized by $\partial V_x/\partial z \leq 0$ at $z = z_{dw}$ for all time. These investigators observed experimentally that a wake region was present for all time downstream of just such a point. For some distance upstream of the fixed x-station (x_d) for which equation [5] holds at a given mean Reynolds number, frequency and amplitude of flow oscillation, one finds wall locations, x_i, where the instantaneous radial gradient of the axial velocity, $[\partial V_x/\partial z]_i$, i = specific wall station, passes routinely through zero during a portion of the flow cycle. At other times, this gradient assumes positive or negative values in response to the driving pressure gradient. However, if one continues to move upstream, a wall station, x_u, is eventually reached where $[\partial V_x/\partial z]_{x = x_u} \geq 0$ for all time. We shall define this point ($x = x_u$) as the beginning of the pre-wake region. Upstream of this region, $[\partial V_x/\partial z]_i > 0$ at the wall for all time and all axial stations. In summary, then, one can examine three specific regions (to the point of flow reattachment) associated with pulsatile laminar flow separation:

Region I: Attached flow, $x_i < x_u$:

$[\partial V_x/\partial z]_{x_{i,t}} > 0$, for all x_i and t;

Region II: Pre-Wake Region, $x_u < x_i < x_d$:

$[\partial V_x/\partial z]_{x_{i,t}} > 0$ or < 0, depending on t at each x_i; and

Region III: Separated Wake Region, $x_i > x_d$:

$[\partial V_x/\partial z]_{x_{i,t}} \leq 0$ for all t until the flow reattaches.

Figure 6 shows schematic velocity profiles which exist in these three regions. Note that in the pre-wake region transient flow reversal occurs near the wall for a portion of the flow cycle. Moreover, if the period of oscillation is very long, these "transient phenomena" may be present for time intervals significantly longer than what might normally be referred to as "transient".

Figures 5 and 8 show the locus of Prandtl's separation criterion for steady flows ($\alpha = 0$) having volumetric flow rates corresponding instantaneously to those given by the relation $Q(t) = 120\ (1 + 1/3 \cos\omega\ t)$. That is, the wall velocity gradient is equal to zero at x = 8.89 cm when $Q_{steady} = Q_{max\ pulsatile} = 160$ ml/min, and it is located at x = 10.41 cm for $Q_{steady} = Q_{min\ pulsatile} = 80$ ml/min. For $Q_{min} < Q_{steady} < Q_{max}$, the separation point is located between $10.41 > x > 8.89$ cm.

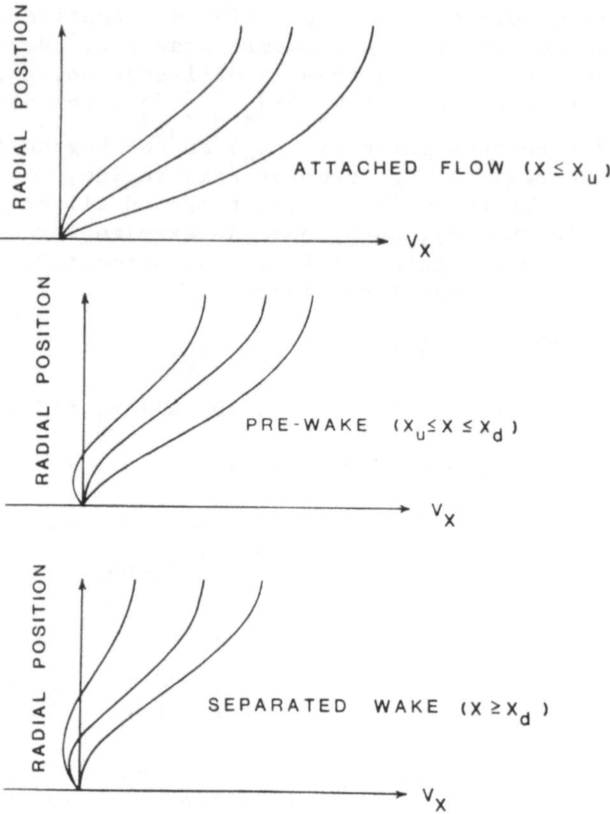

Figure 6 Schematic Velocity Profiles Which Exist in the Attached Flow Region, the Pre-Wake Region and the Separated Wake Region.

Consider what may be occurring when the flow begins to oscillate very slowly, such that $\alpha << 1$. For this particular situation, it is reasonable to suspect that the fluid will behave as if the flow were "quasi-steady". That is to say, for each instantaneous Reynolds number, Re $=V_j R_o/\nu$, corresponding to specific times, t_j, during the flow cycle when the instantaneous mean flow velocity is $V_i = Q(t_j)/A_j$, events change so slowly with respect to time that there are no significant manifestations of phase lags, secondary streaming phenomena, or other effects attributable to the influence of unsteady inertial forces.

The fluid therefore adjusts spontaneously to each instantaneous change in flow and, with no appreciable delay, shows behavior that corresponds precisely to that which would be observed if the flow were, in fact, kept essentially steady at $Q(t_j)$ for each t_j. Thus, at a particular time during the flow cycle, the point of $\partial V_x/\partial z = 0$ at $z = z_{dw}$ is located exactly where it is predicted to be based on steady flow criteria for that corresonding flow rate (or instaneous Reynolds number). That is, for a flow oscillating very slowly between $162.7 \leq Re \leq 325.4$, when $Re = 162.7$, the radial gradient of the axial velocity, evaluated at $z = z_{dw}$, is equal to zero at $x = 10.41$ cm, and as the Re gradually increases to 325.4, this point moves upstream to $x = 8.89$ cm. Completing the cycle, as Re returns to 162.7, the zero-gradient criterion moves back downstream to $x = 10.41$ cm. This is illustrated in Figure 7A, where x_{ss} is the steady state separation point (i.e., for $\alpha = 0$) corresponding to the mean Reynolds number of 244. The amplitude, d, of the locus of $\partial V_x/\partial z = 0$ at $z = z_{dw}$ depends on the amplitude of flow oscillation so that the latter actually controls the trajectory for very low frequencies. It might be inferred that this quasi-steady behavior of the flow is due to the lack of any substantial secondary streaming for small α, especially considering what happens as α is gradually increased.

As the oscillation frequency is increased so that $0 < \alpha < 0.52$, the flow begins to show effects attributable to unsteady inertia and non-linear streaming. Recall that the latter secondary, steady-state phenomenon acts to retard fluid motion near the wall surface. The manifestation of this effect is an upstream displacement of the lower extreme of the point $\partial V_x/\partial z \leq 0$ at $z = z_{dw}$, i.e., Despard and Miller's separation point, x_d, shown in Figure 5 (and again in Figure 7). In addition, characteristic instantaneous velocity profiles at $x = 8.89$ cm now shows backflow near the wall where there was none when $\alpha << 1$. This implies that the upper extremum x_u has also moved upstream as α was increased from 0. Figure 7B illustrates this situation, where both extremes are shown upstream of their positions in Figure 7A.

By the time $\alpha = 0.52$, the point x_d has moved upstream to coincide with its location for the mean Reynolds number equal to 244 (Figure 7C). Again, based on an observed increase in backflow near the wall the point x_u appears to have been also displaced further upstream. However, since the rate of increase of instantaneous backflow at x = 8.89 cm slows as α increases, it may be deduced that x_u does not move upstream quite as fast as x_d does, i.e., that there is a tendency for the extrema to approach each other. In other words, the amplitude of flow oscillation appears to have less of an effect on d at higher oscillation frequencies than it does at lower oscillation frequencies. At any rate, this behavior continues for $0.52 < \alpha < 0.82$ (Figure 7D) until, for $0.82 < \alpha < 1.00$ (maximum steady streaming effect) the maximum upstream displacement for both x_d and x_u is reached (Figure 7E). Furthermore, it seems that the extrema continue to approach each other.

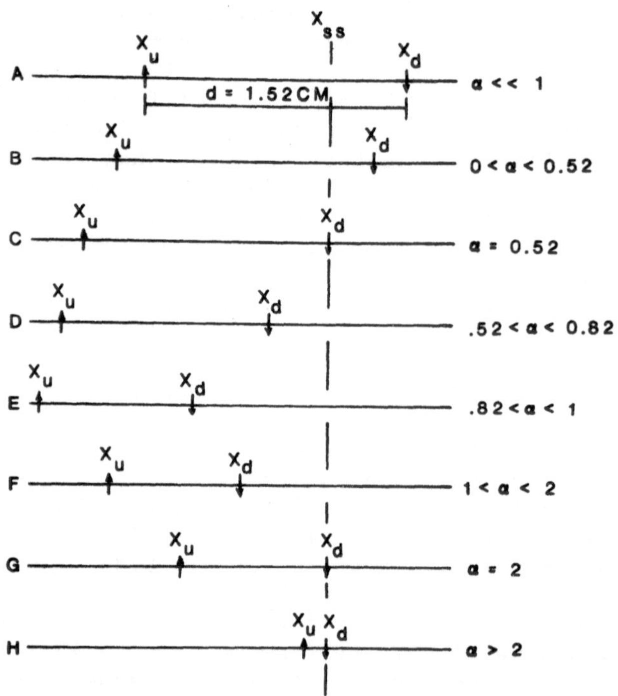

Figure 7 Schematic of Despard and Miller's Separation Point, x_d, and the Beginning of the Pre-Wake Region, x_u, for Several Values of α.

As the frequency of oscillation is increased so that $\alpha > 1.00$, effects of steady streaming begin to disappear and the entire pre-wake region begins to move back downstream (Figure 7F). By the time $\alpha = 2$, x_d is again coincident with its steady state position (Figure 7G). The point x_u has also moved downstream and, while the amplitude, d, of the displacement is smaller than it was, say, for $\alpha < 1$, there is still a relatively large transient pre-wake region. For $\alpha > 2$ (Figure 7H), both theory and experiment indicate that x_d remains at its mean (Re = 244) steady state position but, the point x_u apparently continues to approach x_d so that the two eventually become coincident and totally independent of α. Indeed, just recently evidence has been reported to support this view.

In an experimental investigation of external unsteady flow separation, it has been reported that the point of $\partial V_x/\partial z = 0$ at $z = z_{dw}$ in a steady flow is "dragged" downstream whan a step impulse is applied to the flow rate (29). Hence, it was concluded that a fast acceleration of the flow will sweep the wake region downstream. This also occurs in a high frequency oscillating flow when the instantaneous flow rate changes rapidly from Q_{min} to Q_{max}. Hence, as α increases and the flow begins to experience fast accelerations, it seems entirely likely that a similar mechanism causes x_u to progressively approach x_d.

The description depicted in Figure 7 is summarized and shown more graphically in Figure 8, which is an extrapolation of the experimental findings of Figure 7. Here the axial loci of zero wall velocity gradients are shown as a function of α. Upstream of the curve representing the locus of x_u, the flow is fully attached for all time. Downstream of the curve representing the locus of x_d there is a wake region for all time regardless of the value of α. Note that this curve is the locus of Despard and Miller's criterion for unsteady flow separation. The area between these two curves represents the pre-wake region. Observe that the axial extent of the pre-wake region becomes progressively smaller as α increases and x_u approaches x_d. Particularly interesting, however, is the implication that, for the range of α in which the streaming effect is significant (i.e., $0 < \alpha \leq 2$), the pre-wake region is relatively large. There is even a considerable pre-wake region at $\alpha = 2$, by which time the steady streaming has all but disappeared and x_d is virtually coincident with its mean steady state location. This implies that, if there is a single wall station which divides a fully separated region of flow from one which is attached at all times in a pulsatile flow, it occurs for $\alpha > 2$ and is located at the same position as that predicted by Prandtl's criterion for a corresopnding steady flow having the value Q_{mean}.

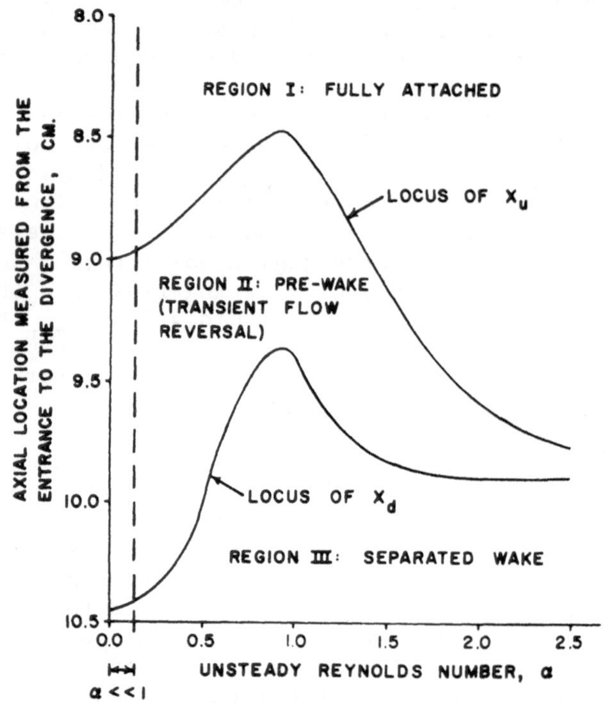

Figure 8 Characteristics of the Fully Attached, Pre-Wake and Separated Wake Regions as a Function of the Unsteady Reynolds Number, α.

Such a point, justifiably, could be called the "point" of onset of flow separation in the sense that downstream of this point there would be a wake region in existence for all time and upstream, the flow would remain attached for all time. For smaller α, this behavior simply does not exist and so here one cannot define the axial location where $\partial V_x/\partial z < 0$ at $z = z_{dw}$ for all time as the "point" of flow separation. The fact that a wake region with transient flow reversals extends upstream of Despard and Miller's separation point at various times during the flow cycle is not negligible. Nor are the effects attributable to such extensive wake regions. This study suggests that perhaps pulsatile flow separation is characterized better by defining the location to be just downstream of $\partial V_x/\partial z \geq 0$ at $z = z_{dw}$ for all time, at least for the case when the volumetric flow rate Q(t) stays positive for all time. The rationale for suggesting this is that the point so defined really locates the beginning of a process which ultimately results in a streamline displacing region of wake flow. On the other hand, it is certainly possible (if not likely) that some totally different mathematical criterion (involving, perhaps, velocity correlations, energy spectra, turbulence intensity, etc.) may be more able to accurately describe pulsatile flow separation. In any case, the greatest difficulty in trying to describe this phenomenon accurately is that, even today, very little quantitative information exists to define unsteady flow separation.

REFERENCES

1. Prandtl, L., "Uber Flussigkeitsbewegung bei sehr kleiner Reibung," Verhandlung des III Intern. Math. - Kongresses, Heidelberg, zur Hydrodynamik u. Aerodynamik, Gottingen, 1927, pp. 1-8, and Edwards Bros., Ann Arbor, Mich., 1943.

2. Blasius, H. Von., "Laminare Stromung in Kanalen Wechselnder Breite," Zeitschr. F. Math. u. Phys., 58: 225-233, 1910.

3. Patterson, G.N., "Flow Forms in a Channel of Small Exponential Divergence," Canad. J. Res., 11: 770-779, 1934.

4. Patterson, G.N., "Viscosity Effects in a Channel of Small Exponential Divergence," Canad. J. Res., 12: 676-685, 1935.

5. Gutstein, W.H., and Schneck, D.J., "In Vitro Boundary Layer Studies of Blood Flow in Branched Tubes," J. Atheroscler. Res., 7 (No. 3): 295-299, June, 1967.

6. Forrester, J.H., and Young, D.F., "Flow Through a Converging-Diverging Tube and its Implications in Occlusive Vascular Disease - I. Theoretical Development," J. Biomechanics, 3: 297-305, 1970.

7. Bentz, J.C., and Evans, N.A., "Hemodynamic Flow in the Region of a Simulated Stenosis," ASME Paper Number 75-WA/Bio-10, December, 1975.

8. Manton, M.J., "Low Reynolds Number Flow in Slowly Varying Axisymmetric Tubes," J. Fluid Mech., 49: 451-459, 1971.

9. Kandarpa, K., and Davids, N., "Analysis of the Fluid Dynamic Effects on Atherogenesis at Branching Sites," J. Biomechanics, 9: 735-741, 1976.

10. Lee, J.S., and Fung, Y.C., "Flow in Locally Constricted Tubes at Low Reynolds Numbers," J. App. Mech., 37: 9-16, 1970.

11. Young, D.F., and Tsai, F.Y., "Flow Characteristics in Models of Arterial Stenoses - I. Steady Flow," J. Biomech., 6: 395-410, 1973.

12. Young, D.F., "Effect of a Time-Dependent Stenosis on Flow Through a Tube," J. Engng. Ind. Trans. ASME, 90: 248-254, 1968.

13. Rott, N., "Unsteady Viscous Flow in the Vicinity of Stagnation Point," Quart. Appl. Math., 13: 444-451, 1956.

14. Sears, W.R., "Some Recent Developments in Airfoil Theory," J. Aero. Sci., 23: 490-499, 1956.

15. Moore, F.K., "On the Separation of the Unsteady Laminar Boundary Layer," in: Gortler, J., (ed.), Boundary Layer Research, Proc. Symp. of Int. Union of Theoret. and Appl. Mech., Berlin, Springer, 1958, pp. 296-311.

16. Sears, W.R., and Telionis, D.P., "Boundary-Layer Separation in Unsteady Flow," SIAM J. Appl. Math., 28 (no. 1): 215-235, January, 1975.

17. Telionis, D.P., "Calculations of Time-Dependent Boundary Layers," in: Kinney, R.B., (ed.), Unsteady Aerodynamics, Proc. Symp. Univ. Ariz., The Arizona Board of Regents, 1975, pp. 155-190.

18. Despard, R.A. and Miller, J.A., "Separation in Oscillating Laminar Boundary-Layer Flows," J. Fluid Mech., 47 (Part 1): 21-31, 1971.

19. Schneck, D.J., and Ostrach, S., "Pulsatile Blood Flow in a Channel of Small Exponential Divergence -- I. The Linear Approximation for Low Mean Reynolds Number," ASME Paper Number 74-WA/Bio-14, Trans. ASME, J. Fluids Eng., 97 (Ser. 1, No. 3): 353-360, Sept., 1975.

20. Schneck, D.J., and Walburn, F.J., "Pulsatile Blood Flow in a Channel of Small Exponential Divergence -- II. Steady Streaming Due to the Interaction of Viscous Effects With Convected Inertia," Trans. ASME, J. Fluids Eng., 98: 707-714, December, 1976.

21. Schneck, D.J., "Pulsatile Blood Flow in a Channel of Small Exponential Divergence -- III. Unsteady Flow Separation," Trans. ASME, J. Fluids Eng., 99: 333-338, June, 1977.

22. Schneck, D.J., and Ostrach, S., "Pulsatile Blood Flow in a Diverging Circular Channel," Case Western Reserve University, Division of Fluid, Thermal and Aerospace Sciences, Technical Report Number FTAS/TR-73-86, Cleveland, Ohio, January, 1973.

23. Schneck, D.J., and Ostrach, S., "Oscillating Blood Flow in a Cylindrical Channel of Small Exponential Divergence," Proc. 3rd Ann. Meeting Biomed. Eng. Soc., p. 40, April, 1972.

24. Schneck, D.J., and Ostrach, S., "Dependence of Unsteady Flow Separation on Frequency of Oscillation," Proc. 26th Ann. Conf. Eng. Med. Biol., 15: 309, October, 1973.

25. Tsahalis, D.T., "Unsteady Boundary Layers and Separation," Ph.D. Thesis, Virginia Polytechnic Institute and State University, Blacksburg, Virginia, August, 1974.

26. Tsahalis, D.T., and Telionis, D.P.," Oscillating Laminar Boundary Layers and Unsteady Separation," AIAA Journal, 12 (#11): 1469-1476, November, 1974.

27. Walburn, F.J., and Schneck, D.J., "An Experimental Technique for Quantifying Unsteady Flow Separation in Diverging Circular Channels. Schneck, D.J. (ed.), Proc. First Mid-Atlantic Conf. on Bio-Fluid Mech., Blacksburg, Virginia, Virginia Polytechnic Institute and State University, 1978, pp. 161-170.

28. Schneck, D.J., and Walburn, F.J., "Unsteady Laminar-Flow Separation in Tubes -- II. The Effect of Variatiions in the Frequency and Amplitude of Flow Oscillations", Virginia Polytechnic Institute and State University Technical Report #VPI-E-79-21, June, 1979.

29. Koromilas, C., "Experimental Investigation of Unsteady Separation," Ph.D. Thesis, Virginia Polytechnic Institute and State University, Blacksburg, Virginia, August, 1978.

APPENDIX Prandtl's Criterion for Flow Separation

The criterion used to determine the onset of laminar flow separation in steady flow was that proposed by Prandtl (equation [1]). Strictly speaking, Prandtl's criterion is a zero shear requirement, i.e.,

$$\tau_{xz} = \mu \left[\partial V_x/\partial z + \partial V_z/\partial x\right] = 0 \text{ at } z = z_{dw} \qquad [A.1]$$

while the other two components are identically zero due to the assumption of axial symmetry and no swirling. However, there are special circumstances (shown by nondimensionalization of equation [A.1] when the zero shear requirement can be replaced by a zero wall velocity gradient requirement. That is, the dependent variables, V_x and V_z, and the independent variables, x and z, are first nondimensionalized by introducing the following new variables:

$$u = V_z/V \qquad w = V_x/U$$
$$z' = z/z_o \qquad x' = x/x_o \qquad [A.2]$$

where the reference quantities, V, U, x_o and z_o are to be chosen in such a manner as to make the dimensionless quantities u, w, x' and z' (and their derivatives), of unit order. Substituting the relations [A.2] into the continuity equation:

$$\partial(zV_z)/\partial z + \partial(zV_x)/\partial x = 0 \qquad [A.3]$$

and rearranging yields:

$$\frac{x_oV}{z_oU}\,\frac{\partial(z'u)}{\partial z'} + \frac{\partial(z'w)}{\partial x'} = 0. \qquad [A.4]$$

If the nondimensionalization is performed correctly, we assume that the derivatives in equation [A.4] are both of unit order. Since the second term is of unit order, it follows that the coefficient of the first term must also be of unit order if the equation is to be valid. It may thus be concluded that:

$$V = Uz_o/x_o. \qquad [A.5]$$

A final substitution of $V = V_z/u$ into equation [A.5] yields

$$V_z = Uz_o u/x_o. \qquad [A.6]$$

Now equation [A.1] may be nondimensionalized with respect to the variables introduced in equation [A.2]. However, the form of V_z given by equation [A.6] will be used instead of that given in equation [A.2]. Substituting the variables into equation [A.1] yields:

$$\tau_{xz} = \mu\left[\frac{U}{z_o}\frac{\partial w}{\partial z'} + \frac{Uz_o}{x_o^2}\frac{\partial u}{\partial x'}\right]. \qquad [A.7]$$

With some rearranging, equation [A.7] becomes:

$$\tau_{xz} = \frac{\mu U}{z_o}\left[\frac{\partial w}{\partial z'} + \frac{z_o^2}{x_o^2}\frac{\partial u}{\partial x'}\right]. \qquad [A.8]$$

In this form, the term $\partial w/\partial z'$ (the radial gradient of the axial velocity) is of unit order while the term $\partial u/\partial x'$ (the axial gradient of the radial velocity) is of order z_o^2/x_o^2. When $(z_o^2/x_o^2) << 1$, the latter term becomes negligible compared with the former. In terms of the geometry of the test section, z_o^2/x_o^2 is on the order of 10^{-4} cm. Thus, the axial gradient term in equation [A.1] may be neglected and Prandtl's criterion becomes:

$$\partial V_x/\partial_z = 0 \text{ at } z = z_{dw}, \qquad [A.9]$$

where V_x is the velocity component in the x-direction and z_{dw} defines the z coordinates of the distal wall of the diverging test section for any arbitrary value of x. This simplification is particularly fortuitous because in the studies reported herein, we had use of only a single channel laser Doppler anemometer system, which made it impossible to measure any more than one velocity component (V_x) at any time or tube location.

THE DYNAMICS OF UNSTEADY BIFURCATION FLOWS

R.J.Liou[*], M.E.Clark[*], J.M.Robertson[*], and L.C.Cheng[+]

*Dept.T.A.M.Univ. of Illinois, Urbana, IL 61801

+Dept.M.E.Wichita State Univ.Wichita, KS 67208

ABSTRACT

Since arterial bifurcations are susceptible to disease, it is important to understand the hemodynamics associated with this common cardiovascular non-uniformity. This paper seeks to extend our understanding of bifurcation flows by using a symmetrical geometry and comparing balanced and unbalanced flows in the branches. Finite difference methods are employed in the vorticity transport-stream function formulation and are aided by a coordinate transform. The basic characteristics of unsteady flow are amply portrayed using the simple oscillatory forcing function. The particular flow division ratios used in this study were QR = 1/2 and 2/3; these values were held constant throughout the period of oscillation. The Karman number and Stokes number of the trunk flow, the basic flow similarity parameters, were taken as 1000 and 10π, respectively. Kinematic results are presented and compared in terms of stream function and vorticity contour plots. Kinetic results are summarized in the form of shear distributions over the regions of interest both in time and space.

INTRODUCTION

Both kinematic and kinetic fluid mechanic events have been coupled to the atherosclerotic process through circumstantial evidence. For example, atherosclerotic plaques have a predilection to occur at arterial bifurcations or in any region of sharp arterial curvature where the fluid dynamic activity becomes abnormal -- either where curvature increases the local velocity or where it causes separation and decreased velocity. Current theories used to explain atherogenesis implicate the development of either high shear rates near the arterial walls (where the endothelial cells are actually damaged mechanically by the shear) or regions of stasis or separation (where the balance of mass transport rates of cholesterol can be upset).

In the arterial system, the blood is delivered to the peripheral cells through a highly-branched system which includes arteries, arterioles, and capillaries. The bifurcation of one vessel into two thus becomes an important feature of the system and inherently involves flow curvatures. One, if not the most important, bifurcation site is the carotid junction where cerebral blood flowing in the common carotid divides into the external carotid (which feeds the face and scalp) and the internal carotid (which carries the major portion of the blood to the brain). Disease often strikes either at this important junction or downstream of it but events that occur at the junction cause the downstream trauma. Atherosclerotic plaque formation can occur near the carotid junction in any or all of the vessels. These blockages themselves can distort the normal flow patterns and cause a loss of blood supply to that most important organ -- the brain.

In the atherosclerotic process, there is a proliferation of collagen cells beneath the endothelial surface as well as a congregation of fatty debris and cholesterol. As time proceeds, the plaque grows larger and intrudes into the arterial passage. As it matures, it can ulcerate and expose the blood platelets to the collagen initiating the formation of a thrombus at the plaque. Such a clot can grow and occlude the artery at the site or a portion of it can break off and move downstream as an embolus. At a downstream junction where the vessels are smaller, the embolus can become lodged and occlude the vessels there. If a significant part of the brain is thus deprived of its blood supply, a stroke occurs.

In view of this sequence of events, it is clear why bifurcations have occupied the attention of hemodynamicists during the past several decades. More information is needed regarding the details of the flow patterns as well as of the shear and pressure patterns in both normal bifurcations and those with stenotic lesions in them (1). A large array of variables are needed to describe a given bifurcation geometry and flow situation. Is the junction itself symmetrical or is the lumen of one efferent larger than the other? At what angles do the efferent vessels leave the trunk? Is the flow in the branches balanced according to the flow area or is it unbalanced with one efferent carrying relatively more of the flow? Does this balanced or unbalanced flow division maintain itself throughout the period of motion or does it change? In this paper the question of flow balance is studied and the effect of the flow division ratio (QR) on flow and shear patterns is delineated for one set of parameters describing an oscillatory flow.

Branching flow studies have been predominately in vitro or numerical studies and often have been limited to steady flow. Early investigations by Stebbins (2), Wesolowski et al. (3), Attinger (4), and Krovitz (5) were in vitro studies using dye streams or fluid birefringence to observe the overall nature of the flow in the branches. More recent flow experimentalists -- i.e., Brech and Bellhouse (6), Mark et al. (7), El Masry et al. (8) -- have looked at the flow patterns more closely in terms of velocity distributions and separation regions and also, for vessels of circular section, the secondary flows. A few unsteady flow studies have been reported: thus, Mark et al. (7) compared the flow field of a two-dimensional bifurcation for simple oscillatory flow with that found by numerical calculation while Talukder (9) and Rodkiewicz and Roussel (10) observed flow patterns (as velocity profiles, separation regions and secondary currents) and wall shears for bifurcations in circular vessels over a range in bifurcation angles, flow-division ratios and Reynolds numbers as well as for pulsatile flow conditions.

Numerical branch flow calculations have principally been made for plane two-dimensional geometries. Steady branch flow calculations have been reported by Lynn et al. (11), Greenfield and Kolff (12), Friedman et al. (13), Cheng (14), Fernandez et al. (15) and Cheng et al. (16). Also, Gokhale et al. (17) calculated steady flow patterns and resultant wall shears of the steady flow through a sequential pair of bifurcations. Calculations of unidirectional pulsatile flow through plane two-dimensional bifurcations have been reported by Hung and Naff (18), Ehrlich et al. (19), Friedman et al. (20), and O'Brien et al. (21). Cheng et al. (16) treated pure oscillatory flow. Additionally,

Clark et al. (1) have reported on unsteady flow calculations delineating the effects of flow division for a bifurcation with a proximal stenosis. In view of the large number of both geometric and fluid-dynamic factors specifying a bifurcation and its flow, there is much yet to be learned about these flows.

THE METHOD

The arterial bifurcation is here modeled in two-dimensions. Where the branches leave the trunk, both the outer wall junctions and the front of the inner wall are rounded. The resultant cusp-like geometry of the inner wall is to some extent physiologically unreal but the exact shape is deemed to be of secondary import. The general arterial bifurcation is asymmetrical both with regard to flow division and geometry. In modeling such a case, the flow in the full trunk and both branches must be determined. Even for the geometrically symmetrical bifurcation (as considered here, see Fig. 1), the unbalanced flow division requires calculation of the full flow field, as contrasted to the bifurcation symmetrical both in geometry and in flow (as also considered here) where only half the field need be calculated. In the more general case (not considered here), a balanced flow through an unsymmetrical bifurcation is defined as one in which the branch flows are proportioned according to their flow areas.

The numerical finite-difference technique is simplified by use of a non-orthogonal transformation which shifts the calculation to a field of square meshes, as has been detailed elsewhere (22). With $X = x/D$, $Y = y/D$ as the dimensionless coordinates of the real flow plane and ξ, η the coordinates of the transformed calculation plane, the geometric transformation is expressed by the relations

$$\begin{aligned} &\text{for trunk,} && \eta_1 = \frac{Y - Y_\ell}{Y_u - Y_\ell} \\ &\text{for lower branch,} && \eta_2 = \eta_0 \frac{Y - Y_\ell}{Y_{m1} - Y_\ell} \\ &\text{for upper branch,} && \eta_3 = \eta_0 + \frac{Y - Y_{m2}}{Y_u - Y_{m2}}(1 - \eta_0) \end{aligned} \tag{1}$$

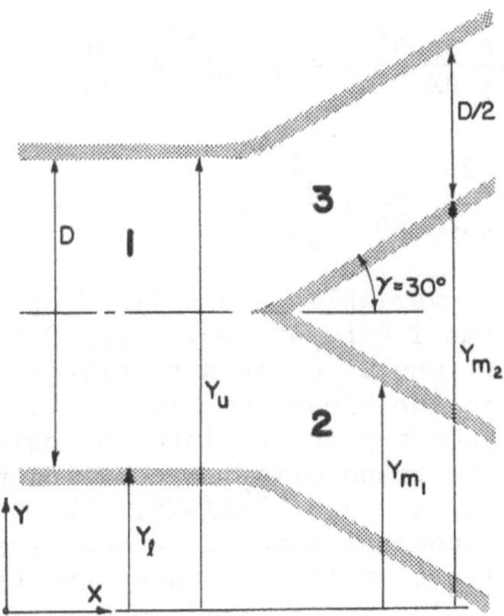

Fig. 1 Definition Sketch for the Bifurcation

where the subscripts ℓ, u, m1 and m2 identify the lower, upper, middle lower and middle upper boundaries, respectively, (see Fig. 1) and where $\eta_0 = (Y_{m1} - Y_\ell)/(Y_u - Y_\ell)$ defines the middle boundary η at the start of the branches. These transformations involve discontinuities in the first and second derivatives at the wall junctions if the straight walls join without transitions. Since the physiological bifurcations are being modeled, the discontinuities resulting from the sharp junctions must be numerically smoothed, as indicated in Fig. 1. For the $\xi - \eta$ plane, the conservative dimensionless vorticity-transport equations of the unsteady plane flow analysis take the form

$$\frac{\partial\Omega}{\partial T} + \frac{\partial\xi}{\partial X}\frac{\partial(U\Omega)}{\partial\xi} + \frac{\partial\eta}{\partial X}\frac{\partial(U\Omega)}{\partial\eta} + \frac{\partial\eta}{\partial Y}\frac{\partial(V\Omega)}{\partial\eta} = \left(\frac{\partial\xi}{\partial X}\right)^2\frac{\partial^2\Omega}{\partial\xi^2} + \\ 2\frac{\partial\xi}{\partial X}\frac{\partial\eta}{\partial X}\frac{\partial^2\Omega}{\partial\xi\partial\eta} + \left[\left(\frac{\partial\eta}{\partial X}\right)^2 + \left(\frac{\partial\eta}{\partial Y}\right)^2\right]\frac{\partial^2\Omega}{\partial\eta^2} + \left[\frac{\partial^2\eta}{\partial x^2} + \frac{\partial^2\eta}{\partial Y^2}\right]\frac{\partial\Omega}{\partial\eta} + \frac{\partial^2\xi}{\partial X^2}\frac{\partial\Omega}{\partial\xi} \qquad (2)$$

and

$$\left(\frac{\partial \xi}{\partial X}\right)^2 \frac{\partial^2 \Psi}{\partial \xi^2} + 2\,\frac{\partial \xi}{\partial X}\,\frac{\partial \eta}{\partial X}\,\frac{\partial^2 \Psi}{\partial \xi \partial \eta} + \left[\left(\frac{\partial \eta}{\partial X}\right)^2 + \left(\frac{\partial \eta}{\partial Y}\right)^2\right] \frac{\partial^2 \Psi}{\partial \eta^2}$$
$$+ \left[\frac{\partial^2 \eta}{\partial X^2} + \frac{\partial^2 \eta}{\partial Y^2}\right] \frac{\partial \Psi}{\partial \eta} + \frac{\partial^2 \xi}{\partial X^2}\,\frac{\partial \Psi}{\partial \xi} = -\,\Omega \tag{3}$$

The dimensionless variables are defined as: velocity $U = uD/\nu$ and $V = vD/\nu$; time $T = t\nu/D^2$, vorticity $\Omega = D^2/\nu$, and stream function $\Psi = \psi/\nu$, where D is a reference dimension (trunk width) and ν is the kinematic viscosity. The effect of the flow-field geometry enters into the calculation through the multiplicative transformation coefficients $\partial\xi/\partial X$, $\partial\eta/\partial X$, $\partial\eta/\partial Y$, $\partial^2\eta/\partial X$ and $\partial^2\xi/\partial X^2$. The nature of these factors depends upon the specific geometry and the manner in which the walls transition between the trunk and the branches.

The flow field is prescribed by the wall boundary conditions on Ψ and Ω and by the inflow and outflow conditions. The inlet and outlet are assumed to be positioned far enough upstream or downstream of the bifurcation to be free of non-uniform flow effects. For cardiovascular fluid flow, the dimensionless Karman number ($\mathbb{K} - (dp/dx)_{amp} D^3/\rho\nu^2$) is frequently used instead of Reynolds number ($\mathbb{R}$) as the parameter fixing the strength of the flow. The dimensionless Stokes number ($S = 2\pi f\, D^2/\nu$) is used to fix the frequency of oscillation. For unsteady flow, the amplitude of the pressure gradient driving the flow is a function of time; for the oscillatory flow used here, $\mathbb{K}$ varies sinusoidally with time. The Ψ and Ω values at the trunk entrance are taken as those of plane oscillatory viscous flow as given by Cheng, et al. (23). For the branch outflows, the same methods are used to calculate the Ψ and Ω profiles but the Karman and Stokes numbers and phase angle differ due to the change in dimensions. However, the assumption that the flow division is time-invariant and the condition that the total outflow always equals the inflow permits definition of these branch flow parameters.

In this plane flow study, the bifurcation is symmetrical with respect to the centerline of the trunk (thus, $\eta_0 = 1/2$), the branch angle γ is 30 degrees, and the ratio of branch areas to trunk area is $\cos\gamma = 0.866$. A five point parabolic transition is used between the trunk and branch walls and for the cusp at the front of the dividing wall. Division of the flow between the branches is specified by the flow division ratio QR, defined as the fraction of trunk flow going to the lower branch. This ratio is taken as invariant during the unsteady flow cycle, a pragmatic but not necessarily physiologically correct assumption. The flow through the symmetrical bifurcation was calculated for two QR values: the balanced flow of $QR = \eta_0 = 1/2$ and the unbalanced one of $QR = 2/3$. The calculations were made for an oscillatory trunk flow of maximum strength prescribed by $\mathbb{K}_t = 1000$ and frequency prescribed by $S_t = 10\pi$. In order for the branch flows to be compatible with the trunk flow, the sum of their flow rates at any time must always equal the instantaneous trunk flow. This requires separate specification of branch Karman and Stokes numbers as well as phase angles; the necessary relations are found at the maximum positive (trunk to branch) flow rates. The frequency of oscillation for trunk and branch flows must be the same: thus, in terms of Stokes numbers

$$\left.\frac{S\nu}{2\pi D^2}\right|_t = f = \left.\frac{S\nu}{2\pi D^2}\right|_{\ell b} = \left.\frac{S\nu}{2\pi D^2}\right|_{ub} \tag{4}$$

where subscripts t, ℓb and ub indicate trunk, lower and upper branches, respectively. Then, since the branch heights are equal to half that of the trunk, their widths (dimension normal to walls) are $(D \cos \gamma)/2$, and

$$S_{\ell b} = S_t \eta_0^2 \cos^2\gamma \quad \text{and} \quad S_{ub} = S_t (1-\eta_0)^2 \cos^2\gamma \tag{5}$$

For this present case of $\eta_0 = 1/2$ and $\gamma = 30°$, the Stokes numbers are $S_{\ell b} = S_{ub} = 0.1875\ S_t = 4.8905$. The flow rates for trunk and branches must be in phase and, specifically, their maximums must occur at the same time. By maximizing the relation for flow rate (23) with respect to $(ST + \phi)$, where ϕ is an arbitrary phase angle, the $(ST + \phi)$ angle for maximum flow can be determined for a given S. Since $ST_t = ST_b$, the difference in $(ST + \phi)$ angles between trunk and branch is just $\Delta\phi = \phi_t - \phi_b$. For the Stokes numbers involved here, $\Delta\phi_{\ell b} = \Delta\phi_{ub} = 0.6989$ radians. Now, $Q = (\mathbb{K}/S) \cdot \text{func}\ (ST + \phi)$ so that $G = (QS/\mathbb{K})_b / QS/\mathbb{K})_t$ is a constant given by the maximum flow conditions, then

$$\mathbb{K}_{\ell b} = G \cdot QR \frac{S_{\ell b}}{S_t} \mathbb{K}_t \quad \text{and} \quad \mathbb{K}_{ub} = G \cdot (1 - QR) \frac{S_{\ell b}}{S_t} \mathbb{K}_t \qquad (6)$$

For the $\eta_0 = 1/2$ cases considered here, $G = 0.53095$ so that for the balanced flow case of $QR = 1/2$, $\mathbb{K}_{\ell b} = \mathbb{K}_{ub} = 176.6$ and for the unbalanced one of $QR = 2/3$, $\mathbb{K}_{\ell b} = 235.4$ and $\mathbb{K}_{ub} = 117.7$. These S_b and $\mathbb{K}_b$ values were used to specify the unsteady flows in the branches at a distance from the junction. The resultant temporal variation in the vessel flows are as indicated in Fig. 2 where $\Delta\phi$ as well as $\Delta\theta$ are shown. The latter angle being 1.239 radians, the phase lag between the pressure gradient and the flow rate.

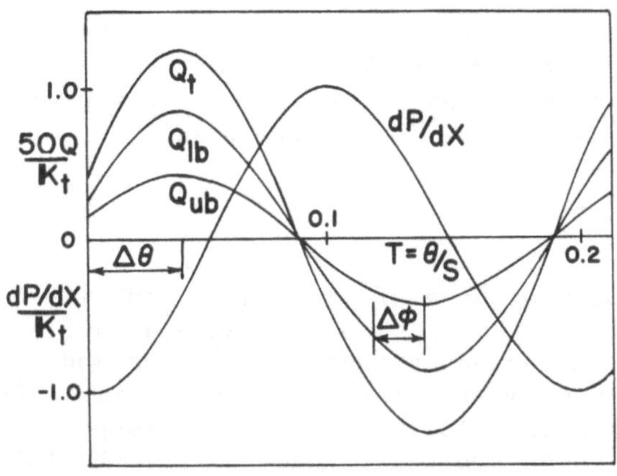

Fig. 2 Temporal Variation in Trunk Pressure Gradient and Trunk and Branch Flows for Unbalanced Flow in a Symmetrical Bifurcation ($\mathbb{K}_t = 1000$, $S_t = 10$, $\eta_0 = 1/2$, $QR = 2/3$).

In the simple rectangular computation field produced by the transform, the finite difference approach is easily applied using central differencing of Eq. 2 and 3. The $2\Omega_{ij}^{N}$ term, which appears when differencing the second derivative, has been replaced by $(\Omega_{ij}^{N+1} + \Omega_{ij}^{N-1})$ to avoid the Richardson instability (N is the time index). Explicit solution of the difference form of Eq. 2 gives the vorticity at the advanced time (N+1) for all interior mesh points. Implicit iterative solution of the difference form of Eq. 3 then yields the stream function at interior points.

The basic mesh size H is taken as D/36 or D/40; these values have been generally satisfactory in terms of solution detail, but at large values of $\mathbf{K}$, a finer mesh near the wall or a refined wall-vorticity relation is needed. Convergence of the Poisson equation (Eq. 3) for the stream function is assessed by the following cease criterion

$$\left|\Psi_{ij}^{k+1} - \Psi_{ij}^{k}\right| < 10^{-3} \tag{7}$$

where k is the number of the iteration. For the unsteady-flow calculations, an alternating sweep direction was used in the trunk with non-alterating inward sweep directions in the branches so as not to sweep from the singularity at the front of the middle wall.

The wall boundary conditions on Ψ are provided by the prescribed inflow conditions, but the boundary wall vorticity Ω is not and must be found from the flow solution. This important information is evaluated from Eq. 3 using $\partial^2\Psi/\partial\eta^2$ as obtained from numerical approximations using values of Ψ adjacent to the wall. Both first and second order expressions have been used for this calculation (24). In this paper an average of the first and second order expressions (as a "1.5 order" form) is used which can be expressed as

$$\Omega_{boundary} = [3\,\Psi_{b+1} - (11/4)\Psi_{b} - (1/4)\Psi_{b+2}]/H^2 \tag{8}$$

where H is the mesh spacing and the subscripts imply boundary and near-boundary positions.

The pressure field is determined using a Poisson equation derived by taking the divergence of the Navier-Stokes equations of motion, i.e.,

$$\nabla^2 P = s = -2\left(\frac{\partial U}{\partial X}\right)^2 - 2\,\frac{\partial U}{\partial Y}\frac{\partial V}{\partial X} \tag{9}$$

where the source term s can be evaluated using the Ψ fields already calculated and converged to the criterion of Eq. 7. In terms of the transformation, the form of the LHS of this equation is the same as Eq. 2 -- the Poisson equation for Ψ. The solution of this equation is subject to Neumann boundary conditions, as follows

$$\left.\frac{\partial P}{\partial \eta}\right|_{wall} = \left(\frac{\partial \eta}{\partial y}\right)^{-1}\left[\frac{\partial \xi}{\partial x}\frac{\partial \Omega}{\partial \xi} + \frac{\partial \eta}{\partial x}\frac{\partial \Omega}{\partial \eta}\right]\Bigg|_{wall} \tag{10}$$

The Poisson pressure equation is written in finite difference form for iterative solution of the pressure at any interior field point more than one node removed from the boundary as

$$\begin{aligned} P^{K+1}_{I,J} = {} & \left[\frac{2\xi_X^{\,2}}{\Delta\xi^2} + \frac{2(\eta_X^2+\eta_Y^2)}{\Delta\eta^2}\right]^{-1} \Big\{ \frac{\xi_X^2}{\Delta\xi^2}\left[P^K_{I+1,J} + P^{K+1}_{I-1,J}\right] + \frac{(\eta_X+\eta_Y^2)}{\Delta\eta^2} \\ & \left[P^K_{I,J+1} + P^{K+1}_{I,J-1}\right] + \frac{(\eta_{XX}+\eta_{YY})}{2\Delta\eta}\left[P^K_{I,J+1} - P^{K+1}_{I,J-1}\right] + \frac{\xi_{XX}}{2\Delta\xi}\left[P^K_{I+1,J} - \right. \\ & \left. P^{K+1}_{I-1,J}\right] + \frac{\xi_X\eta_X}{2\Delta\xi\Delta\eta}\left[P^K_{I+1,J+1} - P^K_{I+1,J-1} - P^{K+1}_{I-1,J+1} + P^{K+1}_{I-1,J-1}\right] - s_{IJ}\Big\} \end{aligned} \tag{11}$$

where K is the iteration step and subscripts on η imply derivatives. The pressure nodes are placed, for convenience, a half mesh away from those for ψ and Ω. In order to solve for the pressure points adjacent to the boundary, the normal derivative condition is incorporated directly into Eq. 11. For example, at the lower wall

$$P^{K+1}_{I,0} = P^{K+1}_{I,1} - \Delta P_{I,1} = P^{K+1}_{I,1} - \frac{\Delta\eta}{\eta_Y}\left[\xi_X\frac{\partial\Omega}{\partial\xi} + \eta_X\frac{\partial\Omega}{\partial\eta}\right]_{I,1} \tag{12}$$

where J = 0 locates a fictitious node a half mesh outside the field and J = 1 the first real node inside the boundary. The revised Poisson pressure equation for J = 1 calculations, simplified by making $\Delta\xi = \Delta\eta = h$ (the mesh spacing), becomes

$$P^{K+1}_{I,1} = [2\xi^2_X + \eta^2_X + \eta^2_Y + \frac{h}{2}(\eta_{XX} + \eta_{YY})]^{-1}\{\xi^2_X(P^K_{I+1,1} + P^{K+1}_{I-1,1}) + (\eta^2 + \eta^2_Y)$$

$$(P^K_{I,2} - \Delta P_{I,1}) + \frac{h}{2}(\eta_{XX} + \eta_{YY})(P^K_{I,2} + \Delta P_{I,1}) + \frac{\eta_X}{2}$$

$$\xi_X(P^K_{I+1,2} - P^K_{I+1,1} + \Delta P_{I+1,1} - P^{K+1}_{I-1,2} + P^{K+1}_{I-1,1} -$$

$$\Delta P_{I-1,1}) + \frac{h}{2}\xi_{XX}[P^K_{I+1,1} - P^{K+1}_{I-1,1}] - h^2 s_{I,1}\} \qquad (13)$$

The field is swept from inlet to outlet and alternately from the lower to the upper boundary, updating values of P as soon as they are calculated.

Since any linear function is a solution of $\nabla^2 P = 0$, the initial pressure field is set so that the proper pressure gradient is imposed on each uniform flow section of the field. With differing Karman numbers in trunk and branch at various times during the cycle, this specification is necessary to avoid excessive iteration. If $P_{I,J}$ is a solution to Eq. 9, then $P_{I,J} + C$ (where C is a constant) is also a solution. The particular solution is achieved by fixing the pressure at one point in the field which is at the center of the trunk inlet for this problem. The outlet branch pressure can be calculated by linear extrapolation.

RESULTS

The temporal variation in the flow fields of the bifurcation is indicated via contour plots of stream function and vorticity at various times in Fig. 3 and 4 for the balanced flow of QR = 1/2. Similar plots for the unbalanced flow of QR = 2/3 have been presented earlier (16). Only a part of the full cycle (period = 0.2) has been selected for presentation from the large mass of data that was generated. The full field was calculated even though the upper and lower halves of the flow in Fig. 3 and 4 should be identical, i.e., no mirror imaging was employed. Some small asymmetries can be detected in the plots but, overall, the field halves appear to closely mirror each other. Half the calculation time could have been saved on these runs if heed had been taken of symmetry but the program for the full field was available and the additional effort that would have been required to program the changes was not deemed to be worth the savings.

Since no separation occurs downstream of the junction, there is little change in the flow patterns near peak oscillatory flow shown in Fig. 3. The method of normalizing the contour values contributes to the constancy of the patterns - ψ_{ref} and Ω_{ref} being taken in each plot as the value at the lower wall at the trunk inlet. In Fig. 4, for times near flow reversal, the patterns show a more complex field

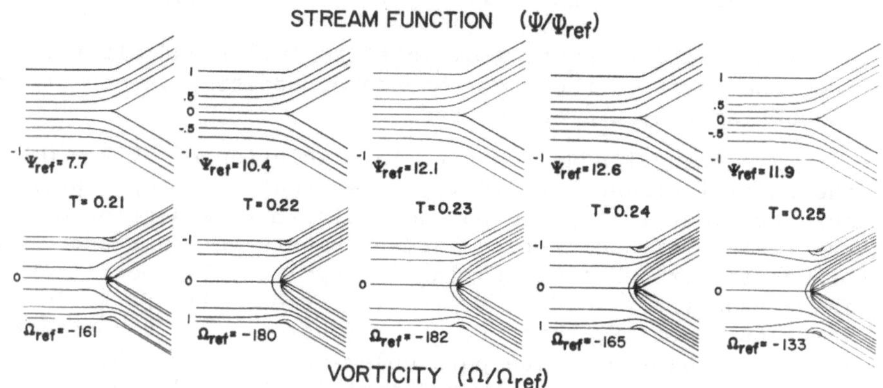

Fig. 3 Temporal Sequence Near Peak Flow of Kinematic Fields for Balanced Flow ($\mathbb{K}_t = 1000$, $S_t = 10\pi$, QR = $\eta_0 = 1/2$)

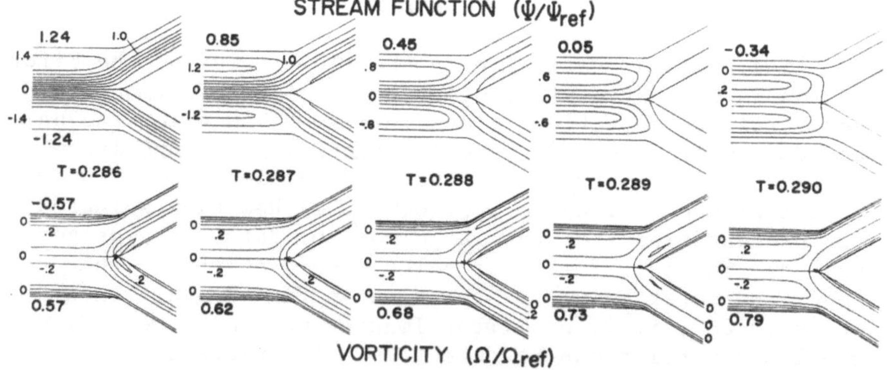

Fig. 4 Temporal Sequence Near Flow Reversal of Kinematic Fields for Balanced Flow ($\mathbb{K}_t = 1000$, $S_t = 10\pi$, QR = $\eta_0 = 1/2$, $\Psi_{ref} = \mathbb{K}_t/1000$, $\Omega_{ref} = \mathbb{K}_t/10$)

at much lower levels of Ψ and Ω. The normalization method for this figure points up the fact that at $T = 0.289$, there is a minimum of activity in the field. Here, both ψ and Ω are normalized by fractions of the Karman number as indicated in Fig. 4. Large zones of circulating flow occupy the major portion of both trunk and branch areas.

The flow fields (as Ψ and Ω contour plots) of the two flow division cases are compared in Fig. 5 for 2 different times -- one, at T = 0.189, where the flow direction was changing from out of the branches into the trunk to out of the trunk into the branches and the other, at T = .24, for flow near maximum from trunk to branches. The fields

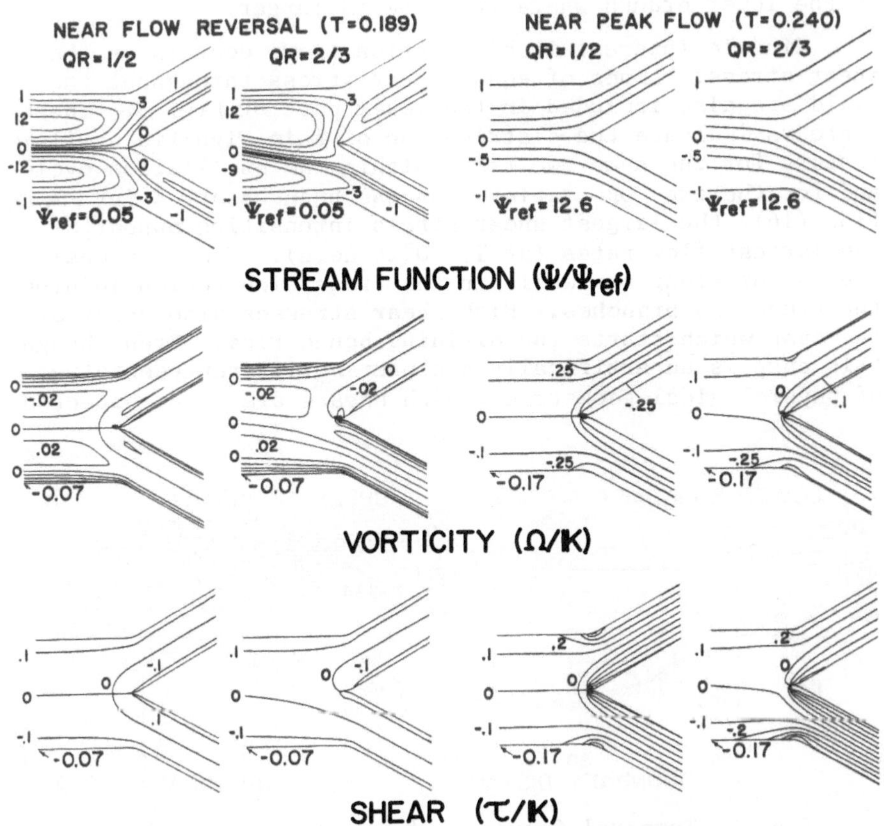

Fig. 5 Comparison of Kinematic and Dynamic Parameters Between Balanced and Unbalanced Flows. ($\mathbb{K}_t = 1000$, $S_t = 10\pi$)

are much more complex near reversal than at peak flow since the inflows and outflows involve complex velocity profiles with both positive and negative velocities. These are evidenced by the appearance of many streamlines which reverse in the same section of the vessel. Contrariwise, the streamline patterns at T = 0.24 are quite straightforward. In each case, the effect of the flow unbalance for the QR = 2/3 flow is clearly evidenced by the distortion of the streamlines near and under the cusp at the start of the branches. The vorticity contour plots of Fig. 5 are more complex with several lines of zero vorticity coursing through the bifurcation. Again, deviation from the balanced case of QR = 1/2 is evidenced for QR = 2/3 by the asymmetrical distortions near the cusp and a higher concentration of contour lines in the lower branch where the flow is larger.

Kinetic interest in bifurcation flows centers on the shear stress. Lines of equal shear stress throughout the field are also included in the lower part of Fig. 5. These contours indicate the distribution of this significant flow feature for the same two times within the oscillation cycle and for the two QR ratios. As has been observed for steady flow (16), the largest shear stress intensities appear at the largest flow rates (at T = 0.24 here). These intensities occur along the walls in the transition region joining the trunk and branches. High shear stresses also occur at the cusp which starts the dividing boundaries. Even though this cusp is an artifically not necessarily representative of physiological geometries, high shears are to be expected

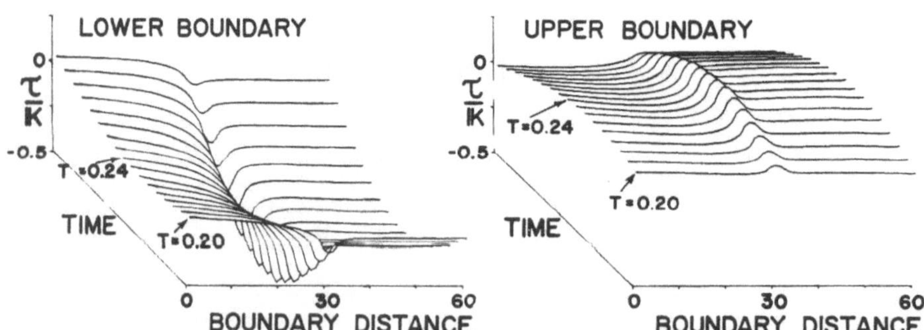

Fig. 6 Temporal Sequence of Lower and Upper Boundary Shear for Part of a Cycle of Unbalanced Flow. (K_t = 1000, S_t = 10 , QR = 2/3)

in this region. Again, it is apparent that unbalanced flow leads to higher shear stresses near the start of the branch due to the larger flow that is shifted to the lower branch. The variation of the boundary shear at both upper and lower boundaries of the field are presented in Fig. 6 in a temporal sequence for part of the cycle for the QR = 2/3 flow. The gradation of the low values of shear at the junction at flow reversal to the high values near maximum forward flow is clearly revealed in this type of plot.

Calculations of the other kinetic feature of bifurcation flow -- pressure field -- have not quite reached the stage of reliability. Tentative results of the pressure calculations are presented in Fig. 7 in the form of pressure contours for times near flow reversal (T = 0.189) and near peak flow (T = .240) for the balanced case. For T = 0.189, there are few deviations from the near uniform increase in pressure from trunk to branch. The gradient in the trunk is somewhat less than that in the branches as should be the case for the local Karman numbers that occur (i.e., those calculated on the basis of the local width, not the trunk width D). The cusp region is devoid of stagnation activity as might be expected since the flow is reversing from out of the branches into the trunk to out of

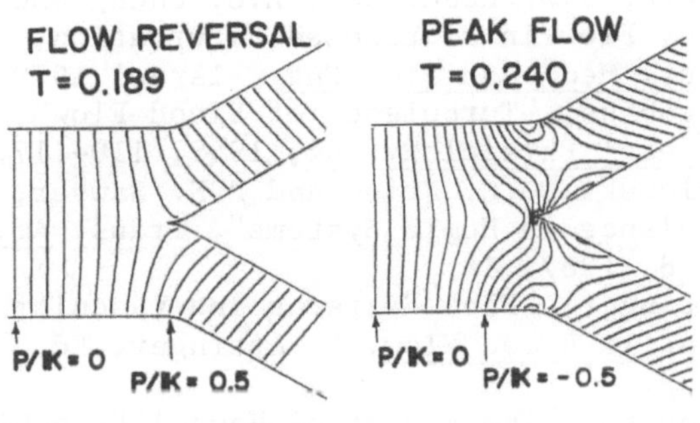

Fig. 7 Pressure Contours at Flow Reversal and at Peak Flow for a Balanced Flow ($\mathbb{K}_t$ = 1000, $S_t = 10\pi$, QR = η_0 = 1/2, Contour Interval = 0.05)

the trunk into the branches. For T = 0.240, however, there is considerable pressure variation in the junction region with local rises at the cusp and local drops at the wall transitions. One disturbing feature of the plot at T = 0.240 (also present to a lesser degree at T = 0.189) is the fact that the contours do not run normal to the walls in the supposedly uniform flow regions in the branches. Experience with the evaluation of pressure fields in other flows indicates that once the pressure is disturbed at a vessel nonuniformity much longer distances are required for it to return to normal values than other variables require. Hence, the size of the flow field calculated here may not have been large enough. The nature of the pressure field is of general interest but it has not yet at least been accorded a prominent place in the list of hemodynamic events of great significance.

ACKNOWLEDGEMENTS

This work was supported by the National Science Foundation (Grant No. Eng. 76-84528, UIUC, and No. Eng. 78-01962, WSU), by the Research Board at the University of Illinois, and by the Computer Center at Wichita State University.

REFERENCES

1. M.E. Clark, J.M. Robertson, L.C. Cheng and S.P. Girrens, "Unsteady Flow in a Stenosed Bifurcation," Digest XII Int. Conf. Med. and Bio. Eng., Israel,1979,Part IV,88.8.
2. W.E. Stebbins, "Turbulence of Blood Flow", Quart J. Experimental Physiology, 44, 1959, 110-117.
3. A. Wesolowski, C.C. Fries and P.N. Sawyer, "Development of Turbulence in Hemic Systems", Trans. Am. Soc. Artif. Organs, 8, 1967, 18.
4. E.D. Attinger, "Flow Patterns and Vascular Geometry", in Pulsatile Blood Flow, E. Attinger, Ed., McGraw-Hill, 1966, 179.
5. L.J. Krovitz, "The Effect of Vessel Branching of Hemodynamic Stability", Phys.Med.Biol., 10, 1965, 417-427.
6. R. Brech and B.J. Bellhouse, "Flow in Branching Vessels", Cardiovascular Res., 1, 1973, 593-600.
7. F.F. Mark, C.B. Bargeron, O.J. Deters and M.H. Friedman, "Experimental Investigation of Steady and Pulsatile Laminar Flow in a 90° Branch", ASME J.Appl.Mech.,44,1977,372-377.

8. O.A. El Masry, I.A. Feurstein and G.F. Round, "Experimental Evaluation of Streamline Patterns and Separated Flows in a Series of Branching Vessels with Implications for Atherosclerosis and Thrombosis", Circulation Res. 43, 197, 608-618.
9. N. Talukder, "An Investigation on the Flow Characteristics in Arterial Branchings", ASME paper 75-APMB-4, April 1975.
10. G.M. Rodkicwicz and C.L. Roussel, "Fluid Mechanics in a Large Arterial Bifurcation", ASME J. Fluids Eng. 95, 1973, 108-112.
11. N.S. Lynn, V.E. Fox and L.W. Ross, "Computation of Fluid Dynamical Contributions to Atherosclerosis at Arterial Bifurcations", Biorheol. 9, 1972, 61-66.
12. H. Greenfield and W. Kolff, "The Prosthetic Heart Valve and Computer Graphics", J. Amer. Med. Assn., 219, 1972, 69-74.
13. M.H. Friedman, V. O'Brien and L.W. Ehrlich, "Wall Shear and Separation in Pulsatile Flow Through a Branch", Proc. 26th ACEMB, 1972, 305.
14. R.T. Cheng, "On the Study of Convective Dispersion Equation", in Finite Elements in Flow Problems, T.J. Oden et al., Ed., UAH Press 1974, 29-47.
15. R.C. Fernandez, K.J. DeWitt and M.R. Botwin, "Pulsatile Flow Through a Bifurcation with Applications to Arterial Disease", J. Biomech. 9, 1976, 575-580.
16. L.C. Cheng, M.E. Clark, J.M. Robertson and N.A. Chao, "Effects of Flow Division on Bifurcation Flow Characteristics, Proc. 1st Mid-Atlantic Conf. on Bio-Fluid Mech., VPI, Blacksburg, 1978, 151-160.
17. V.V. Gokhale, R.I. Tanner and K.B. Bischoff, "Finite Element Solution of the Navier-Stokes Equations for Two-Dimensional Steady Flow Through a Section of a Canine Aorta Model", J. Biomech., 11, 1978, 241-249.
18. T.C. Kung and S.A. Naff, "A Mathematical Model of Systolic Blood Flow Through a Bifurcation", Proc. 8th Int. Conf. Med. and Bio. Eng., 1969, 20-11.
19. L.W. Ehrlich, M. Friedman and V. O'Brien, "Digital Simulation of Periodic Flow in a Bifurcation", Proc. 25th ACEMB, 1972, 215.
20. M.H. Friedman, V. O'Brien and L.W. Ehrlich, "Wall Shear and Separation in Pulsatile Flow Through a Branch", Proc. 20th ACEMB, 1973, 305.
21. V. O'Brien, L.W. Ehrlich and M.H. Friedman, "Unsteady Flow in a Branch", J. Fluid Mech., 75, 1976, 315-336.

22. M.E. Clark and J.M. Robertson and L.C. Cheng, "Interactive and Non-Uniform Unsteady Physiological Flows by Finite Difference Transforms", Proc. Symp. Computer Methods in Eng., USC, Aug. 1977, 1, 497-506.
23. L.C. Cheng, M.E. Clark and J.M. Robertson, "Numerical Calculations of Oscillating Flow in the Vicinity of Square Wall Obstacles in Plane Conduits", J. Biomech., 5, 1972, 467-484.
24. P.J. Roache, Computational Fluid Mechanics, Hermosa Publ., 1972, Albuquerque, 143.

STEADY FLOW AT THE CAROTID BIFURCATION

K. Balasubramanian, D.P. Giddens, R.F. Mabon

School of Aerospace Engineering

Georgia Institute of Technology
Atlanta, GA. 30332

INTRODUCTION

Despite the fact that deaths from heart and blood vessel disease have been declining since 1968, these diseases still account for over 50 percent of the deaths in the United States each year. Atherosclerosis is a major form of arterial disease, and its genesis, proliferation and detection are complex and unsolved problems. Although exact mechanisms for lipid accumulation in the arterial wall are not fully established, the tendency for plaques to occur at preferred sites in the vasculature has stimulated the hypothesis that fluid mechanical factors play a causative role in atherogenesis and may, additionally, be involved in proliferation of early lesions. The passage of blood components into the initima and media, accretion of fibrin and other circulating cellular elements on the intimal surface, the rate of proliferation of intimal cells, and the extent of accumulation of fibrous elements within the subendothelial region may all be dependent, to some extent, on hemodynamic factors. Furthermore, as early disease progresses towards advanced obstruction, plaque encroachment upon the vessel lumen can create localized flow disturbances long before a hemodynamically significant stenosis is present. For these reasons the role of fluid mechanics in the genesis, proliferation and detection of atherosclerosis has received increasing attention.

Because of the tendency for atherosclerosis to develop at sites of vessel branching and bifurcation, numerous investigators have studied branching flows. A complete list of references and results would be rather lengthy to include here; however, among the conclusions from these studies are:

(i) strong secondary flows may be encountered in the branches

(ii) the critical Reynolds number for turbulence in branches is below 2000 and decreases with branch angle

(iii) in steady flow a separation region is formed at the branch with the lower mean velocity, and separation persists even in pulsatile flow

(iv) increasing the branch angle and making the apex blunter tend to cause increased flow disturbances

(v) wall shear stress near the apex is considerably higher than at the outer walls of the branches.

These observations give an indication of general features of the flow field. However, the sensitivity of the flow to branch angle, flow division, Reynolds number, and pulsatility demonstrate that selection of a particular anatomical site for detailed study is necessary for a more complete understanding of the relationship between fluid dynamics and atherosclerosis. For this reason and because of the importance of its involvement in extracranial vascular disease, the carotid bifurcation was selected as the vascular branching for extensive hemodynamic study. Furthermore, the importance of stroke as a major health problem and the accessibility of the carotids to study with ultrasound techniques add to the motivation for investigation of this particular site in the vasculature. The results presented here arise from the first phase of our research and are, consequently, limited in scope. This paper contains a brief description of the model employed and the results of flow visualization and velocity measurement for steady flow at a Reynolds number representative of the mean value found normally in the human subject. Additional studies at other Reynolds numbers are in progress and future experiments will incorporate a pulsatile flow. However, the results to date demonstrate several interesting features, some of which are peculiar to the carotid arteries, which may have bearing on both the puzzle of atherogensis and the problem of early detection of localized plaque development.

METHODS OF EXPERIMENT

Model and Flow Conditions

The design of the carotid model and the flow conditions employed for the experiments were guided by the requirement of relevance to human physiology. Initially, it was necessary to decide on whether to model the carotid artery of a specific individual or to construct a model representative of an "average" configuration. There are advantages and disadvantages associated with either approach; however, in view of the fact that this initial phase was to involve only fluid dynamic measurements and not the pathology of the arterial wall itself, it was decided to take the approach of constructing a model whose geometry was derived by averaging the carotid configuration from many subjects. The bulk of the information used to develop the model was taken from angiograms made available by two local hospitals. Although large numbers of these were available, there are several disadvantages in their use. Since each x-ray photograph is a planar projection, it is necessary to use bi-planar studies to obtain an accurate geometry; and thus anterio-posterior and lateral views were employed in determining the branch angle formed by the internal and external carotid arteries. Additionally, there was usually no absolute length scale present on the angiograms so that all length dimensions were rendered dimensionless by the apparent diameter of the common carotid. In the case of older adults, arterial disease was usually already present so that the outline of a healthy vessel could not be obtained from these subjects, making it often impossible to delineate a normal sinus. As expected, it was found that angiograms of young adults, which would perhaps form the most appropriate data base, were exceedingly scarce. Although angiograms of children were available, the fact that they were yet in a growing stage reduced the value of measurements obtained from their arteriograms. However, despite these shortcomings 67 angiograms from 50 subjects less than 18 years of age and 57 angiograms from 22 adults between the ages of 34 and 77 years were employed in this study.

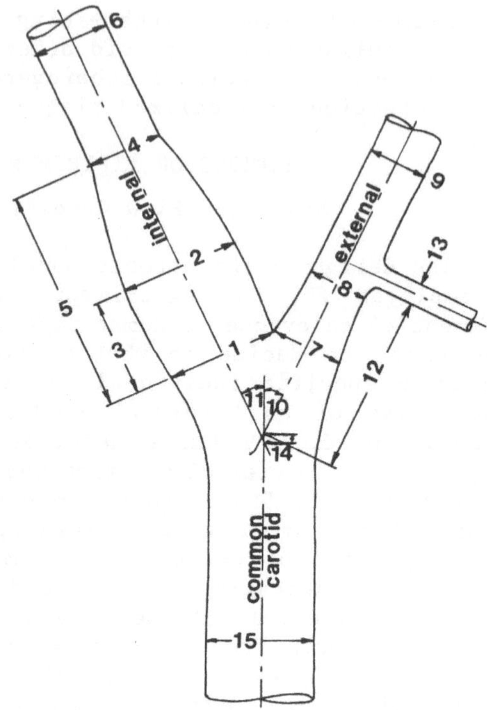

Figure 1. Measurements made from angiograms to obtain carotid bifurcation geometry. (Figure is drawn to scale based on mean dimensions).

In each angiographic outline, the centerlines corresponding to the common, internal and external carotid arteries (CCA, ICA, and ECA, respectively) at the bifurcation, as well as the superior thyroid artery, were identified; and the lengths and angles were measured at 15 locations carefully chosen to describe the geometry of the carotid bifurcation, as indicated in Figure 1. All lengths were obtained using a scale graduated in hundredths of an inch, and nondimensionalised with the diameter of the common carotid artery. The following assumptions were made to relate the measurements obtained from the angiograms to the actual lengths and angles in the human subject.

(i) the CCA is parallel to the major anatomical structures of the neck, and hence to the plane of the angiogram in both the lateral and anterio-posterior projections;

(ii) the arteries and the carotid sinus are circular in cross-section;

(iii) the parent and daughter vessels are in the same plane at the bifurcation; and

(iv) all regions of the bifurcation being studied are magnified equally in any given angiogram.

Granting these assumptions, the diameters of the arteries and the carotid sinus are equally magnified, and their dimensionless values in the angiograms are the same as the corresponding nondimensional values in the physiologic state. This is also true of the length along the common carotid artery. However, the branch angle and lengths along the internal and external carotid arteries are smaller in the angiograms due to the distortion produced by the projection of the three-dimensional bifurcation within the body upon a two-dimensional plate of film. These factors were taken into account by utilizing pairs of biplanar angiograms. Table 1 gives the resulting values and standard deviations for the fifteen different measurements. Additional details and a more complete discussion of the difficulties encountered are given in Ref. 1. The degree of bluntness of the apex cannot be ascertained from the angiograms. Therefore, material from cadavers and from arteries excised during endarterectomy were studied with regard to this aspect. These specimens indicated a rather acute apex angle which was blunted only slightly by a rounded leading edge.

With all lengths being nondimensionalized by the common carotid diameter, the geometric description afforded by Table 1 is complete once an average value for this reference length is known. A study of several sources revealed that a value of 8 mm for the internal diameter of the common carotid is reasonably representative of adults, and this value was employed in the final geometric scaling.

The study demonstrates that there is no "typical" carotid bifurcation. There is considerable variability among subjects in a given age group, and the geometry for a given individual undergoes changes with age. However, the data collected serve in defining an "average" configuration for purposes of constructing a model whose flow field

TABLE 1
Dimensions of the Bifurcation
Obtained from Angiograms
(Nondimensional Values)

	Location*	Sample Size	Mean Valve	Std. Dev.
Internal	1	40	1.043	.206
Carotid	2	25	1.110	.138
	3	11	.908	.308
	4	25	.718	.120
	5	13	2.139	.447
	6	53	.694	.097
External	7	36	.694	.165
Carotid	8	13	.689	.125
	9	40	.581	.096
Angles	10	23	25.143^{o}	11.022
	11	23	25.362^{o}	10.422
Other	12	11	1.860	.812
	13	15	.233	.040
	14	51	-.207	1.192

*See Figure 1.

characteristics offer a representation of the types of phenomena to be expected in an actual carotid bifurcation.

Two models were constructed. The first was a glass-blown model to be used in flow visualization studies, while the second was machined from plexiglass and employed in experiments to obtain velocity data. Glass provides a better quality material for photographic work; and in view of the fact that the visualization studies are qualitative in nature, it was deemed satisfactory to have the model constructed by an expert glassblower. Slight differences exist between this model and the very accurately machined plexiglass bifurcation, and these are documented in Ref. 1.

In the initial phase of the research it was decided to concentrate upon steady, rather than pulsatile, flow. Little was known about the flow behavior at the outset, and the added complication of pulsatility seemed prohibitively difficult as a starting point. A typical velocity waveform for the common carotid artery yields a frequency parameter ($\alpha = R\sqrt{\omega/\nu}$ where R is the radius, ω is the frequency of oscillation, and ν is the kinematic viscosity) of approximately 5 if the period of the heartbeat is used and 8 if the systolic period is taken as the effective frequency. Although for these values of α a quasi-steady assumption is not expected to be quantitatively accurate, trends occurring in steady flow would very likely be present when flow is pulsatile. Furthermore, since the normal carotid maintains an almost constant positive flow rate following systole, for much of the cardiac cycle a quasi-steady assumption is probably very accurate.

The medical literature was explored to obtain information on typical mean flow rates in the common carotid and its branches; and, although considerable variation was found, the values of 500, 300 and 150 ml/min were selected as representative for common, internal, and external carotid arteries, respectively. No information on flow rate through the thyroid branch from the external carotid was found. Due to the extensive variability in this branch location and the fact that visualization studies showed little effect of varying flow through the thyroid vessel, this branch was omitted from the plexiglass model.

Employing 8 mm as the common carotid diameter, 500 ml/min as the mean flow rate, and 0.035 cm^2/sec as representative of blood viscosity gives a mean Reynolds number of 380. Thus, a value of 400 was selected for extensive initial study, and values as high as 1200 are currently being investigated. The average flow division ratio of percentage flow rate into the internal carotid to that in the external carotid was taken to be 70:30. As perturbations about this, however, values of 60:40 and 80:20 were also employed during the research.

The actual model was scaled upward in size to allow better visualization and velocity resolution. It was convenient to employ standard glass tubes in the model construction, and tubes with internal diameters of 31, 22 and

17.6 mm were used to represent the common carotid and its branches. The resulting diameter ratios are within 0.02 of the values in Table 1. The model test section was connected to a constant head reservoir by a 2.72 m glass tube to insure fully developed flow; and electromagnetic flowmeter probes were connected between the outflow planes of the branches in the model and tubes leading to the downstream reservoir, allowing measurement of individual branch flow rates. Water/glycerin mixtures were employed (viscosity = 0.12 gm/cm sec) as the working fluid at a Reynolds number of 400. Since geometric similarity to the human carotid bifurcation was preserved and dynamic similarity was obtained with the Reynolds number, Dean number similarity for flow through curves tubes was automatically satisfied.

Instrumentation

Flow Visualization. Hydrogen bubble and dye injection methods were employed to visualize flow patterns. A cathode wire of 0.002 inch (50.8 μ m) diameter, stretched across the diameter of the parent tube, is placed in the coupling connecting the model to the long glass tube and is arranged such that it can be rotated about the axis of the parent vessel. This enables positioning of the wire at any desired orientation and the entire cross-section of the tube can be covered. The disturbance produced by the wire itself is negligible, as the Reynolds number based on wire diameter is at the most 4.0 at the center of the tube, when the upstream Reynolds number in the parent vessel is 1200. The anode is a brass foil placed inside the settling chamber to prevent the oxygen bubbles from entering the flowfield. Sodium sulfate was added to the solution to enhance uniform bubble generation. The hydrogen bubbles are illuminated by four powerful (250 W) spotlights from the side, and the flow patterns thus revealed are photographed from above on Kodak 2475 recording film, using a Nikkormat EL single lens reflex camera with a 55 mm close up lens, against a dark background provided by black velvet cloth.

To inject dye into the model a long flexible narrow tube is inserted from the downstream end of the internal carotid artery. Since the main purpose of these studies is to investigate flow near the walls of the bifurcation, the dye is injected rapidly and allowed to mix with the upstream fluid. The flexible tube is withdrawn quickly from the bifurcation to keep the flow field clear of any external influences. Sufficient time is allowed for the disturbances produced by the injection process and subsequent withdrawal of the tube to subside before recording the data. Black ink is used for the injections, and water employed as the liquid since the low velocities are conducive to the study of dye patterns. The water is periodically replaced as it becomes dark due to the ink added.

The illumination for this part of the visualization is provided by the same lamps used for the hydrogen bubble technique. However, they are placed in line with the camera and shielded by a white velvet cloth to provide a uniformly bright background against which the flow patterns are photographed. The same camera and lens as before are used, but with Kodak Tri-X Pan film.

Velocity Measurement. A DISA 55L Mark II Laser Doppler Anemometer (LDA) system was employed for velocity measurements in the plexiglass model. The system consists of a 15 milliwatt helium-neon laser, a transducer-optics package which includes a Bragg cell, a photomultiplier and receiving optics, and a frequency tracker for processing the Doppler signal. The laser and optics package are mounted on a support that allows translation in all three directions and rotation about a vertical axis. The least count for translations is 0.001 inch (0.025 mm) and 0.5 degree for rotation. Silicon carbide particles with a diameter of 1.5 μm are used for seeding the flow. Because of the complex geometry of the model used, movement of the laser mount does not always bear a linear relationship to the movement of the sampling volume in the liquid medium; and optical corrections are required as discussed in detail in Ref. 1.

A Hewlett-Packard 5451B Fourier Analyzer was employed to process the analog output of the frequency tracker. Averaging times from 25 to 50 seconds were employed in calculating the velocity at a point in the flow. No turbulence was observed at a Reynolds number of 400.

RESULTS

Flow Visualization

The visualization studies revealed a bifurcation field that is dominated by the presence of secondary flows.

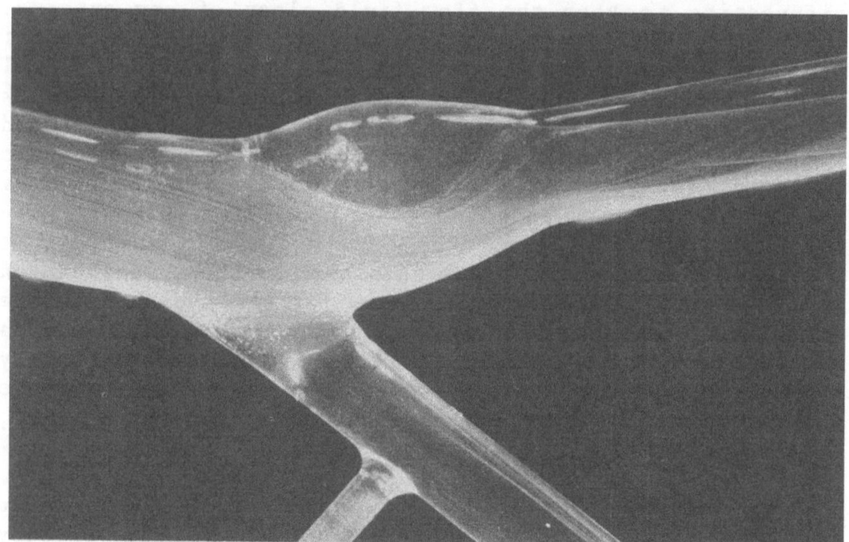

Figure 2. Steady flow through the glass model as visualized with hydrogen bubbles illustrating flow separation (Re = 400, 70% flow through the ICA, hydrogen generating wire is in the plane of bifurcation).

Figure 2 is taken from the hydrogen bubble experiments with the bubble-generating wire oriented in the plane of symmetry of the bifurcation (that is, in a horizontal plane if the camera lens is assumed to be viewing vertically downward) and illustrates the large zone of flow separation which is encountered as flow enters the carotid sinus. The Reynolds number is 400, and 70 percent of the flow is through the internal carotid. Figures 3 and 4 show views when the wire is oriented normal to the previous direction, and the flow conditions are the same. The complex helical nature of flow in the sinus is clearly seen in these figures. The bubble streaklines appear to be more or less unidirectional in the immediate neighborhood of the apex along the inner wall of the sinus. However, careful scrutiny reveals streaklines which are directed circumferentially from the apex towards the outer wall and are then entrapped within the helical patterns of the sinus. These patterns are suggestive of higher shear rates near the apex and of shear stresses along regions of the sinus wall which

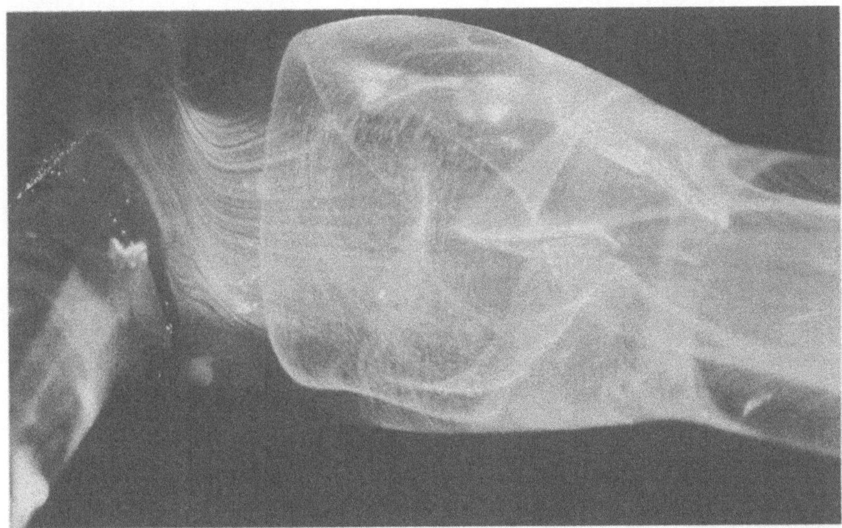

Figure 3. Complex flow field in the sinus as seen with hydrogen bubbles (Re = 400, 70% flow through the ICA, hydrogen wire is perpendicular to the plane of bifurcation).

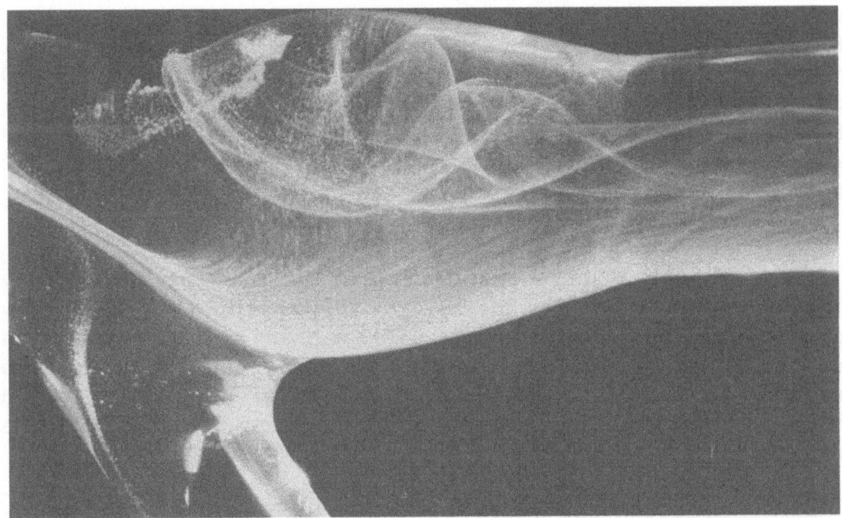

Figure 4. Flow field in the sinus seen along the plane of bifurcation. (Conditions same as for Figure 3).

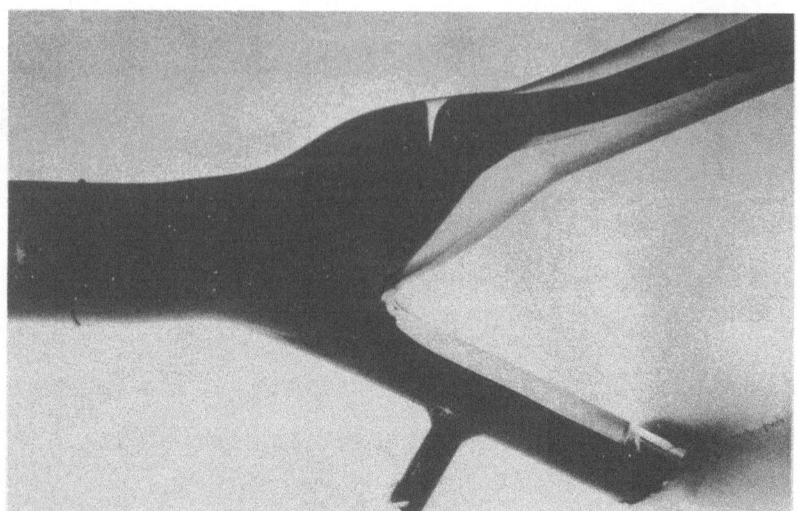

Figure 5. Rapid clearing of dye near the apex and at the walls of the internal carotid (Re = 400, 70% flow through the ICA).

are oriented at substantial angles to the axial direction. Figure 5 gives a different perspective of the regions of high shear stress near the apex, again for Re=400 and 70 percent flow in the internal carotid. This photograph was taken at a time subsequent to the complete filling of the model with dye. The clear zones along the inside walls of the branches, distal to the apex, arise due to higher velocities transporting the dye downstream.

Figure 6 illustrates the stagnation of flow along the outer wall as it enters the sinus for the same flow conditions. Dye has remained within this region while much of the remainder of the bifurcation has been cleared. Also evident in this photograph is a circumferential stagnation band just at the bifurcation of the common carotid into the branches.

Although the separated flow zone has negative velocities present, to term it a recirculation region is a misnomer since fluid is not entrapped and recirculated within a closed surface. Rather, it is a region of complex

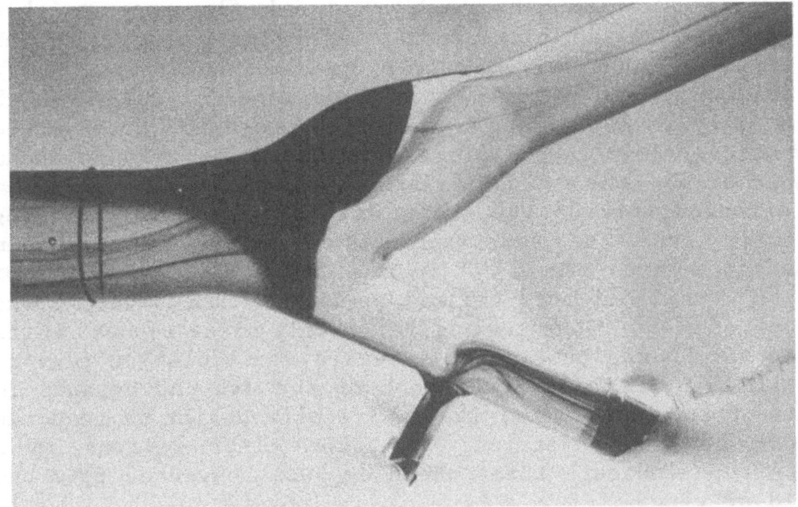

Figure 6. Accumulation of dye at the entrance to bifurcation. (Re = 400, 70% flow through the ICA).

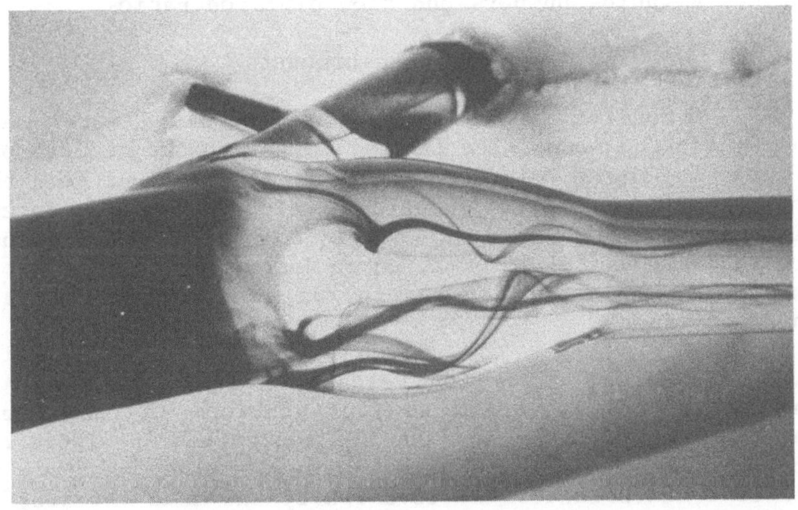

Figure 7. Helical flow in the carotid sinus (Re = 800, 60% flow through the ICA). Note the absence of dye along the outer wall of sinus indicating no recirculation.

secondary flow within which reversed flow occurs, but streamlines do not close upon themselves. A particle which enters this zone may encounter upstream velocities for a period, but it will follow a well-defined trajectory which eventually leads it into a helical pattern to be convected downstream. This is illustrated in Figure 7 for a Reynolds number of 800 and a flow division ratio of 60:40. The view is directed towards the outer wall of the sinus, looking directly into the separated flow region. The upstream separation line forms a "C" in this figure where the ends of the "C" trail off into helical patterns. Immediately downstream of the "C" the zone is relatively clear of dye, which would not be the case if this were a recirculation region. It should be recalled that each of the dye photographs is taken at some time after the entire bifurcation is rendered opaque by ink injection, and thus clear regions in a photograph indicate zones where dye was convected from the field of view.

The figures shown here are merely a sampling from the hundreds of photographs taken. Reference 1 contains numerous other results of the flow visualization study at various Reynolds numbers and flow division ratios.

Velocity Measurements

Extensive measurements of velocity were made throughout the model. The LDA is limited to the measurement of only a single velocity component and as seen from the previous figures, the velocity field is clearly three-dimensional. However, the single-component measurements provide very useful information in describing flow development, locating zones of separation, and estimating wall shear stresses.

Figure 8 gives the velocity profiles measured in the plane of bifurcation at four stations in the common carotid artery. (The stations are identified in Figure 10) for Re=400 and a 70:30 flow division. The velocity is rendered dimensionless by division by the mean velocity in the CCA. Sections CC1 and CC2 are in the constant area region of the common carotid, section CC3 is just distal to the entrance plane of the bifurcation, and section CC4 is approximately 0.5 diameters distal to section CC3. Little upstream influence of the bifurcation is seen in the pro-

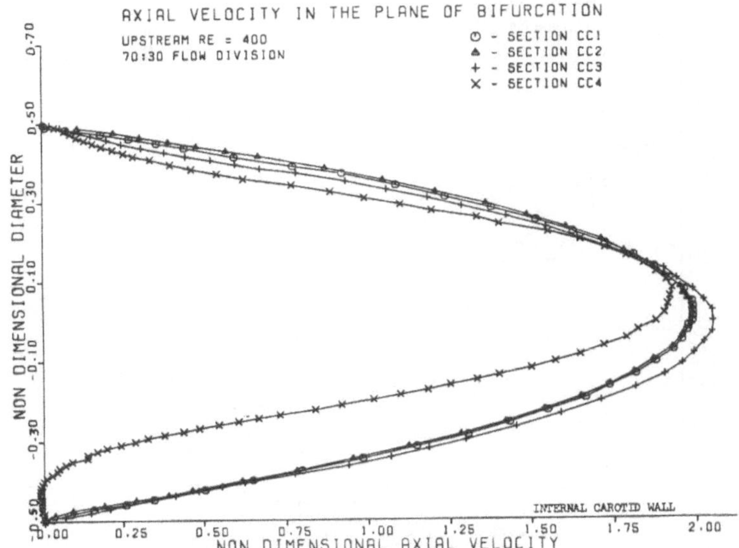

Figure 8. Inplane velocity profiles in the common carotid artery. (The sections are shown in Figure 10).

files for the first two stations, although there is a slight shift of the peak velocity towards the external carotid. At section CC3 the peak shifts towards the internal carotid, and the velocity gradient at the wall near the common-external junction is reduced. At section CC4 reversed flow can be seen near the carotid sinus wall and a further reduced velocity gradient occurs at the external carotid wall. However, no flow separation in the ECA was detected for these conditions. The velocity component graphed in Figure 8 is that directed along the axis of the common carotid artery.

Figure 9 gives results of velocity measured along the centerline of the ICA. The origin of the axial distance begins at the intersection of the CCA and ICA axes, and the velocity component is that along the ICA axis. Three curves are given, corresponding to Re=400 and flow division ratios of 60:40, 70:30 and 80:20. The sharp drop in axial velocity is a consequence of the secondary flow field which creates a skewing of the axial velocity profile towards the apex and a flow separation along the outer wall. The measurements

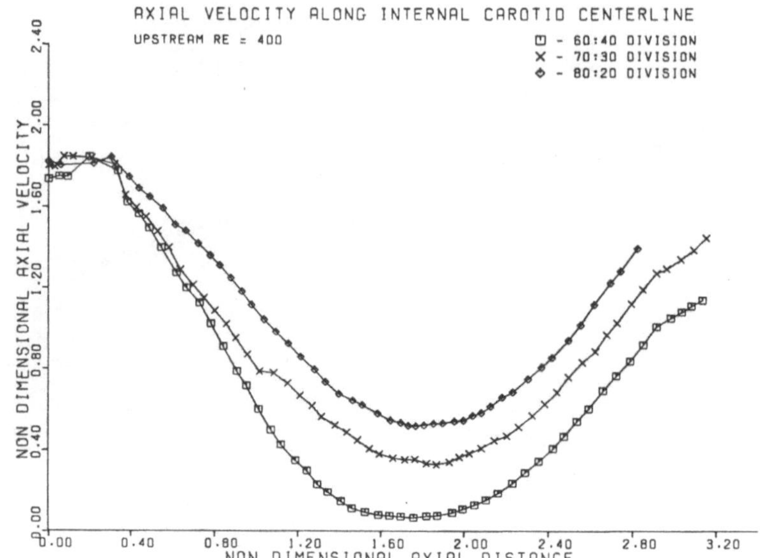

Figure 9. Distribution of axial velocity along the centerline of the ICA showing the effect of changing the division of flow between branches.

indicate that decreasing the flow division ratio at a fixed Reynolds number creates stronger secondary flow effects.

A composite illustration of the development of axial velocity profiles is given in Figure 10 for Re=400 and a 70:30 flow division. The profiles were all taken in the plane of bifurcation. The tendency towards low, or even negative, shear rates at the outer wall of the CCA-ICA junction is clearly demonstrated, as is the existence of high shear rates along the wall near the apex. Interestingly, however, the shear rates along the outer sinus wall are relatively high once flow has passed the separated flow zone. Also, velocity profile measurements in a plane perpendicular to the bifurcation plane and through the sinus axis indicate relatively high shear rates at what might be termed the top and bottom walls (Ref. 1). Thus, the sinus wall was found to encounter low shear rates only in the neighborhood of the outer corner of the CCA-ICA junction.

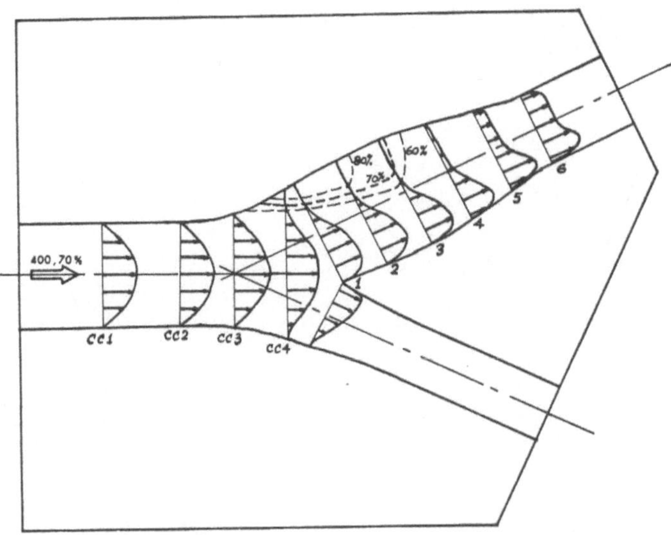

Figure 10. Velocity profiles in the plane of bifurcation when Re = 400 with 70% flow through the ICA. (Dotted lines in sinus denote boundary of reversed flow zones in the plane of bifurcation).

The change in size of the separated flow region with variations in flow division is also depicted in Figure 10. The outlines of these zones were determined from axial velocity profiles in the bifurcation plane. They are not intended to represent dividing streamlines but are simply the locus of points of zero velocity in the profiles.

Because of various theories regarding the role of fluid mechanics in atherogenesis (e.g. Refs. 2-4), there has been considerable interest in values for wall shear stress. Although this quantity was not measured directly, the velocity profiles were employed to estimate shear rate at the wall and a Newtonian relation employed to calculate wall shear stress. The procedure employed was to measure the velocity at three radial points, beginning as near to the wall as LDA signal quality would allow and moving outward for a distance of 0.2 mm. (The length of the LDA sample volume is approximately 1.1 mm in the fluid). A least-squares straight line was fit to these three velocity values, and its slope was taken as the value of shear rate

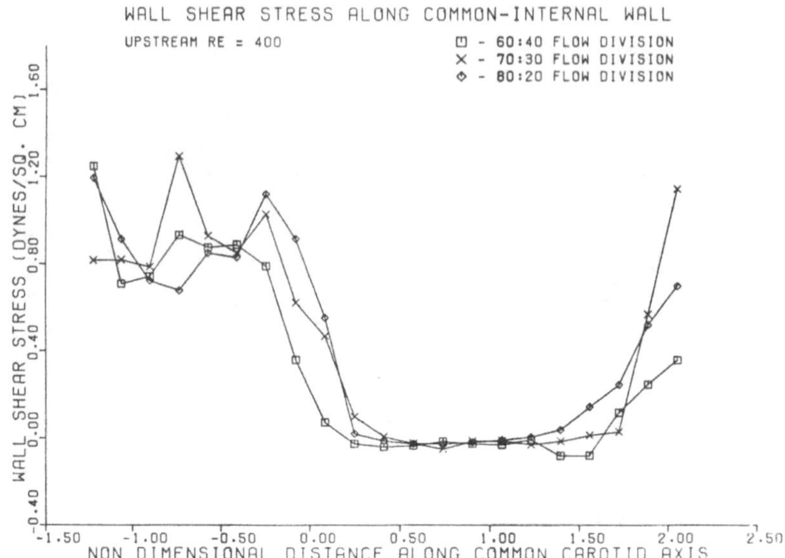

Figure 11. Distribution of shear stress along the CCA-ICA wall when RE = 400. (Shear Stress Values are Dimensionless. Disregard Units.)

at the wall. Employing this procedure at upstream stations in the CCA gave wall shear rates within 10 percent of the Poiseuille value. In high shear rate regions this accuracy would be compromised and additional experiments are required to define better the experimental error. However, the results obtained for these measurements along the CCA-ICA outer wall are graphed in Figure 11 for Re=400 and three flow division ratios, assuming a Newtonian form for the constituitive equation. The negative values corresponding to reversed flow may be seen, and the extent of flow reversal is shown to increase with decreasing flow division into the ICA. Also, the rapid increase in wall shear stress distal to the reversed flow zone is readily apparent. The shear stress values have been made dimensionless by dividing by the upstream CCA Poiseuille value, which is 5.83 dynes/cm^2 for a blood viscosity of 0.035 cm^2/sec. when the Reynolds number is 400. Table 2 gives wall shear stress values at the inner and outer sinus walls assuming a blood model. Again, the outer wall experiences a relatively low shear stress while the inner wall stress is considerably larger.

TABLE 2

Wall Shear Stress in the Human Carotid Sinus in the Plane of Bifurcation.

Upstream Re = 400; Flow through ICA is 70%.

Sinus Section*	Shear Stress in dynes/cm^2 Inner Wall	Outer Wall
1	--	-.19
2	14.1	-.08
3	18.2	-.05
4	25.2	2.72
5	28.8	15.8
6	30.5	24.3

*See Figure 10 for section locations.

DISCUSSION

The results presented here represent only the initial phase of a long-range study of the carotid bifurcation and, consequently, are not proclaimed as comprehensive. Steady flow is at best a rough approximation, considering the frequency parameters typical of the carotid artery; and limiting the investigations to a single Reynolds number is restrictive. However, this initial phase has provided a sound basis for ongoing research which is continually directed towards more realistic conditions. Also, it has yielded interesting information in its own right.

Both flow visualization and velocity measurements demonstrate the important influence of the carotid sinus on the flow field. The flow is seen to separate near the CCA-ICA junction but not at the CCA-ECA junction under the conditions studied. Regions of low wall shear stress in the ICA are confined to this separation region, while the apex experiences a much higher stress. Additionally, there is a circumferential band of low wall shear stress at the termination of the CCA. The flow visualization studies demonstrate that streaklines near the upper and lower (as opposed to inner and outer) sinus walls have significant tangential components and that these change sharply with flow division and Reynolds number. If this carries over to the pulsatile case, it is expected, then, that wall shear stress vectors in the sinus undergo a continual change of direction during the cardiac cycle. Generally speaking, the secondary flow and wall shear stress behavior seem to be strongly influenced by flow division and, to a lesser degree, by Reynolds number. No turbulence was observed during these studies although it is certain that if the Reynolds number were increased sufficiently, turbulence would occur.

The contrasts in this three-dimensional study and previous theoretical investigations in idealized branching flows are noteworthy. Friedman et al (Ref. 5) have made calculations of wall shear stress along the outer wall of a symmetric two-dimensional bifurcation with sharp corners. Their results indicate an increasing wall shear stress in the parent vessel as the corner is approached. The present data for a three-dimensional model with rounded corners show a decreasing wall shear stress followed by separation. And, of course, numerical studies of two-dimensional bifurcations do not predict secondary flows. Hence, wall shear stress directions and values at upper and lower walls cannot be predicted.

The existence of separation is very likely highly dependent upon the geometry and flow conditions. In a study of flow in a model of a common iliac bifurcation (Ref. 6) no separation near the corners was detected for an equal division of flow.

Given the limited scope of the present results it would be improper to extract far-reaching conclusions regarding the role of hemodynamics in atherogenesis. There is a tendency, recognized in the medical literature, for atherosclerotic plaque to develop near the outer wall of the CCA-ICA junction. The wall shear stress in this region is relatively low, which might be construed as supporting Caro's theory of atherogenesis (Ref. 2). However, the separated flow region does not appear to be a recirculation zone since it is cleared away of dye by the secondary flow field. Thus, although the stress is low in this zone, it would not be accurately described as a stagnation region. In addition, the configuration of this region changes, even in steady flow, with flow division and Reynolds number.

The existence of higher, unidirectional wall shear stress in the neighborhood of the apex has been hypothesized by Fry (Ref. 4) to be a factor in the relative immunity of this site to advanced atherosclerotic involvement, provided the magnitude of the stress is below some acute damage level. The present results do not demonstrate excessively high shear stress values at the apex and do indicate a unidirectional nature under all flow conditions studied.

The possibility of variation in wall shear stress direction leading to impairment in the protective endothelial barrier was raised by Fry (Ref. 4). The present experiments show that the wall shear stress in the sinus region has greatly varying direction with regard to spatial distribution; but until pulsatile flow studies are completed, it is premature to state that shear stress direction at a point on the outer sinus wall varies significantly with time.

Despite the simplifications involved the present study has served to provide new knowedge of the fluid dynamics of a perticularly important anatomical site, the carotid bifurcation. However, it must be followed by research in which conditions closer to the physiologic state are considered and, most importantly, by a direct investigation of the interaction between the flow environment and the pathology of the arterial wall.

REFERENCES

1. Balasubramanian, K. - An experimental investigation of steady flow at an arterial bifurcation; Ph.D. Thesis, Georgia Institute of Technology, Atlanta, 1980.

2. Caro, C.G. - Transport of material between blood and walls in arteries (in) Atherogenesis: Initiating Factors; A Ciba Foundation Symposium (Vol. 12), New Series, Elsevier Excerpta Medica, Amsterdam, North Holland, 1973.

3. Fry, D.L. - Certain histological and chemical responses of the vascular interface to acutely induced mechanical stress in the aorta of the dog; Circulation Research (24): 93, 1969.

4. Fry, D.L. - Response of the arterial wall to certain physical factors (in) Atherogenesis: Initiating Factors; A Ciba Foundation Symposium (Vol. 12), New Series, Elsevier Excerpta Medica, Amsterdam, North Holland, 1973.

5. Friedman, M.H., et al - Calculations of pulsatile flow through a branch; Circulation Research (36): 227, 1975.

6. Deters, O.J., et al - Velocities in unsteady flow through casts of human arteries; 32nd Annual Conference of Engineering in Medicine and Biology (21): 101, 1979.

ACKNOWLEDGEMENTS

This research was supported by grants from the National Science Foundation (ENG 76-23876) and the National Institutes of Health (HI 20835).

PULSATILE FLOW THROUGH A CONSTRICTED ARTERY

V. O'Brien and L. W. Ehrlich

Applied Physics Laboratory
Johns Hopkins University
Laurel, Maryland 20810

INTRODUCTION

As part of our long standing interest in the etiology of atherosclerosis we are presently concentrating on an isolated moderate stenosis in a straight artery. The attack is three pronged, chronic *in vivo* experiments with healthy dogs, physical experiments in transparent models and approximate theoretical solutions. By studying the fluid dynamic field and the endothelial response we hope to establish the normal physiological state and thus be better able to define the disease mechanisms leading to deposition of atherosclerotic plaque.

This paper will deal only with the latter two prongs of the program, the theoretical pulsatile predictions via numerical approximate solutions and our efforts to establish physical flow data to verify the method. Previous experiments[1]* and theory[2] dealing with steady flow in stenosis geometries have been successful in showing some degree of mutual confirmation. We expect even better verification for the unsteady flows in spite of the difficulty of measurements which require special experimental techniques.

A brief mathematical section sets forth the basic dynamic parameters and defines a "simple pulsatile flow". The stenosis shapes chosen for the approximate finite-difference flow solutions are given by an uncomplicated analytic expression which facilitates the mapping of a median physical flow

*References are listed at the end.

plane to a more convenient computational strip. Examples of results for steady and simple pulsatile flows are shown, which reveal time-dependent wall shearing stress details and transient recirculation regions that may be important in atherogenesis or plaque development. A few difficulties are mentioned.

The measurement of periodic flows through a stenosis geometry is outlined in the experimental section. The heart of the flow facility is the transparent test section containing a parallel inlet section and the stenosis. Centerline data, taken by means of laser Doppler velocimetry, in the parallel inlet flow set the unsteady boundary conditions for a numerical simulation. Off-center velocity data allows a check of the consistency of the experiments and theoretical predictions for the parallel unsteady inflow. A number of features can then be measured for comparison to the more interesting predictions in the stenosis vicinity and downstream.

MATHEMATICAL BACKGROUND

Blood is taken as a Newtonian incompressible fluid. Motion of the arterial walls is neglected. The geometry is axisymmetric, a single constriction in a long straight round tube of radius a. Steady flow depends upon a single dynamic parameter, the Reynolds number $Re \equiv (2Ua/\nu)$, but periodic unsteady flow depends upon a reduced frequency (Stokes number) as well, $f_R \equiv (2\pi a^2/\nu T)$. (Here U is the space-average velocity in the inlet tube, ν is the kinematic viscosity of the fluid and T is the period of the pulsatile flow.) We call the periodic waveform simple when it consists of a steady portion and a single harmonic $\cos \omega t$, where $\omega = 2\pi/T$. The pulsatile flow depends on amplitude ε when the normalized streamfunction (flux) is

$$\psi = 1 + \varepsilon \cos \omega t \qquad \text{(simple pulse wave)}, \qquad (1)$$

where ε can vary from nearly zero to values greater than one. When the periodic signal is rich in harmonics, all major harmonic coefficients are important.

Combining the continuity equation and the curl of the momentum (Navier-Stokes) equation, the normalized flow equations can be written as the coupled pair

$$\Omega = -\frac{1}{r} D\psi \tag{2}$$

$$f_R \frac{\partial\Omega}{\partial\tau} + \frac{1}{2}\frac{\overline{Re}}{r}\left[J(\psi,\Omega) + \frac{\Omega}{r}\frac{\partial\psi}{\partial x}\right] = \frac{1}{r} D(r\Omega)$$

where the cylindrical coordinate system (r,x) has

$$D \equiv \frac{\partial^2}{\partial r^2} - \frac{1}{r}\frac{\partial}{\partial r} + \frac{\partial^2}{\partial x^2}$$

the Jacobian

$J(\psi,\Omega) = \dfrac{\partial(\psi,\Omega)}{\partial(r,x)}$; and $\overline{Re}$ is the time-mean Reynolds number.

These equations, with appropriate viscous boundary conditions, are solved in stenosis geometries by finite-difference approximation; see Ref. 3 for a brief discussion of methods.

NUMERICAL SIMULATIONS

Previous experiments and steady flow solutions have shown the departure from parallel flow depends upon the radial extent of the constriction (or the severity of the stenosis, expressed as per cent reduction in flow area by clinicians). The length of the constriction has a lesser influence, and the exact contour even less. Our computations were done with stenosis shapes given by a simple two parameter conformal transformation:

$$\zeta = U + iV = A\eta + B \tanh \frac{\pi}{2}\eta, \text{ where } \eta = x + ir. \tag{3}$$

This allows selection of a throat ratio and a more, or less, gentle approach to the throat; see Fig. 1a. Any curve of the set can be used, e.g., either darkened curve can be used as the lumen boundary, both providing about the same throat diameter ratio, d_R, but the inner curve reaches farther up and downstream for the same per cent local area change. The advantage of the conformal transformation is two-fold. It provides an *exact* mapping to a more convenient rectangular computation (U,V) plane, and also gives the least complication in the transformed (U,V) partial differential equations. Moreover, it provides a standard to test more arbitrary numerical approximate mappings. All the mappings send the curved boundary into a straight line allowing the consistent application of the viscous nonslip boundary con-

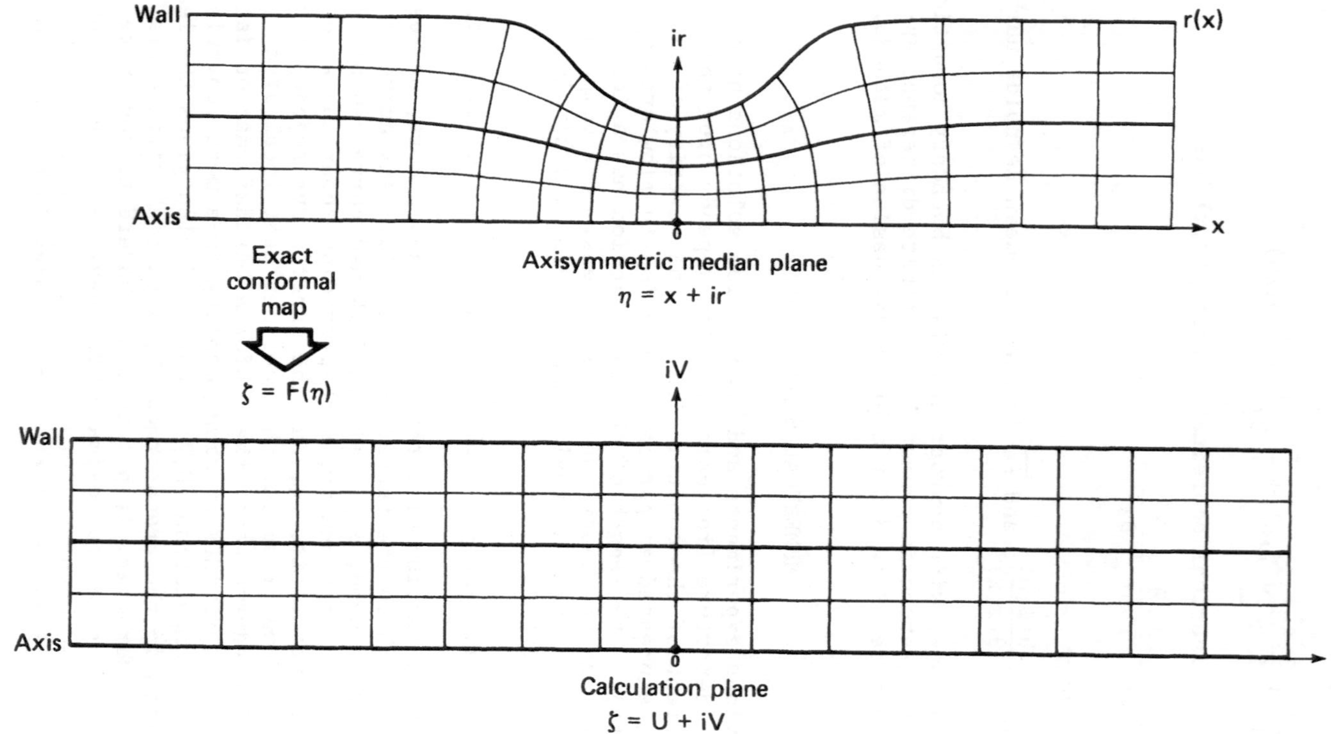

Figure 1. Mapping the stenosis to a computation plane.

dition to $O(h^2)$, where h is the finite-difference mesh. This is chosen in the transformed plane (Fig. 1b) and provides a finer resolution just where it is desired, in the throat region, in the original physical plane.

In principle, any smooth constriction in a parallel tube corresponds to an exact analytic expression that conformally maps the physical median flow plane into the U, V computational strip, but generally the $\zeta(\eta)$ is not known beforehand. It can, however, be approximated by semi-analytical techniques. Alternatively, one can do the whole mapping as a numerical approximation but the buildup of errors may degrade the numerical stability of the solution methods (see Refs. 4, 5).

A variety of solution methods were tried for a steady test case (Ref. 5) providing mutual verification of the predicted ψ, Ω values. However, all the pulsatile flows discussed here were obtained by explicit time-marching with a small enough time step to assure numerical stability. For ease of presentation and ready comparison to experimental results,[1,6] the periodic flux wave is simple, Eq. 1.

Fortunately, the shape of the flux wave is not too important to the numerical technique. Fine time-steps allow faithful representation of the flux wave and our experience has been that an arbitrary wave of the same peak value can be handled with the same relaxation parameters as the simple wave.[7]

Inflow (Outflow)

The exact parallel inflow and outflow streamfunction,[8] and the vorticity $\Omega(r)$, as $x \rightarrow \pm\infty$ are known as a function of f_R (or "Womersley's parameter" $\alpha = \sqrt{f_R}$). A value of $f_R = 100$ is taken as typical for a realistic model. In the computations the parallel flow values are assigned at finite distances up and downstream of the throat. This introduces little effect upstream, provided the inflow condition is imposed where the area differs very little from πa^2. But downstream the "end-condition" must be placed farther away as Re increases to avoid influencing the throat solution. (This can only be established by cut-and-try; but the artifices introduced by too short an outflow length are now clearly recognized). It is possible to reduce the semi-infinite rectangular (U,V) domain to a finite one by another mapping $\phi(\zeta)$ which affects only U, but of course this complicates the p.d.e.'s some more. Selecting a uniform mesh in the latter transformed plane introduces greater variability in the mesh spacing in the physical plane.

Solutions

Results have been obtained for a range of Re and ε at the fixed value $f_R = 100$. These include moderate and severe stenoses. When ε is zero (steady flow), the flow details near the throat are qualitatively in agreement with those shown previously by other workers.[1,2,4,9] Like the previous theory[2,4] our calculations show an increase in the magnitude of the wall vorticity peak with Re (Fig. 2). Separation in the median plane advances slightly toward the throat ($x = 0$) with Re increase, but the major change in recirculation length occurs in the travel of the reattachment point. Of course, changes in the recirculation pattern are accompanied by changes in velocity profiles and pressure gradients. An example is shown in Fig. 3 where some velocity data[9] is also included. The axial and radial velocity components at the throat are also shown, because the formation of a central jet is well established even at low Re for fairly severe stenoses. The radial velocities change sign as the jet expands to the wall downstream.

Results for the pulsatile flow cases computed are more difficult to illustrate. Not that the information is not available, but there is not enough space to show all the time-dependent details of a single case, much less how they depend on the geometric and dynamic parameters. For convenience, we concentrate on the wall vorticity distributions. With the simple pulsatile waveform, we select the quarter-cycle times of maximum, minimum, and mean flux. The wall vorticity (Ω_W) patterns are illustrated for $\varepsilon = 1.0$ in Fig. 4. First note the distributions at the mean flux ($\tau = \omega t = \pi/2 + n\pi$) are not the same (and both differ from the corresponding steady one). The shape of the wall vorticity distribution curve at peak flux is qualitatively similar to steady flow at the same *instantaneous* Reynolds number. But the distribution at minimum flux is far different from that of the corresponding steady Reynolds number (exactly zero here).

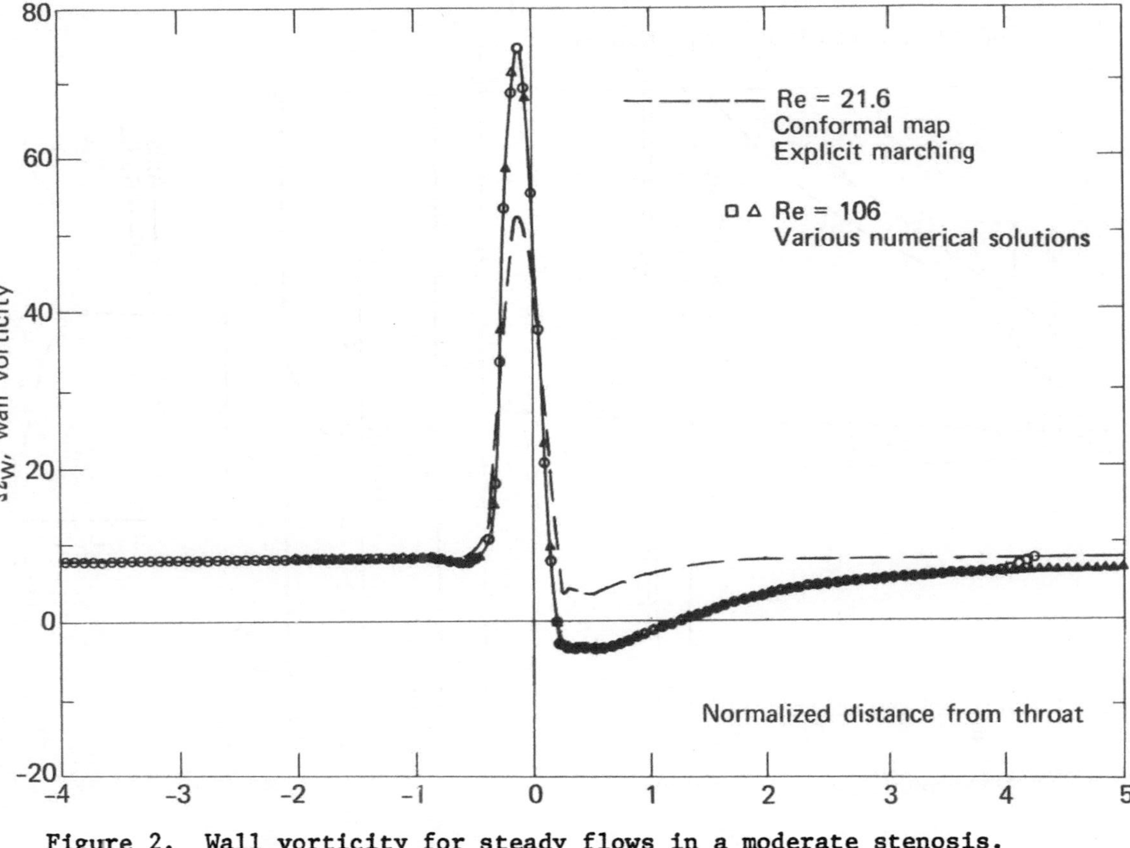

Figure 2. Wall vorticity for steady flows in a moderate stenosis.

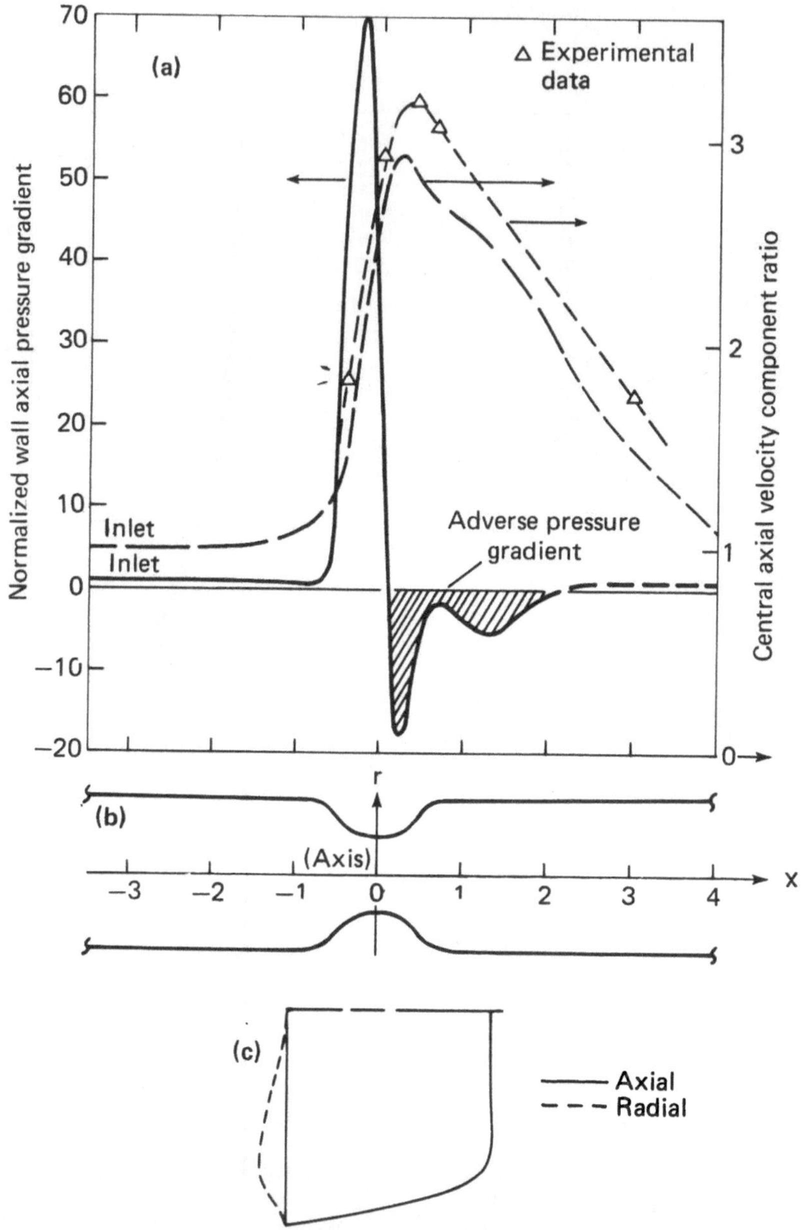

Figure 3. Flow details for Re = 21 flow: a) axial centerline velocities and pressure gradients, b) geometry and c) throat u and v.

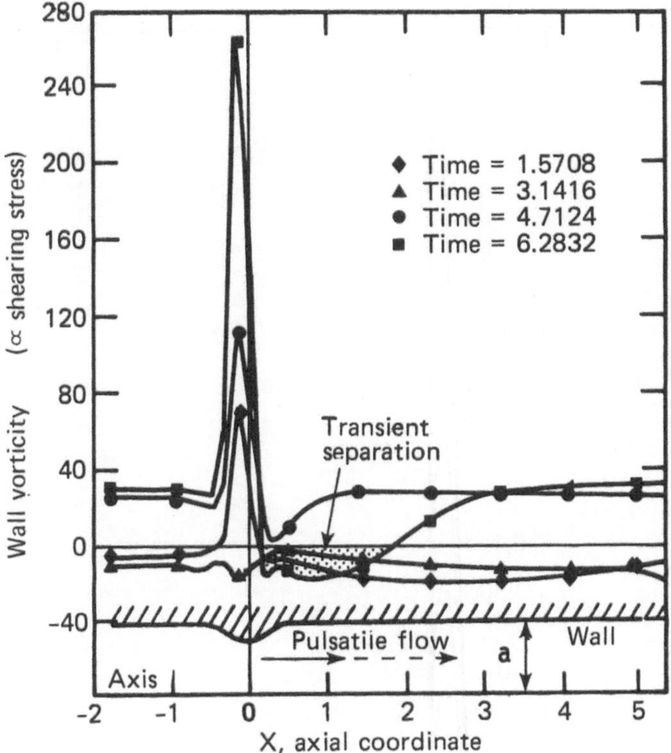

Figure 4. Distributions of wall vorticity Ω_W at quarter cycle times during the simple pulse wave (stenosis area = 0.56, $\overline{Re}$ = 213, f_R = 100, ε = 1.0).

These differences between the instantaneous flow patterns and the corresponding steady ones are accentuated as ε increases. Only when ε is very small (0.1) does the pattern tend to be near the quasi-steady one. However, all the spatial distributions of wall vorticity at peak flux tend to have the greatest magnitude at nearly the same location just before the throat, Fig. 5. Likewise the separation point is nearly the same regardless of ε but the reattachment location is significantly different.

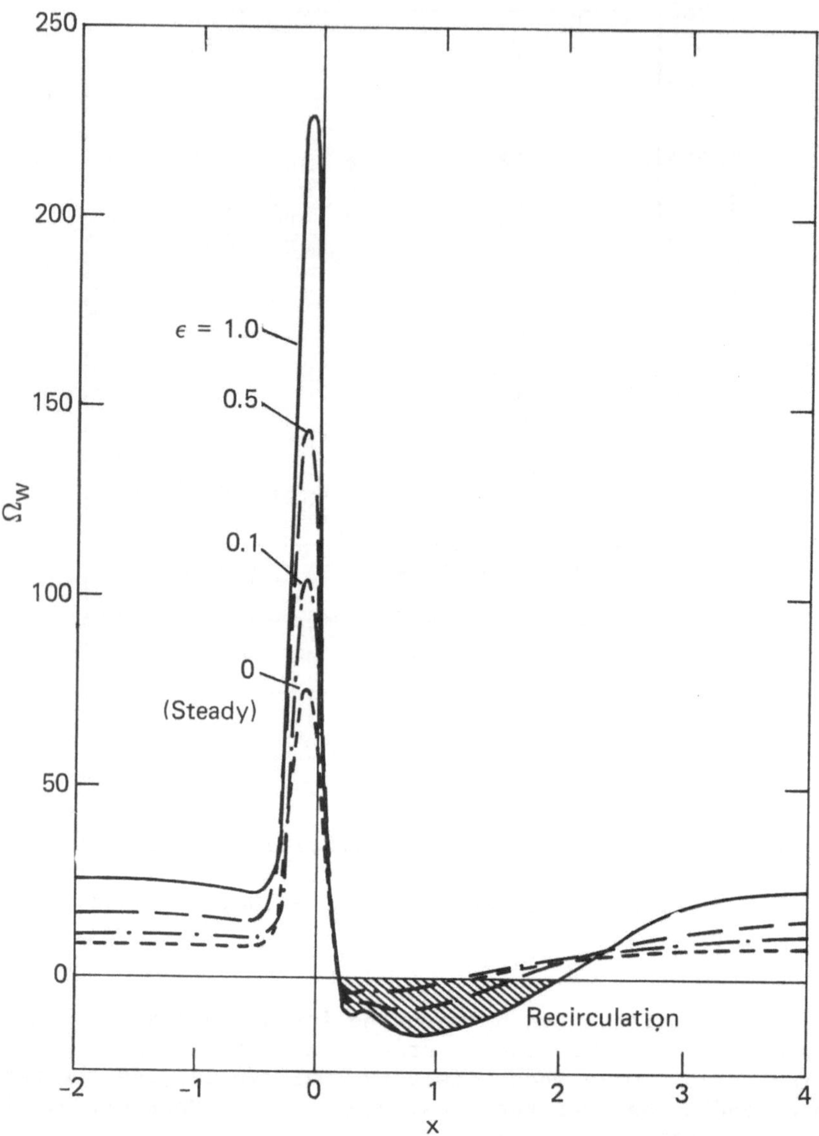

Figure 5. Distribution of wall vorticity at the time of peak flux for three pulsatile flows with the same mean Reynolds number, Re = 213, but different values of ε (stenosis area = 0.56, f_R = 100).

Some of the differences of unsteady wall vorticity distributions from the steady mean one are due simply to the changes within the parallel inflow as it varies through the cycle. The relative magnitude of the wall vorticity of the oscillatory portion in simple pulsatile flow is *greater* than ε (relative magnitude of flux). This means the total (oscillatory plus steady) wall vorticity can be transiently negative during part of the cycle for moderate ε. (The value of ε which just causes zero Ω_W at its low point in the cycle is dependent on f_R; see Appendix A in Ref. 11). There is a phase shift also (dependent on f_R) so at the two instants when the flux is equal to the steady (mean) value, the velocity and vorticity profiles across the lumen are different from each other (and the mean). One has to distinguish accelerating and decelerating flow.

Because some hypotheses on atherogenesis invoke recirculation regions,[10] there may be more interest in the wall extent of such regions during the pulsatile flow cycle. At each end of a bounded recirculation there is a stagnation point on the wall, which is given by a zero of Ω_W. So a space-time plot of wall vorticity zeroes, Fig. 6, graphically illustrates the changing recirculation regions. Time ($\tau = \omega t$) goes upward so the bottom portion ($\tau < \pi$) represents decelerating flow. The deceleration recirculation region enlarges with time until it disappears completely when the total wall shear reverses sign. However for a short time there is a small pocket of positive wall shear about the throat bounded by negative inflow/outflow values. As the volume flow accelerates from its minimum at π (zero in this example) the inflow wall shear changes back to positive. As the flow accelerates more strongly all the wall experiences positive wall shear until the point of incipient recirculation behind the throat is reached. This region grows larger not only through the rest of the acceleration phase, but continues to grow larger still during the deceleration until reverse flow occurs all along the walls as the cycle repeats.

Difficulties

Converged pulsatile numerical results for $\overline{Re}$ up through 200 with $\varepsilon = 1$ have been obtained for a moderate stenosis (area reduction to 50%), but not this high for a severe stenosis (area reduction to 25%). The latter case undoubtedly

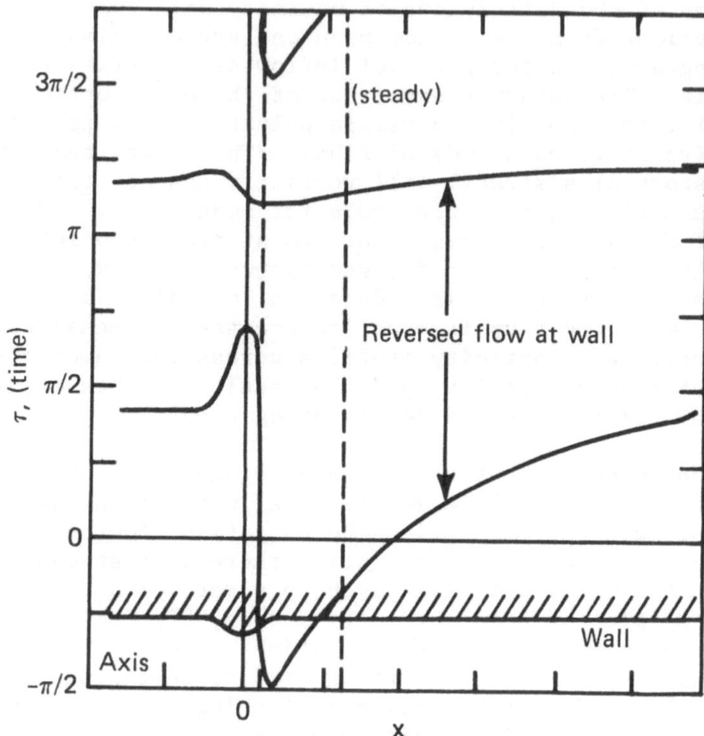

Figure 6. Locations of wall stagnation points with time = zeroes of wall vorticity (stenosis area = 0.56, Re = 106, f_R = 100, ε = 1.0).

involves steeper shear gradients at a given Re and the lack of convergence is likely due to too coarse a mesh. A finer mesh, though more costly for a computation, would improve stability, provide more accurate extremum values of Ω_W and better recirculation region resolution.

The time-marching techniques, either implicit or explicit, are likely to run into numerical instability for much smaller values of f_R. Much higher values than 100 are going to suffer from steep gradient difficulties, as above for the higher Re.

MEASURING STENOSIS FLOWS

The basic flow facility, see Fig. 7, consists of a head tank (1), a connecting long straight inlet section (2) to the transparent test section (3) and an outlet section (4) which directs the flow through a rotating valve (5) before it enters the dump tank (6). The fluid is returned to the head tank by means of an adjustable flow pump. The test section is made by polymerizing a styrene monomer about a machined mandrel. The outer surface of the casting is highly polished. A photograph, Fig. 8, shows the 50% stenosis, air-filled. The flow system is filled with a clear organic fluid (Eugenol) which closely matches the index of refraction of the cast plastic so the effects of wall curvature upon a laser beam at the fluid-solid interface are minimized.

The periodic flow is produced by rotating the valve (5) at a constant frequency by an electric motor to a desired fixed reduced frequency. Eugenol viscosity at room temperature is about 6 centistokes. The average volume flow rate (and mean Reynolds number) is determined by the heights of liquid in the head and dump tanks, which are maintained at a constant differential by the return pump. It can be measured and monitored in the parallel inlet section, see Fig. 9.

Only the test section is transparent for that is where the laser Doppler velocimetry (LDV) is carried out. The one-component optical system is similar to that described previously for pulsatile flow measurements in a two-dimensional branch.[12] So that small and negative velocities can be measured, the center frequency is shifted by a rotating diffraction grating disc. For a single (say axial) velocity component the first order diffracted beams on either side of the central maximum are selected from the line of beams emerging from the rotating disc. These are focussed at the measuring "point" in the flow with the beam plane aligned to measure axial velocities. The scattered light from the small moving particles with which the Eugenol is seeded is collected on the face of a photomultiplier tube. The time-dependent output signal is filtered, amplified and then autocorrelated by a probability and correlation analyzer (Honeywell SAICOR SAI-43A) to yield the Doppler shift, proportional to velocity. The unsteady signals are averaged over at least 100 cycles with a signal enhancer, externally triggered by a periodic pulse from the rotating valve.

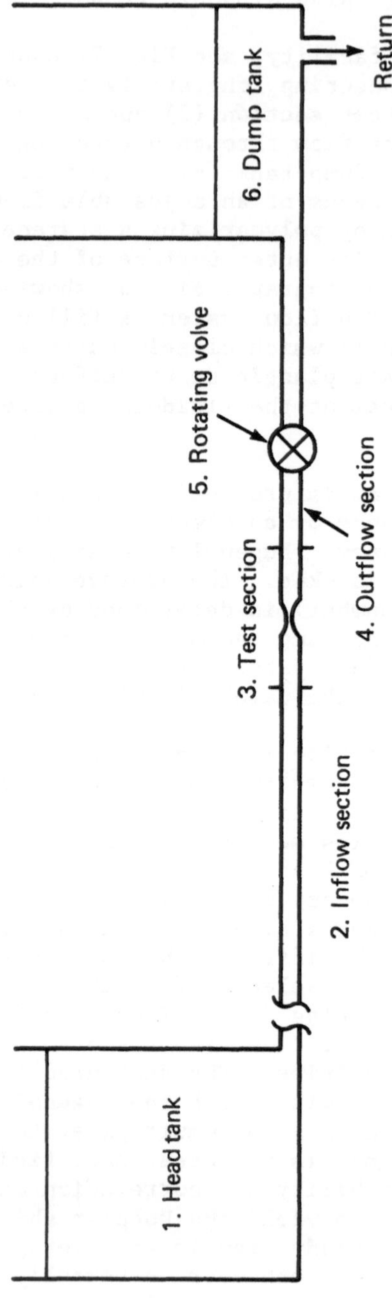

Figure 7. Schematic of the stenosis flow facility in the Biodynamics Laboratory.

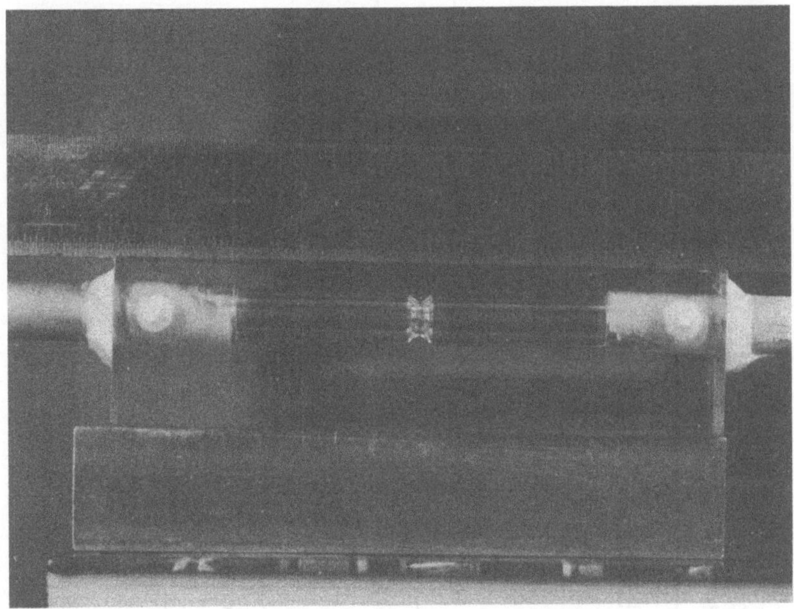

Figure 8. A photograph of the cosine test section (stenosis area = 0.50).

Since the flow in the stenosis vicinity contains major deviations from axial velocity vectors, we aim to measure these non-axial components by a novel approach. This uses the three brightest beams from the line of beams from the rotating diffraction grating (all of different frequency).* The center (maximum intensity) beam of frequency f_c, is translated sideward by a prism, and the triad of beams is all focussed together by the same lens at the measuring point, Fig. 10a. Each pair of focussed beams defines a plane and a vector component to be measured. Taking two pairs in turn, two components can be measured and reduced to orthogonal velocity components u, v. (The third pair is redundant but allows an accuracy check; see Fig. 10b).

The LDV provides accurate velocities and good spatial resolution. Time-dependent velocity data on the centerline

*U.S. Patent No. 4,148,585 (1979).

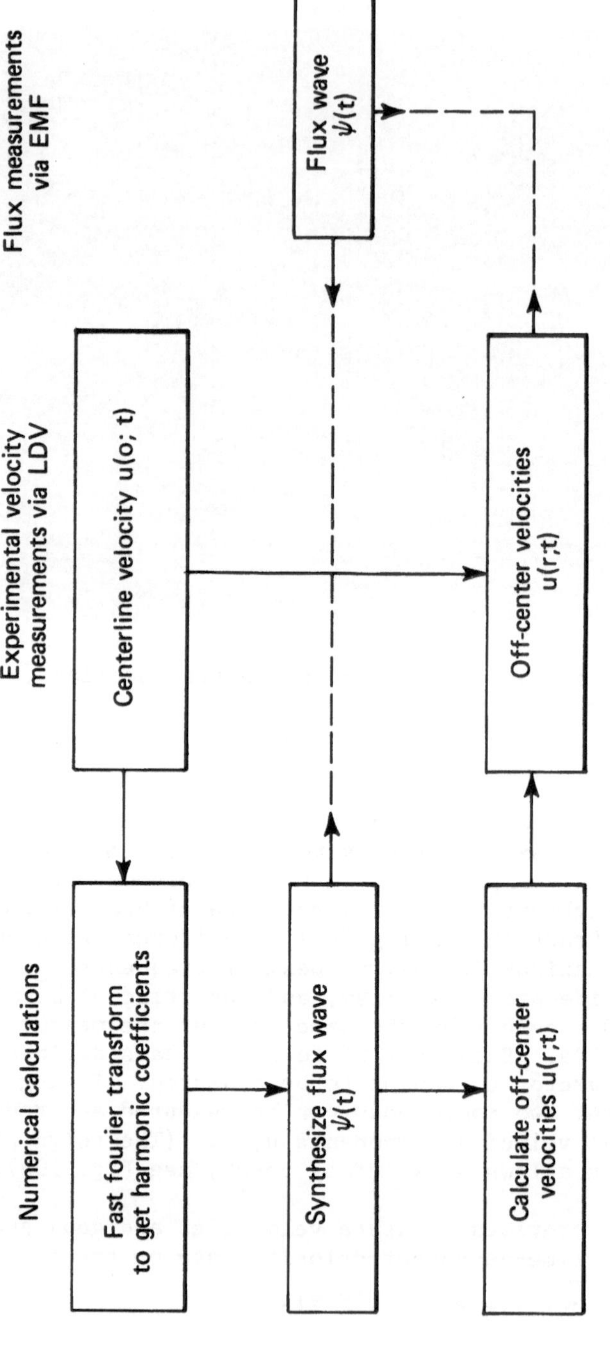

Figure 9. Parallel (inflow) procedure for pulsatile flow.

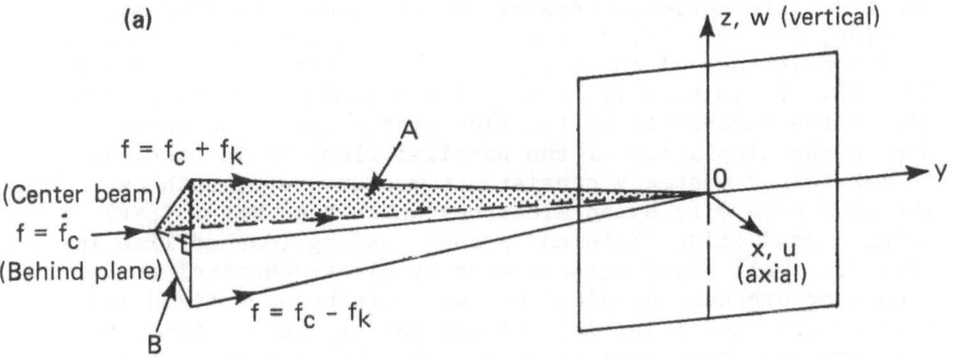

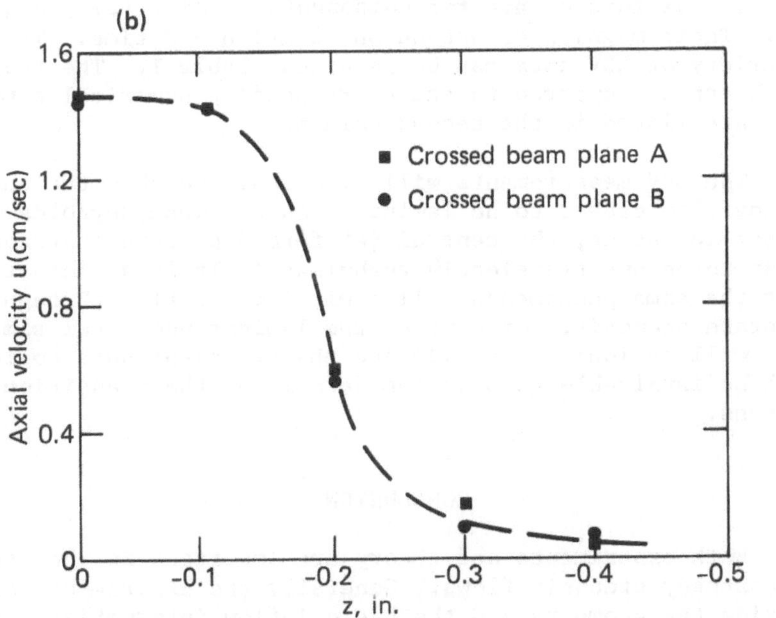

Figure 10. The laser beam configuration for two velocity components and a test of its operation.
(a) Triad beam geometry for u measurement.
(b) Axial velocity profile measurements on a jet vertical centerplane

within the parallel inflow *before* the stenosis provides all the flow information necessary for the numerical flow simulation, Fig. 9. LDV also provides the mutual check on the self-consistency of the total parallel flow; see Fig. 9 and Ref. 12. It is readily seen that any laminar periodic flow within the capability of the flow system can be handled. Due to the simplicity of the parallel flow, it is possible to make other unsteady consistency measurements, such as the flux wave $\psi(t)$ by an electromagnetic flow meter (EMF) using a conducting fluid and proper scaling (dashed line in Fig. 9). Wall shear measurements by electrochemical techniques or pressure gradient measurements between fixed locations are also feasible, but neither of these methods have the time and space resolution of LDV. The velocity data (and time or space averages) provide all the parallel flow information, as does the numerical solution.

Within the stenosis itself and immediately downstream the flow vectors $\vec{q}$ have two components, radial (v) and axial (u). Their magnitudes depend on location and time. Thus a variety of LDV data can be recorded, Table I. The features which can be compared to the corresponding numerical simulation are listed in the second column.

The LDV measurements will also indicate when the unsteady flow ceases to be laminar. As the mean Reynolds number increases, the central jet formed by severe stenoses tends to become transiently turbulent.[1] It is anticipated that the same phenomenon will limit laminar flow through a moderate stenosis. Of course, the laminar numerical simulation will no longer be valid and the transient data collected will be invaluable to describe details of the transition process.

CONCLUSION

Both experiments and theory are important in the study of unsteady stenosis flows. Generally the experiments must provide the geometry and the basic inflow information. Then the numerical theory can give the complete picture, whose accuracy can be assessed from a relatively few measurements in the vicinity of the stenosis. Qualitatively the higher the mean Reynolds number, the more severe the occlusion and the higher the peak-to-mean flux ratio, the higher the peak

Table 1. Types of velocity measurements within the stenosis R(x).

Data	Features
Centerline u(x,0;t)	Median velocity profile
Off-centerline u(x,r;t)	Axial velocity profile at $x = x_i$
Near wall u(x,R-δ;t)	Transient negative velocities (δ << 1)
Radial component v(x,r;t)	Radial velocity profiles at $x = x_i$
Near wall $\vec{q}_p(R-\delta;t)$	Wall shear, $\Delta q_p/\delta$ (p ∿ parallel)
Stagnation point locations $x_s(t)$	Ends of transient recirculation regions

wall shearing stress in the constriction and the larger the extent of the recirculation region. Application to *in vivo* flows is obvious; when the laminar inflow to a particular axisymmetric stenosis is known, any desirable feature of the flow field can be simulated.

Acknowledgments

This work is supported by NIH under Grant HL23291. The expert experimentalists who do the LDV measurements are O. J. Deters and F. F. Mark.

REFERENCES

1. D. F. Young and F. Y. Tsai, "Flow characteristics in models of arterial stenosis - I. Steady Flow and II. Unsteady Flow," J. Biomech. 6, 395-410 (1973).

2. M. D. Deshpande, D. P. Giddens, and R. F. Mabon, "Steady laminar flow through modelled vascular stenoses," J. Biomech. 9, 165-174 (1976).

3. V. O'Brien and L. W. Ehrlich, "Pulsatile flow through stenosed arteries," ASME 1977 Biomechanics Symp. (ASME AMD Vol. 23, pp. 113-116).

4. J. S. Lee and Y. C. Fung, "Flow in locally constricted tubes at low Reynolds numbers," J. Appl. Mech. 37, 9-16 (1970).

5. L. W. Ehrlich, "The numerical solution of a Navier-Stokes problem in a stenosed tube: a danger in boundary approximations of implicit marching schemes," Comp. & Fl. 7, 247-256 (1979).

6. D. J. Schneck and F. J. Walburn, "Unsteady laminar-flow separation in tubes - II. The effect of variations in the frequency and amplitude of flow oscillations," Rpt. VPI-E-79-21, June 1979.

7. V. O'Brien, L. W. Ehrlich and M. H. Friedman, "Unsteady flow in a branch," J. Fl. Mech. 75, 315-336 (1976).

8. S. F. Grace, "Oscillatory motion of a viscous liquid in a long straight tube," Phil. Mag. 5, 933-939 (1928).

9. J. C. Bentz and N. A. Evans, "Hemodynamic flow in the region of a simulated stenosis," ASME paper 75-WA/Bio-10, 1975 (also Bentz Ph.D. thesis, Univ. Penna., 1974).

10. F. B. Gessner, "Hemodynamic theories of atherogenesis," Circ. Res. 33, 259-266 (1973).

11. W. McMichael, "Interphase mass and heat transfer in pulsatile flow," Ph.D. thesis, Rice Univ. (1973).

12. F. F. Mark, C. B. Bargeron, O. J. Deters, and M. H. Friedman, "Experimental investigations of steady and pulsatile laminar flow in a 90° branch," Trans. ASME (JAM) 99E, 372-377 (1977); see also Proc. 28th ACEMB, 1975.

INDEX